Environmental Footprints of Products and Processes

C000256912

Series Editor

Subramanian Senthilkannan Muthu, Head of Sustainability - SgT Group and API, Hong Kong, Kowloon, Hong Kong

Indexed by Scopus

This series aims to broadly cover all the aspects related to environmental assessment of products, development of environmental and ecological indicators and eco-design of various products and processes. Below are the areas fall under the aims and scope of this series, but not limited to: Environmental Life Cycle Assessment; Social Life Cycle Assessment; Organizational and Product Carbon Footprints; Ecological, Energy and Water Footprints; Life cycle costing; Environmental and sustainable indicators; Environmental impact assessment methods and tools; Eco-design (sustainable design) aspects and tools; Biodegradation studies; Recycling; Solid waste management; Environmental and social audits; Green Purchasing and tools; Product environmental footprints; Environmental management standards and regulations; Eco-labels; Green Claims and green washing; Assessment of sustainability aspects.

More information about this series at https://link.springer.com/bookseries/13340

Felipe Luis Palombini ·
Subramanian Senthilkannan Muthu
Editors

Bionics and Sustainable Design

 Springer

Editors
Felipe Luis Palombini
Federal University of Rio Grande do Sul
Porto Alegre, Brazil

Subramanian Senthilkannan Muthu
SgT Group and API
Hong Kong, Kowloon, Hong Kong

ISSN 2345-7651   ISSN 2345-766X (electronic)
Environmental Footprints and Eco-design of Products and Processes
ISBN 978-981-19-1814-8   ISBN 978-981-19-1812-4 (eBook)
https://doi.org/10.1007/978-981-19-1812-4

This Springer imprint is published by the registered company Springer Nature Singapore Pte Ltd.
The registered company address is: 152 Beach Road, #21-01/04 Gateway East, Singapore 189721,
Singapore

Contents

About the Editors

Felipe Luis Palombini is a Post-Doctoral Researcher in the Graduate Program of Botany (PPGBot) at the Federal University of Rio Grande do Sul (UFRGS), in Porto Alegre (Brazil). He received his M.Sc. and his Ph.D. in the Graduate Program in Design (PGDesign), in the School of Engineering (UFRGS). His main research topics are focused on bionics, sustainability, design, and materials selection and characterization via finite element analysis and X-ray microtomography. He is associated with the Design and Computer Simulation Research Group (DSC), and the Plant Anatomy Laboratory (LAVeg), at UFRGS.

Dr. Subramanian Senthilkannan Muthu currently works for SgT Group as Head of Sustainability and is based out of Hong Kong. He earned his Ph.D. from The Hong Kong Polytechnic University, and is a renowned expert in the areas of Environmental Sustainability in Textiles & Clothing Supply Chain, Product Life Cycle Assessment (LCA), and Product Carbon Footprint Assessment (PCF) in various industrial sectors. He has 5 years of industrial experience in textile manufacturing, research and development, and textile testing and 7 years of experience in life cycle assessment (LCA), carbon and ecological footprints assessment of various consumer products. He has published more than 100 research publications, written numerous book chapters, and authored/edited over 100 scientific books in the areas of Carbon Footprint, Recycling, Environmental Assessment, and Environmental Sustainability.

Sustainability in the Biom*

Nancy E. Landrum and Taryn Mead

Abstract Bioinspired innovation is a growing field and is often attributed to sustainable design outcomes. After reviewing existing literature in bionics, biomimetics, and biomimicry, a comparative analysis was used to compare and contrast these subdisciplines. This theoretical analysis aims to reveal differences between bioinspired design approaches to show that each is distinct and to position bioinspired design approaches along the sustainability spectrum. This research contributes to the conceptualization of sustainability within bioinspired innovation and advances nuanced perspectives for scholars and practitioners in this field.

Keywords Biomimicry · Biomimetics · Bionics · Bioinspired design · Sustainable design · Eco-design · Responsible design · Strong sustainability

1 Introduction

Early design work encouraged consumption which contributed to environmental degradation through the extraction of raw materials and accumulation of waste [11, 29, 44]. It was the critical work of Papanek [44] that called for improvements in the design profession. Early responses to this call focused on reducing the environmental impact of products but each approach, such as green design and ecodesign, had shortcomings [8]. A later wave of design approaches for sustainability turned to nature-inspired design [8].

This research did not receive any specific grant from funding agencies in the public, commercial, or not-for-profit sectors

N. E. Landrum (✉)
Independent Scholar, Leipzig, Germany
e-mail: nancyelandrum@gmail.com

T. Mead
Western Colorado University, 1 Western Way, Gunnison, CO 81231, USA
e-mail: tmead@western.edu

Nature-inspired or biologically inspired designs are now present in many fields, including architecture, industrial design, engineering, medicine, urban planning, materials science, management, agriculture, and more. Research has shown that exposure to biological examples increased novelty in design ideas while exposure to human-engineered examples decreased variety in design ideas [64]. Nonetheless, it has also been shown that designers do not always make analogous use of biological phenomena [9] and that a better understanding of biological phenomena could help designers improve the application of biologically inspired, or bioinspired, design.

There are three bioinspired design approaches that are commonly referenced interchangeably: bionics, biomimetics, and biomimicry. Researchers and practitioners alike have called for research that offers a clearer delineation between these approaches based on factors such as methodology, origins, and connections to sustainability [24, 25, 36, 53]. Using comparative analysis of the existing literature, the aim of this chapter is to reveal differences and similarities between these three bioinspired design approaches to illustrate the ways in which each is distinct and represents varying degrees of sustainability, thus responding to the call of Papanek [44] to make design more responsible. While it is widely recognized that the aforementioned terms are used interchangeably in many contexts, we believe that additional analysis is necessary to develop and define the unique, nuanced trajectory of each discipline in its respective field and toward sustainability. We propose to advance the relationship between bioinspired disciplines and sustainability and further differentiate between the dichotomous categorization of "weak vs. strong" biomimicry categorization [7] by applying a phased model of sustainability. The rest of this chapter is organized as follows. In Sect. 2, we review the three bioinspired design approaches of bionics, biomimetics, and biomimicry. A brief history of each design approach is presented and practical examples are offered. In Sect. 3, we introduce strong sustainability theory, the sustainability spectrum, and a corresponding five-stage model of sustainability. In Sect. 4, we place each of the design approaches along the sustainability spectrum and explain how these design approaches can be positioned from less sustainable to more sustainable and showing that, ultimately, only one of these design approaches is sustainable. Section 5 discusses the practical value of this comparative analysis for the field of design studies. Section 6 suggests the limitations of our work and future directions for sustainable design research. Finally, Sect. 7 offers our conclusion.

2   Bioinspired Design

Bionics, biomimetics, and biomimicry are biologically informed disciplines that lead to bioinspired design solutions [24]. All three share a common approach toward design: (1) observation of biological models, (2) translation of biological principles or strategies, and (3) application of biological principles or strategies to design [19, 63]. Despite their similarities, we will demonstrate that each term has unique aspects.

2.1 Bionics

Jack Steele, formally trained as a medical doctor, is credited with coining the term bionics in 1958, a combination of biology and technics. His work on bionics and cybernetics was the inspiration for science fiction books and television shows of the 1970s. Werner [41] and others further advanced bionics into a field of study. It is also noteworthy that the term "bionik" is the German language interpretation of the concept overall, with Nachtigall relying on this as his first language.

Experts place bionics within the fields of biology, medicine, and engineering, viewing bionics as a sub-field of engineering that is focused on robotics and mechanics [24]. Bionics is the Anglo-Saxon term used in medicine, however, it is also the term used for the overall discipline in German-speaking countries [54]. Bionics employs the principles of physics for creative problem-solving, is practiced by functional biologists and engineers in the fields of medicine and cybernetics, specializes in technical complexity and innovation, and seeks mechanically innovative solutions [24]. Practitioners study mechanics and processes in nature and replicate those processes in engineering and computing design. Bionics focuses on nature's mechanical abilities or technology without regard for ecology [24] and is a "prediction and control approach to learning from nature" ([61], p. 292). The practice of bionics fails to consider sustainability, ecology, or society and can ultimately lead to unsustainable solutions [24, 61]. The intent of bionics is to extract physics principles found in nature and apply them to solutions [24] by using nature only for the inspiration it can provide in developing technical design [41]. Thus, bionics extracts physics principles to create technical solutions that perform the same function as nature but does not imitate those principles in the same way as found in nature. The innovation culture and narrative of those who adopt bionics don't have a clear motivation for sustainability beyond learning from nature [34]. Rather, the emphasis in developing the innovation is on immediate return on investment and taking advantage of strategic business opportunities [34].

Examples of bionics include researchers who are studying the echolocation of bats to integrate similar mechanical functions into drones for improved navigation [15, 39]. Popular applications of bionics are those in which robotic movements mimic those of humans, such as the prosthetic arm and hand system developed for veterans [47]. Another bionics example might be a building developed by architects and designers that is inspired by the form of termite mounds [54].

2.2 Biomimetics

Otto Schmitt, a biophysicist formally trained as an engineer, is largely credited with launching this field of study in the 1950s; in 1969, he coined the term biomimetics from the roots bios (life) and mimesis (imitate) [59]. Schmitt (founder of biomimetics) and Steele (founder of bionics) were colleagues in the 1960s at

Wright-Patterson Air Force Base, but each took his work in a different direction, thus separating biomimetics from bionics.

Experts place biomimetics within the discipline of engineering [24]. Practitioners study and imitate nature in design [8] and engineering [20], the focus is predominantly on the mechanical capabilities of the structure–function relationship [2, 24]. Biomimetics employs principles of biology [2], is practiced by biologists, engineers, and designers in the fields of medicine, information technology, economics, and systems science; specializes in mechanical abilities, technical complexity, and innovation; and seeks to imitate nature without a focus on sustainability [24]. Biomimetics not only draws inspiration from nature's designs, like bionics, but focuses on the application or replication of nature's design, unlike bionics. Research into the process of biomimetic innovation frequently focuses on the analogical transfer of biological strategies (e.g., [9, 21, 55]), rather than broader reaching metaphorical inspirations that are frequently applied in biomimicry. And while there is great "promise" that biomimetics will produce more sustainable results, these results are frequently called into question [17].

Examples of biomimetics can be found in stronger fibers modeled after spider webs, multifunctional materials modeled after nature's efficient creation of materials that form multiple functions, and superior robots that mimic both shape and performance of biological creatures [2]. Continuing with our earlier bionics example of a building with constant temperature, while bionics would mimic the structure, biomimetics would mimic the structure and the function of the termite mound chimneys to help vent hot air out and keep the interior temperature constant. This design was mimicked in the Eastgate Centre shopping mall in Harare, Zimbabwe, and maintains a comfortable inside temperature year-round without the use of traditional heating and cooling systems [16].

2.3 Biomimicry

The term biomimicry is generally acknowledged to have first appeared in a doctoral thesis in 1982 [38] but was conceived in its modern interpretation by Janine Benyus in 1997. Benyus [3] defines biomimicry as "the conscious emulation of life's genius" (p. 2) and a science that imitates nature's models, uses an ecological standard to measure what is sustainable, and values nature as a mentor for learning. Though biomimicry was likely practiced throughout history prior to the industrial revolution [8], Benyus was the first to connect bioinspired solutions to the evolutionary sustainability of the human species.

The terms biomimetics and biomimicry are often used interchangeably [3, 7, 31, 57, 58]. In fact, some refer to biomimetics as "reductive biomimicry" [3, 49]. However, biomimicry goes one step further than biomimetics in that biomimicry uses nature as a model and mentor but adds the dimension of "measure" to determine what is sustainable [3, 12, 13]. Biomimicry doesn't focus only on the mechanics or process but also includes mimicking form or shape and as well as interactions within and between systems. Therefore, not only do we learn from and imitate nature, but biomimicry designs create conditions that are conducive to life. This means pursuing efficiency in using benign materials and renewable energy in a closed-loop system without producing waste [3, 42] and also designing with regeneration in mind [61]. The nature-focused ethos and the realization that humans are part of nature is what separates biomimicry from the other terms [24]. The intent of biomimicry is not to just extract biology principles found in nature but to learn from nature and develop solutions that are life-sustaining [3, 13, 24, 50]. Biomimicry has been described as an approach for more holistic systems design [1, 6], suggesting a more encompassing perspective than the simple analogical translation of biological functions.

It is for these reasons that some refer to biomimicry as "sustainable biomimetics" [25] and "holistic biomimicry" [3] although much of what passes for biomimicry today would more accurately be defined as biomimetics. Wahl [61] argues that biomimicry is ecologically informed, more holistic, and simultaneously considers humans, ecosystems, social systems, and economic systems, this would be described as a co-evolutionary level of sustainability [28]. That is, modern interpretation and practice of biomimicry have co-evolved with our sustainability challenges and the ongoing adaptation of the concept of sustainability in ever-changing global realities [14, 30, 48].

Experts place biomimicry within the fields of design, business, architecture, and philosophy [24], though there are exceptions, e.g., in engineering [49] and chemistry [59]. Practitioners mimic complex living systems which are supportive of life on earth to solve design challenges. Biomimicry employs biological and life-sustaining principles, is practiced by ecologists, environmental scientists, designers, architects, economists, and biologists; incorporates a nature-focused ethos with minimal technical complexity; and designs solutions focused on life-sustaining principles, but which might lack real-world applicability [24]. Biomimicry draws inspiration from nature's designs (like bionics and biomimetics), focuses on the application or replication of nature's design (like biomimetics), but unlike either bionics or biomimetics, biomimicry incorporates sustainability into the design as an explicit component of the methodology [5]. Those who adopt biomimicry have innovation cultures and narratives that can be described as "aspirational" in that they are ambitious and seek to "be like nature," sustainability is the purpose and is intrinsically motivated, and it is setting a model for others in its sustainability orientation [34].

An example of biomimicry is the "Factory as Forest" initiative [18]. This work began as a project between consultancy Biomimicry 3.8 and carpet manufacturer Interface and continues with other companies through Project Positive [4]. The goal is to mimic interactions between systems with the explicit goal of achieving sustainability. This work focuses on transforming the built environment to become an active participant in its surrounding ecosystem. Continuing with our example of a building that is self-regulating in temperature, biomimicry designers would ensure the building and its materials are life-sustaining in a way that allows self-sufficiency, self-regulation, zero waste, and participation in its surrounding ecosystem, such as the Factory as Forest concept at Interface.

It is precisely because these three terms are often confused and used interchangeably that we focus on revealing differences between these three bioinspired design approaches. We summarize these differences in Table 1.

Equipped with this knowledge, we can now offer a simplified yet more nuanced definition of the three bioinspired design approaches showing how each adds a progressive layer toward achieving sustainability (Table 2). Bionics gains design inspiration from nature through the utilization of physics principles in technological design but does not mimic nature. Biomimetics gains inspiration from nature and mimics natural design through increased control of mechanics and structure in technical design. Biomimicry gains inspiration from nature, mimics natural design, and uses nature as a measurement against which to define sustainability through a nature-based ethos applied to design.

Table 1 Differences between bioinspired design approaches

	Bionics	Biomimetics	Biomimicry
Intent	Employs principles of physics [24], inspiration from nature to develop technical design [41]	Employs principles of biology [2] for increased prediction, manipulation, and control [61]	Employs biological principles and life-sustaining principles [24] for human adaptation [3]
Disciplines and practitioners	Functional biologists, engineers, medicine, cybernetics [24]	Biologists, engineers, designers, medicine, information technology, economics, systems science [24]	Ecologists, environmental scientists, designers, architects, economists, biologists [24]
Specialization	Technical complexity & innovation [24]	Mechanical abilities with technical complexity & innovation [24]	Nature-focused ethos with minimal technical complexity [24]
Solution	Creates mechanically innovative solutions that lack sustainability [24]	Solutions that imitate nature but lack sustainability [24], though with the aforementioned exceptions	Solutions that adhere to life-sustaining principles but may lack real-world applicability [24]

Adapted from Bar-Cohen [2], Iouguina et al. [24], Nachtigall [41], Wahl [61]

Table 2 Defining characteristics of bioinspired design approaches

	Bionics	Biomimetics	Biomimicry
Gains design inspiration from nature	x	x	x
Nature serves as a model of design		x	x
Nature serves as a mentor for learning and a measure of sustainability			x

3 Bioinspired Design and Sustainability

While intentionality can be a useful indicator of sustainability-oriented innovation, it provides little indication of the systemic impacts of any innovation, bioinspired or otherwise. Some methodologies specifically indicate steps and criteria for sustainable design (e.g., Biomimicry 3.8's Design Toolbox), but this is not a guarantee of better performance when the innovation outcomes are scrutinized in a social and ecological life-cycle analysis. Despite the "Biomimetic Promise" of superior innovation performance [17], assumptions that innovations are inherently sustainable because they somehow emulate a biological strategy result in a naturalistic fallacy that because something is "natural," it is inherently better [7]. It has been shown that bioinspired approaches can generate both positive and negative impacts [40], therefore, these assumptions about sustainability are worthy of further analysis and consideration [8]. Given this wide array of approaches and perspectives [35, 36], proposed that rather than viewing bioinspired approaches in a dichotomous frame of an innovation being *unsustainable* or *sustainable,* it is more useful to gauge the sustainability of a bioinspired innovation along a gradient of *less* to *more sustainable* [33] as shown in Fig. 1.

3.1 Sustainability Spectrum

Sustainability can best be understood as a gradient ranging from less sustainable to more sustainable which leads us to understand that each gradient holds different meanings for different people. As such, the sustainability spectrum proposes a continuum of worldviews within environmentalism to help us understand the varying positions along the gradient. The original spectrum of sustainability included four worldviews: very weak, weak, strong, and very strong [43, 45, 46, 56]. Landrum [28] expanded the model to include an intermediate position.

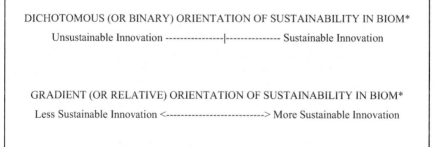

DICHOTOMOUS (OR BINARY) ORIENTATION OF SUSTAINABILITY IN BIOM*

Unsustainable Innovation ----------------|--------------- Sustainable Innovation

GRADIENT (OR RELATIVE) ORIENTATION OF SUSTAINABILITY IN BIOM*

Less Sustainable Innovation <---------------------------> More Sustainable Innovation

Fig. 1 Dichotomous versus gradient orientation. Adapted from Mead et al. [35]

Very weak sustainability. Very weak sustainability is a worldview that sees the natural environment for its instrumental value to humans. The focus is on human-made technocentric solutions to our sustainability challenges that have been extracted from nature (a more exploitative orientation) and a belief in the ability of humans to develop technological solutions that are superior to nature, and which will improve life [43, 45, 46]. In corporate contexts, this is defined by compliance-based decision-making and strongly influenced by external factors that force change [28]. Examples may include using bioinspiration to innovate new materials to replace previously used materials that have been banned.

Weak sustainability. Weak sustainability is a worldview that is less exploitative and has taken a more accommodative orientation. This worldview sees more value, albeit self-serving, in adopting solutions from nature, such as reduced costs, new markets, or improved reputation. However, this stage is still a "manipulative and technocentric position" ([43], p. 88) in that it uses natural resources to develop solutions based upon human ingenuity and technology. Also called business-centered sustainability in relation to corporate environments, this position views sustainability as internally driven by the reduction of costs and eco-efficiency [28]. An example of bioinspiration applied through a weak sustainability lens might be a structural color coating that mimics the Morpho butterfly wing but relies on toxic materials to produce the effect.

Both very weak and weak sustainability worldviews are technocentric in that they view humans as dominant over nature and seek technological solutions to environmental problems. These positions view nature for its instrumental value and allow for human-made solutions that can improve upon nature.

Intermediate sustainability. Landrum [28] posits that there exists an intermediate worldview between weak and strong sustainability that has characteristics of both strong and weak sustainability but is not clearly situated on either side of the spectrum. In this worldview, there is an emphasis on systems-level sustainability that goes beyond a single organization or product. Also referred to as systemic sustainability in a business context, this position looks outside the company and works with others to improve conditions within its sphere of influence [28].

One example of intermediate sustainability can be found in global efforts to reduce single-use plastics. This movement has led designers to create many plastic alternatives. Biobased plastics attempt to create systemic change away from fossil fuels but they also have negative environmental impacts. Eating utensils made from sugar cane, for example, are chemically identical to polyethylene terephthalate (PET) plastic, do not biodegrade, and create the same long-term waste as plastic from fossil fuels [27]. Polylactic acid (PLA) biobased plastic is recyclable, biodegradable, and compostable if commercially composted [27]. However, if they are discarded in the conventional waste stream, they produce the same environmental problems as traditional plastic [27].

Strong sustainability. Strong sustainability is a worldview that is more radical and considers self-sufficiency and cooperation [44]. Landrum [28] describes further, equating strong sustainability with regenerative approaches to innovation, design, and operations. A regenerative approach is characterized by the goal of repairing and restoring human and natural systems. This might be exemplified by Interface's "Net-Works" program which collects, cleans, and upcycles discarded fishing nets from beaches and oceans and remanufactures them into bioinspired carpet tiles [23].

Very strong sustainability. Very strong sustainability sees the intrinsic value of nature and believes in the need for humans to co-evolve alongside nature. This position supports "obedience to natural laws" ([43], p. 91) and the maintenance of life-sustaining conditions. Landrum [28] also refers to this as the co-evolutionary phase of sustainability within corporate contexts where companies view themselves as reintegrating with natural systems using science-based approaches and steady-state economics. An example of this is Biomimicry 3.8's application of Ecological Perfor-mance Standards that redesign operations of factories and cities to use "nature as measure" and systematize integration with ecological systems [18, 62].

Both strong sustainability and very strong sustainability worldviews are ecocentric in that they view humans and nature as co-existing and seek solutions that allow humans and nature to flourish together. These worldviews believe nature has intrinsic value and that man-made solutions are not superior to nature's solutions.

4 Positioning Bioinspired Design Approaches

Drawing from prior work on the sustainability spectrum [43, 45, 46, 56] and the stages of corporate sustainability [28], the concept of "weak" versus "strong" biomimicry has also been proposed as a way to differentiate between those innovations that aim to make some gains in sustainability versus those innovations that have no such intention [7]. When we view nature simply as a source of inspiration for human design, we develop solutions that are technocentric rather than ecocentric [7]. These designs are anthropocentric and focus only on humanity's needs and nature's instrumental value to humans [32]. This "weak" biomimicry allows us to design solutions that serve human needs and continues to advance the notions of human separation from nature and human control over nature, in fact viewing nature's designs as deficient and our technological designs as supplementary [7]. "Strong" biomimicry, on the other hand, enables us to design solutions that are situated within nature and in harmony with ecosystems [7], an approach that positions humans and nonhumans in a bioinclusive relationship where our focus is not to reduce human impact but to have a generative impact on nature recognizing our interconnected life system needs and the intrinsic value of nature [32]. For example, the natural world has been the inspiration and model for military technologies, spacecraft, nanomachines, and even surveillance cameras [31] and has been interpreted as merely another methodology for the enslavement of nature [26]. But replacing a conventional climbing robot with a gecko-inspired climbing robot [37, 51] or developing a painless needle that imitates a mosquito's stinger [10] does little to advance sustainability and may, in fact, create a rebound effect that drives more consumption.

Table 3 Bioinspired approaches and sustainability

	Bionics		Biomimetics	Biomimicry	
Gradient	Less sustainable		Mixed	More sustainable	
Sustainability spectrum	Very weak	Weak	Intermediate	Strong	Very strong
Stage of corporate sustainability	Compliance	Business centered	Systemic	Regenerative	Co-evolutionary
Worldview	Technocentric		Mixed	Ecocentric	

Based upon our analysis, we can position the three bioinspired design approaches along the sustainability continuum (Table 3). Bionics solves human-defined problems by looking to nature for solutions. It is an anthropocentric and technocentric approach that extracts design ideas from nature and applies them to design solutions without necessarily replicating the mechanics of nature. For this reason, we classify bionics within the very weak and weak sustainability worldviews.

Biomimetics extracts design ideas from nature but without consideration of the sustainability of the design. However, biomimetics focuses more on the adoption of nature's mechanics and, for this reason, we classify biomimetics as an intermediate sustainability worldview.

Biomimicry seeks nature-defined solutions to apply to human problems. Biomimicry is ecocentric and is not focused on extraction but rather on learning from nature in terms of form, process, and system and ensuring the sustainability of the design. Therefore, we classify biomimicry within the strong and very strong worldviews.

5   Discussion and Implications

There is growing interest in sustainable design. Bioinspired approaches for sustainable design include bionics, biomimetics, and biomimicry but the terms are often confused and erroneously used interchangeably. Although [62] believes it is of limited use to distinguish between these various approaches because they all contribute a degree of sustainability, experts agree that clearer distinction is needed between these concepts [24, 25].

Contrary to [62], our comparative study shows that these three bioinspired sustainable design approaches do not each contribute a degree of sustainability. Our analysis shows that bionics, biomimetics, and biomimicry are distinct terms regarding intent, disciplines and practitioners, specialization, and solutions. This allows us to identify more nuanced differences and to develop more precise descriptors for each bioinspired approach toward design. Given that sustainability is a much-contested concept [22], we show how [28] five-stage model can be used as a framework to illuminate a path toward sustainable bioinspired design. Using this framework, we can order these three bioinspired design approaches along a gradient of less sustainable to more sustainable following the sustainability spectrum [43, 45, 46, 56]. On the left side of the spectrum, bionics is classified within very weak sustainability and weak sustainability and do not contribute to sustainability. Therefore, this design approach is appropriate when designers are not seeking sustainability but, nonetheless, want a design inspired by nature. Biomimetics is an intermediate position between weak and strong sustainability. This design approach is appropriate when designers create a design inspired by nature and which uses models found in nature but is not made with sustainable materials or in a sustainable way. On the right side of the spectrum, biomimicry is classified within strong sustainability and very strong sustainability and does contribute to sustainability. This design approach is appropriate when designers are seeking a design that is inspired by nature, follows models found in nature, and uses nature's measure of sustainability in methods and materials.

For design and innovation practitioners seeking sustainable solutions, the nuanced distinction of the different phases of sustainability is a necessary inclusion in the design process. Understanding the differences between bionics, biomimetics, and biomimicry approaches can be useful to designers and to the field of design in determining which approach to use depending upon the desired outcome. Designers can choose bionics for novel designs inspired by nature, or they can choose biomimicry for a sustainable design inspired by nature.

Cooper ([11], p. 15) posits that the future of design "offers the significant potential of design to change the world at all levels and to do so in an ethical, trustworthy and collaborative manner." To do this, design must be sustainable. Of all the bioinspired approaches, only biomimicry will lead to sustainable design and only with intention and accountability toward sustainability.

6 Limitations and Future Directions

The authors recognize that several attempts have been made to classify and distinguish between the bioinspired disciplines as they relate to sustainability. While we have attempted to be inclusive in our selection of literature reviewed, we recognize this is a vast and evolving conversation, with both memes and specific words quickly evolving.

Further research is needed to aid in delineating the variety of bioinspiration terms that are often misunderstood or used interchangeably, such as those highlighted here. In addition, future research would benefit from the application of [28] staged sustainability framework to new and existing case studies of bioinspired design to better exemplify how it can be relevant for designers. Finally, the staged sustainability framework [28] can be applied to other sustainable design approaches beyond bioinspired design methods to assess their contribution toward sustainability.

7 Conclusion

Victor Papanek proclaimed that few professions are more harmful than industrial design while simultaneously advocating that design is the most powerful tool for shaping our environment [44]. Papanek called for responsible design that had purpose, served humanity, and protected the environment; design that could change the world [44]. Since that time, numerous sustainability elements have entered the design profession. Of interest here are the bioinspired approaches of bionics, biomimetics, and biomimicry. These concepts are related and often used interchangeably, leading to confusion. Designers have called for better clarity between these terms [24, 25]. This is the goal of the current chapter: to provide clarity.

Our comparative analysis of the terms bionics, biomimetics, and biomimicry uses sustainability theories and frameworks to provide clarity. We define bionics as a design method that gains design inspiration from nature, biomimetics goes one step further to use nature as the design model, while biomimicry extends both concepts and exclusively uses nature as a mentor and measure of sustainability (Table 2). We also show that it can be useful to think of sustainability as being along a gradient of less to more sustainable [35], as shown in Table 3. Using the sustainability spectrum [43, 45, 46, 56], we applied [28] framework of five stages of corporate sustainability to define a placement for each of the three terms, bionics, biomimetics, and biomimicry. From this exercise, we show that bionics is less sustainable, adopts a technocentric worldview, represents weak or very weak sustainability, and is aligned with compliance and business-centered stages of sustainability. Biomimetics represents intermediate sustainability with mixed technocentric and ecocentric worldviews that are somewhere between less and more sustainable, and is aligned with systemic sustainability. Biomimicry is most sustainable, adopts an ecocentric worldview, represents strong or very strong sustainability, and is aligned with regenerative and co-evolutionary stages of sustainability. Improved understanding of these terms reveals that they are distinct concepts with each subsequent approach building upon the other. This understanding can help designers choose appropriate methods suitable to their intended purpose. It is clear that if the design intends to be sustainable, biomimicry is the only solution.

References

1. Baek J, Meroni A, Manzini E (2015) A socio-technical approach to design for community resilience: a framework for analysis and design goal forming. Des Stud 40:60–84. https://doi.org/10.1016/j.destud.2015.06.004
2. Bar-Cohen Y (2006) Biomimetics—Using nature to inspire human innovation. Bioinspiration Biomimetics 1(1):1–12
3. Benyus J (1997) Biomimicry: innovation inspired by nature. Harper Collins, New York
4. Biomimicry 3.8 (2021) Project positive. https://biomimicry.net/project-positive/

5. Biomimicry Institute (2020) The biomimicry design process. https://toolbox.biomimicry.org/methods/process/#:~:text=The%20Biomimicry%20Design%20Process&text=The%20Biomimicry%20Design%20Spiral%20provides,solutions%20to%20a%20design%20challenge

6. Blizzard J, Klotz L (2012) A framework for sustainable whole systems design. Des Stud 33(5):456–479. https://doi.org/10.1016/j.destud.2012.03.001

7. Blok V, Gremmen B (2016) Ecological innovation: biomimicry as a new way of thinking and acting ecologically. J Agric Environ Ethics 29(2):1–15

8. Ceschin F, Gaziulusoy I (2016) Evolution of design for sustainability: from product design to design for system innovations and transitions. Des Stud 47:118–163

9. Cheong H, Shu L (2013) Using templates and mapping strategies to support analogical transfer in biomimetic design. Des Stud 34(6):706–728

10. Cohen D (2002) Painless needle copies mosquito's stinger. NewScientist. https://www.newscientist.com/article/dn2121-painless-needle-copies-mosquitos-stinger/. Accessed 1 Nov 2021

11. Cooper R (2019) Design research—Its 50-year transformation. Des Stud 65:6–17

12. Dicks H (2017) Environmental ethics and biomimetic ethics: nature as object of ethics and nature as source of ethics. J Agric Environ Ethics 30:255–274

13. Dicks H (2019) Being like Gaia: biomimicry and ecological ethics. Environ Values 28(5):601–620

14. Dorst K, Cross N (2001) Creativity in the design process: co-evolution of problem-solution. Des Stud 22(5):425–437

15. Eliakim I, Cohen Z, Kosa G, Yovel Y (2018) A fully autonomous terrestrial bat-like acoustic robot. PLoS Comput Biol. https://doi.org/10.1371/journal.pcbi.1006406

16. Fehrenbacher J (2012) Biomimetic architecture: green building in Zimbabwe modeled after termite mounds. Inhabitat. https://inhabitat.com/building-modelled-on-termites-eastgate-centre-in-zimbabwe/. Accessed 1 Nov 2021

17. Gleich A, von Pade C, Petschow U, Pissarskoi E (2010) Potentials and trends in biomimetics. https://doi.org/10.1007/978-3-642-05246-0

18. Green J (2016) The factory as forest. https://dirt.asla.org/2016/10/18/the-factory-as-forest/. Accessed 1 Nov 2021

19. Helfman Cohen Y, Reich Y (2016) Biomimetic design method for innovation and sustainability. Springer International Publishing, Switzerland

20. Helms M, Vattam S, Goel A (2009) Biologically inspired design: process and products. Des Stud 30(5):606–622. https://doi.org/10.1016/j.destud.2009.04.003

21. Helms M, Goel A (2012) Analogical problem evolution in biologically inspired design. In: Proceedings of the 5th international conference on design computing and cognition. Springer, College Station, Texas, Berlin

22. Imran S, Alam K, Beaumont N (2014) Reinterpreting the definition of sustainable development for a more ecocentric reorientation. Sustain Dev 22(2):134–144. https://doi.org/10.1002/sd.537

23. Interface (2021) The net-works programme. https://www.interface.com/EU/en-GB/about/mission/Net-Works-en_GB

24. Iouguina A, Dawson JW, Hallgrimsson B, Smart G (2014) Biologically informed disciplines: a comparative analysis of terminology within the fields of bionics, biomimetics, biomimicry and bioinspiration, among others. Des Nat VII 9(3):197–205

25. Jacobs S (2014) Biomimetics: a simple foundation will lead to new insight about process. Int J Des Nat Ecodyn 9(2):83–94

26. Johnson E (2011) Reanimating bios: biomimetic science and empire. Doctoral dissertation, University of Minnesota, Minneapolis, MN. https://conservancy.umn.edu/bitstream/handle/11299/117375/1/Johnson_umn_0130E_12236.pdf. Accessed 31 Oct 2021

27. Krieger A (2019) Are bioplastics really better for the environment? Read the fine print. GreenBiz. https://www.greenbiz.com/article/are-bioplastics-really-better-environment-read-fine-print. Accessed 31 Oct 2021

28. Landrum N (2018) Stages of corporate sustainability: integrating the strong sustainability worldview. Organ Environ 31(4):287–313. https://doi.org/10.1177/1086026617717456

29. Lloyd P (2019) You make it and you try it out: seeds of design discipline futures. Des Stud 65:167–181
30. Maher M (2000) A model of co-evolutionary design. Eng Comput 16:195–208
31. Marshall A, Lozeva S (2009) Questioning the theory and practice of biomimicry. Int J Des Nat Ecodyn 4(1):1–10
32. Mathews F (2011) Towards a deeper philosophy of biomimicry. Organ Environ 24(4):364–387
33. McElroy M, Jorna R, van Engelen J (2008) Sustainability quotients and the social footprint. Corp Soc Responsib Environ Manag 15(4):223–234. https://doi.org/10.1002/csr.164
34. Mead T (2017) Factors influencing the adoption of biologically inspired innovation in multinational corporations. Doctoral dissertation, University of Exeter, Exeter, UK. http://hdl.handle.net/10871/30466. Accessed 31 Oct 2021
35. Mead T, Borden DS, Coley D (2020) Navigating the Tower of Babel: the epistemological shift of bioinspired innovation. Biomimetics 5(4):60. https://doi.org/10.3390/biomimetics5040060
36. Mead T, Jeanrenaud S (2017) The elephant in the room: biomimetics and sustainability? Bioinspired Biomimetic Nanobiomater 6(2):113–121
37. Menon C, Murphy M, Sitti M (2004) Gecko inspired surface climbing robots. In: 2004 IEEE international conference on robotics and biomimetics, pp 431–436. https://doi.org/10.1109/ROBIO.2004.1521817
38. Merrill C (1982) Biomimicry of the dioxygen active site in the copper proteins hemocyanin and cytochrome oxidase. Doctoral dissertation, Rice University, Houston, TX. https://scholarship.rice.edu/handle/1911/15707. Accessed 31 Oct 2021
39. Miller M (2018) A drone made for a dark night (or a dark knight). UC Magazine. https://magazine.uc.edu/editors_picks/recent_features/batbot.html. Accessed 1 Nov 2021
40. Montana-Hoyos C, Fiorentino C (2016) Bio-utilization, bio-inspiration, and bio-affiliation in design for sustainability: biotechnology, biomimicry, and biophilic design. Int J Des Objects 10(3):1–18
41. Nachtigall W (1997) Vorbild Natur: Bionik-Design für funktionelles Gestalten. Springer, Berlin
42. Oguntona O, Aigbavboa C (2017) Biomimicry principles as evaluation criteria of sustainability in the construction industry. Energy Procedia 142:2491–2497. https://doi.org/10.1016/j.egypro.2017.12.188
43. O'Riordan T (1989) The challenge for environmentalism. In: Peet R, Thrift N (eds) New models in geography. Unwin Hyman, London, England, pp 77–102
44. Papanek V (1971) Design for the real world: human ecology and social change. Van Nostrand Reinhold, New York
45. Pearce D (1993) Blueprint 3: measuring sustainable development. Earthscan, London, England
46. Pearce D, Turner R (1990) Economics of natural resources and the environment. The Johns Hopkins University Press, Baltimore, MD
47. Pellerin C (2016) DARPA provides groundbreaking bionic arms to Walter Reed. DOD News, U.S. Department of Defense. https://www.defense.gov/Explore/News/Article/Article/1037447/darpa-provides-groundbreaking-bionic-arms-to-walter-reed/. Accessed 31 Oct 2021
48. Poon J, Maher M (1997) Co-evolution and emergence in design. Artif Intell Eng 11(3):319–327
49. Reap J, Baumeister D, Bras B (2005) Holism, biomimicry and sustainable engineering. In: ASME 2005 international mechanical engineering congress and exposition, pp 423–431
50. Rowland R (2017) Biomimicry step-by-step. Bioinspired Biomimetic Nanobiomater 6(2):102–112
51. Schiller L, Seibel A, Schlattmann J (2019) Toward a gecko-inspired, climbing soft robot. Front Neurorobot 13:106. https://doi.org/10.3389/fnbot.2019.00106
52. Schmitt O (1969) Some interesting and useful biomimetic transforms. In: Proceedings of the third international biophysics congress of the international union for pure and applied biophysics. International Union for Pure and Applied Biophysics, Cambridge, MA, p 297
53. Speck O, Speck D, Horn R, Gantner J, Sedlbauer KP (2017) Biomimetic bioinspired biomorph sustainable? An attempt to classify and clarify biology-derived technical developments. Bioinspir Biomim 12:1–15
54. Sugár V, Leczovics P, Horkai A (2017) Bionics in architecture. YBL J Built Environ 5(1):31–42

55. Töre Yargın G, Moroşanu Firth R, Crilly N (2017) User requirements for analogical design support tools: learning from practitioners of bioinspired design. Des Stud. https://doi.org/10.1016/j.destud.2017.11.006

56. Turner R (1993) Sustainability: principles and practice. In: Turner RK (ed) Sustainable environmental economics and management: principles and practice. Belhaven Press, London, pp 3–36

57. Vincent J (2009) Biomimetics—A review. Proc Inst Mech Eng Part J Eng Med 223:919–939

58. Vincent J, Bogatyreva O, Bogatyrev N, Bowyer A, Pahl A (2006) Biomimetics: its practice and theory. J R Soc Interface 3(9):471–482

59. Vincent BB, Bouligand Y, Arribart H, Sanchez C (2002) Chemists and the school of nature. Cent Euro J Chem 1–5. Accessed 1 Nov 2021. https://core.ac.uk/download/pdf/52816959.pdf

60. Volstad N, Boks C (2012) On the use of biomimicry as a useful tool for the industrial designer. Sustain Dev 20(3):189–199

61. Wahl D (2006) Bionics versus biomimicry: from control of nature to sustainable participation in nature. WIT Trans Ecol Environ Des Nat III Compar Des Nat Sci Eng 87:289–298

62. Wahl DC (2016) Designing regenerative cultures. Triarchy Press, Axminster, England

63. Wanieck K, Fayemi P-E, Maranzana N, Zollfrank C, Jacobs S (2017) Biomimetics and its tools. Bioinspired Biomimetic Nanobiomater 6(2):1–14

64. Wilson J, Rosen D, Nelson B, Yen J (2010) The effects of biological examples in idea generation. Des Stud 31(2):169–186

Two-Way Bionics: How Technological Advances for Bioinspired Designs Contribute to the Study of Plant Anatomy and Morphology

Felipe Luis Palombini, Fernanda Mayara Nogueira, Branca Freitas de Oliveira, and Jorge Ernesto de Araujo Mariath

Abstract Bionics is fundamentally based on the development of projects for engineering, design, architecture, and others, which are inspired by the characteristics of a biological model organism. Essentially, bionics is based on a transdisciplinary approach, where teams are composed of researchers trained in a variety of disciplines, aiming to find and adapt characteristics from nature into innovative solutions. One of the key steps in a bioinspired project is the comprehensive study and analysis of biological samples, aiming at the correct understanding of the desired features prior to their application. Among the most sought natural elements for a project to be based on, plants represent a large source of inspiration for bionic designs of structures and products due to their natural efficiency and high mechanical performance at the microscopical level, which reflects into their functional morphology. Therefore, examining their microstructure is crucial to adapt them into bioinspired solutions. In recent years, several new technologies for materials characterization have been developed, such as X-ray Microtomography (μCT) and Finite Element Analysis (FEA), allowing newer possibilities to visualize the fine structure of plants. Combining these technologies also allows that the plant material could be virtually investigated, simulating environmental conditions of interest, and revealing intrinsic

F. L. Palombini · J. E. de Araujo Mariath
Laboratory of Plant Anatomy – LAVeg, Graduate Program in Botany – PPGBot; Institute of Biosciences, Federal University of Rio Grande do Sul – UFRGS, Av. Bento Gonçalves, Porto Alegre, RS 9500, Brazil
e-mail: jorge.mariath@ufrgs.br

F. M. Nogueira (✉)
Laboratory of Algae and Plants of Amazon – LAPAM, Federal University of Oeste do Pará – UFOPA, Campus Oriximiná, Rodovia PA-254, 257, Oriximiná, Brazil
e-mail: fer.m.nogueira@hotmail.com

F. L. Palombini · B. F. de Oliveira
Design and Computer Simulation Group – DSC, School of Engineering, Graduate Program in Design – PGDesign, Federal University of Rio Grande do Sul — UFRGS, Av. Osvaldo Aranha 99/408, Porto Alegre, RS 90035-190, Brazil
e-mail: branca@ufrgs.br

© The Author(s), under exclusive license to Springer Nature Singapore Pte Ltd. 2022
F. L. Palombini and S. S. Muthu (eds.), *Bionics and Sustainable Design*, Environmental Footprints and Eco-design of Products and Processes,
https://doi.org/10.1007/978-981-19-1812-4_2

properties of their internal organization. Conversely to the expected flow of a conventional methodology in bionics—from nature-to-project —besides contributing to the development of innovative designs, these technologies also play an important role in investigations in the plant sciences field. This chapter addresses how investigations in plant samples using those technologies for bionic purposes are reflecting on new pieces of knowledge regarding the biological material itself. An overview of the use of μCT and FEA in recent bionic research is presented, as well as how they are impacting new discoveries for plant anatomy and morphology. The techniques are described, highlighting their potential for biology and bionic studies, and literature case studies are shown. Finally, we present future directions that the potential new technologies have on connecting the gap between project sciences and biodiversity in a way both fields can benefit from them.

Keywords Biomimetics · Biomimicry · X-ray microcomputed tomography · Finite element analysis · Monocotyledons · Bamboo · Bromeliaceae · Micro-CT

1   Introduction

Based on billions of years of life and natural selection, bionics takes advantage of the attributes that made each species successful to this day. This field of applied sciences is fundamentally associated with the development of bioinspired solutions based on a certain aspect or characteristic extracted from the natural world. Such solutions can be used on and applied into projects from a variety of fields [6]: from product design [60, 92], architecture [61, 70, 101], engineering [20, 95], and materials science [31, 144, 145] to biomedicine [53, 128], management [105], and robotics [100]. Having its origins focused on applications in military projects, mainly marked by the development of the SONAR, according to [136], bionics can be defined "as the study of living and life-like systems with the goal to discover new principles, techniques, and processes to be applied in man-made technology". One of the forerunners of this branch of development was the then-Major from US Air Force Jack Ellwood Steele, who coined the term "bionics" in 1958 through the Greek term βίος, from "life", and the suffix ʾικός, as "related to" (or "pertaining to", "in the manner of"), in order to promote it as a new science [42]—the same etymology is also found, for example, in words like "mechanics" (related to machine), "mathematics" (related to knowledge or learning), "dynamics" (related to power), and "aesthetics" (related to the perception of the senses). Besides, the term "Biomimetics" (as what is "related to the imitation of life") was also proposed by the polymath Otto Herbert Schmitt, also in the late 1950s [135]. Either way, despite preferences, both terms are equally important and are known to be referred to the same connections between bioinspiration and applications, aimed at benefiting society:

> Let us consider what bionics has come to mean operationally and what it or some word like it (I prefer bio-mimetics) ought to mean in order to make good use of the technical skills of scientists specializing, or, more accurately, despecializing into this area of research. Presumably our common interest is in examining biological phenomenology in the hope of

gaining insight and inspiration for developing physical or composite bio-physical systems in the image of life. [114]

From that moment onward, it can be emphasized the realization of different scientific events that sought to disseminate these new areas of research—like the first Bionics Symposium, entitled "Living Prototypes—the Key to New Technology", which occurred in Dayton (OH, USA), in 1960 [38]. More recently, in addition to those terminologies' appearance in popular media, more terms have gained attention, particularly in research papers. Biomimicry and bioinspiration are examples of other names that can be emphasized. Therefore, despite those and more popular terminologies, all of them tend to represent the same goals of performing technical analyses of natural elements with aims at their application through multiple technologies for innovative results. Considering new research findings in scientific publications, the growth in bionic-related papers is remarkable. Figure 1 shows the annual number of papers published that contains some of those "bio*-related" terms, from 1990 to 2020 including bionics, biomimetics, biomimicry, bioinspiration, etc. It is noteworthy that in the past decade the average number of bio*-related papers has grown from around 3,000 to 10,000, yearly, according to the Web of Science™ platform. And not only when considering scientific publications, the presence of bio*-related works is noticed. When analyzing the total number of published patents, following data retrieved from the same platform, in 1990, approximately 0.002% of all globally published pieces were related to bionics. As for 2020, that segment increased to about 0.074%. While the annual number of published patents increased about 14 times from 1990 to 2020, the annual number of bionic-related patents increased 395 times, i.e., the growth rate of patents that use nature as a source of inspiration is over 26 times greater than all other areas in the past three decades. That emphasizes the impact R&D in bionics has on innovation.

From trabeculated bone tissue and biomechanics of animals to the cellular arrangement and seed dispersal of plants, nature has come a long way in discovering

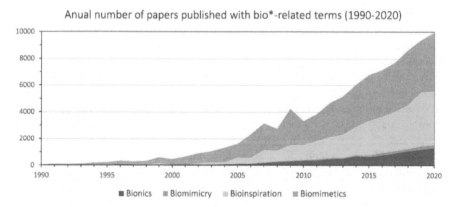

Fig. 1 Total number of papers published containing bio*-related terms, from 1990 to 2020, and registered in Web of Science™ platform

different measures to help to protect from predators, attracting pollinators, reinforcing against weather conditions, moving efficiently, saving energy, among many others. Regarding the characteristics that make biological compounds, materials, structures, arrangements, or systems so interesting for applications, they can be found in all magnitudes of scales: from the biochemical composition of biopolymers (e.g., collagen, melanin, suberin, cellulose, lignin, etc.) [37, 131], to the complex equilibrium of entire ecosystems [110]. Hence, there are equally many ways to investigate a certain natural characteristic and, depending on the level of which it is presented or intended to be inspired on, we have to adopt particular approaches, in terms of techniques, equipment, and research protocols to analyze it properly. If researchers choose wrong or incomplete investigation procedures, there is a chance that some vital details regarding the desired characteristic would remain unresolved or unexplored, impairing the correct understanding of the "hows" or the "whys" of the examined biological sample, and consequently the correct application of it in a bioinspired project. Another key aspect to be considered during this type of investigation is being able to count on a transdisciplinary team. Even if the correct approaches and techniques are followed, the natural object of study would also require the interpretation of a biologist. Much more than assisting on the explanation of some specific analyzed characteristic, a biologist can also provide new directions from which the research can proceed to address the issue at hand. Therefore, finding the most adequate methods and specialists is as important as the objects of study themselves.

One of the most fascinating biological objects of study for their complexity in a multitude of levels of hierarchical organizations is plants. They can be considered fantastic examples of how an organism can adapt to several adversities of external conditions in order to prosper in its environment. For instance, by efficiently depositing and shaping biopolymers, cells, and tissues where they are most mechanically needed. In his work "Plant Biomechanics—An Engineering Approach to Plant Form and Function", Karl [82] emphasized that "plants are the ideal organisms in which to study form-function relations". Their shape is primarily based on the direction and rate of their growth [59], being this immediately related to characteristics of each type of biological component, as well as how they are arranged throughout the individual. Such arrangement influences several physiological features of plants, from molecular and chemical, to physical and mechanical. And it is the complexity of this organization that gives plants their fascinating attributes as materials. Mechanically, for instance, the hierarchy levels of plants, from the cell wall to the organization of supporting tissues, are the foundation of their properties [112]. The gradient distribution of the material in the stem of plants can also be found in multiple scales, from the macroscopic level to the microscopic and molecular ones [120]. One of the main examples of plants with outstanding and highly efficient structural properties is bamboo. Macroscopically, the hollow stem of this monocot is periodically divided into solid cross sections, called the diaphragm, in the nodal regions. These perpendicular reinforcements act like ring stiffeners, increasing the overall flexural strength of the plant, by preventing the stem from failing by ovulation of its cross section during buckling or bending [138]. When a ductile thin cylindrical structure is bent, its failure is usually reached in buckling when part of its length collapses, due to the initially circular section becoming elliptical and unstable [129], i.e., the longitudinal forces of local tension and compression in the tube also tend to ovalize its cross section, thus precipitating the elastic deformation and reducing the flexural stiffness, known as the "Brazier effect" [10]. Galileo Galilei (1564–1642) demonstrated in 1638 that materials applied in the periphery—rather than in the center—of constructions provided more resistance to bending forces, using hollow stacks of grasses to illustrate it [83], which is aligned to the effect described almost 300 years later. Microscopically, when analyzing the stem's cross section, the scattered distribution of the vascular bundles in monocots, named atactostele [29]—in which they are bigger and dispersed in the inside and smaller and clustered in the outside—concentrates more material, with a higher relative density, in the regions under which they are subject to the greatest stresses [68, 92]. The higher the relative density, the higher the elastic modulus, therefore the plant is stiffer on the external side, increasing the moment of inertia of the stem [37]. Despite the sclerenchyma bundles being scattered, the parenchyma ground tissue can act as a matrix, in a composite analogy, by distributing stresses throughout the plant, even if a local pressure is applied [92]. Furthermore, combining the sclerenchyma bundles with the low density, foam-like ground parenchymatic material [22], this gradient distribution in the cross-section direction also contributes to the plant's high efficiency, i.e., considerable stiffness and strength along with reduced weight. Even the actual geometry of each sclerenchyma bundle was verified as having a role in the performance of the stem by increasing the local compressive strength of the fibers [95]. Despite cylindrical structures being considered more efficient in bending when the direction of the lateral load is unknown [37], the different shapes of the vascular bundles of bamboo [69] can be particularly efficient due to their radial orientation in the culm. Given the fact that individual fiber bundles tend to locally bend toward the center, during the universal bending of the stem they contribute by micro-stiffening the structure [95]. Therefore, most failure modes noticed on bamboo culm during bending or compression is due to fiber splitting [117], when the sclerenchyma is detached longitudinally from the parenchyma [41]. And even this failure characteristic is somewhat reduced in the plant by the arranging of the vascular bundles in the nodal region, at which they translate horizontally and interlace themselves in the longitudinal bundles, when connecting to new branches or leaves, and thus preventing them from detaching completely [96, 99, 143]. All those features represent a number of possibilities to be applied on bioinspired solutions. Consequently, once studied, they can be explored, either individually or combined, as a source of inspiration for the development of analogous bionic projects like beams, columns, and thin-walled structures designs, bioinspired in bamboo [18, 33, 57, 71, 146]. And, as previously mentioned, prior from applying the characteristics of a certain biological material in the design of a bionic or bioinspired project, the sample first have to be investigated, by means of observation, analyses, or simulations. In order to benefit from such complex features of plants, engineers and designers have to consider how to approach the investigation, by the selection of the appropriate equipment as well as count with the presence of a specialist from biological sciences—just like in most bamboo-inspired design cases, a botanist was involved.

It is easy to link up that the growing knowledge in plant anatomy and plant morphology was due to technological advances in biological imaging (bioimaging). Considered of primary importance for all lines of research in plant sciences [29], plant anatomy has its first investigations dating back to the seventeenth century [27]. In the middle of that century, with the first developments of early microscopes, Robert Hooke (1641–1712) introduced the term "cell" in his book "Micrographia" [54]—derived from the Latin "*cella*", as a small enclosed space [37, 141]—when observing cork samples, in reference to the small cavities surrounded by walls [27]. A few years later, Nehemiah Grew (1655–1703) first described plant tissues in his book "The Anatomy of Plants" [40], published in English in 1682 [64]. While writing his book, Grew actually searched for the advice of Hooke in explaining how cells expand and why some tissues are stiffer and stronger than others [83]. Grew also identified that plant tissues needed to be conceptualized in 3D because they were composed of microscopic structures with distinct spatial relationships [13]. In 1838, botanist Matthias Schleiden (1804–1881) expanded the characteristics of cells by reporting that all plant tissues consist of organized masses of cells, which was later extended by zoologist Theodor Schwann (1810–1882) to all animal tissues, and these basic units do not differ fundamentally in its structure from each other. Therefore, in 1838, the cell theory of Schleiden and Schwann was proposed for all forms of life [28, 64, 137]. And this was also the formal statement of "cell biology" as a newly created and still unexplored field of science [1]. The next major step—the understanding of its basic principles and functions—occurred when advances in staining, lighting techniques, and optics increased the contrast and resolution view of the cell's internal structures, especially after the introduction of the transmission electron microscopy (TEM) at the beginning of the 1940s [1]. The evolution of photomicrography followed the studies of Thomas Wedgwood (1771–1805), William Talbot (1800–1877), and others, where images were captured with solar-based microscopes (Overney and [88]. In modern times, the digital system and image processing have overcome the human eye limitations, like low brightness and small differences of light intensity in a luminous background, so the modern CCD (charged-coupled device) and CMOS (complementary metal-oxide semiconductor) had become an integrating part of the microscope system. Besides, since the acquisition of images is in digital format, different software became able to correct brightness, contrast, and noise in the images, allowing the reduction of artifacts and limitations of the optical system and the human eye. More recently, advances in bioimaging technologies have pushed the boundaries on new interpretations and insights regarding the structure and morphology of biological samples. In the modern timeline, we can point out the discovery of X-rays in 1895, by German engineer and physicist Wilhelm Röntgen (1845–1923) [109], which rapidly led to the proliferation of this type of equipment in medicine, by allowing a clear visual separation between soft and hard tissues [4]. Later, physicist Ernst Ruska (1906–1988) and electrical engineer Max Knoll (1897–1969) presented the first full design of the electron microscope [63], after years of development of electron-based imaging [39]. In 1936, the first commercial transmission electron microscopy (TEM) was presented in the UK, despite not being completely functional, and only becoming available by commercial companies after the end of World War II [139]. In 1935,

Knoll also presented a prototype of the first scanning electron microscopy (SEM), which was improved by Manfred von Ardenne in 1938, and became functional by Vladimir K Zworykin and his research group at the Radio Corporation of America in 1942 [72]. Even though both TEM and SEM techniques required specific protocols for sample preparation [2], the technology became one of the most important pieces of science equipment for the study of materials in plant sciences and many other fields up to the present day [15]. However, important details of microstructures in TEM and SEM still required trained eyes in order to be reconstructed and volumetrically visualize the observed specimen in 3D [50]. More recently, preliminary works employing Atomic Force Microscopy (AFM) were published in the mid-1990s, mainly focusing on the topography of pollen grains [19, 111]. However, the authors stated that despite the high resolution, the 3D presentation of the surface, and the absence of specific preparation methods, generally AFM was still limited by its depth of field. The work of Minsky searching the three-dimensional structure of neural connections led him to propose in 1957 the principles of confocality, and since then, the basic principles are used on all confocal microscopes. The modern confocal laser scanning microscopy has been a powerful tool in elucidating 3D complex internal structures of cells and tissues, allowing studying dynamic processes visualization in living cells, like rearrangements of cytoskeleton and chromosomes. Despite its benefits, the technique has a limited penetration depth ranging from 150 to 250 μm [23, 89]. It is noteworthy that the methodology choice is an important step to be followed by the research team, considering each one has advantages and limitations owing to the image resolution, sample size, and physical preparation, in addition to the need for sectioning [48]. Simultaneously, resources were applied to the progress of 3D imaging. In this regard, mathematical models were also critical for allowing us a better comprehension of the complexity of the biological world. In 1917, Austrian mathematician Johann Radon (1887–1956) proved [106] that an "infinite" object (N) could be reconstructed based on a "finite" number of projections ($N - 1$) of it [116]. Such mathematical and physical bases, especially with the works of English electrical engineer Godfrey Hounsfield (1919–2004) and South African physicist Allan Cormack (1924–1998), resulted in the development of the first feasible X-ray computed tomography (CT) scanners, in the early 1970s [56].

In addition to the findings in biomedical sciences, the dissemination of new 3D observation technologies added new levels of comprehension in many disciplines, including traditional fields such as plant sciences. Consequently, once enough information from a certain biological sample is gathered, improvements and entirely new concepts on bionic designs can be proposed. However, despite the "biology–technology–bionics" direction being well established, its orientation can be questioned. This chapter addresses how technological advances are contributing not only to innovation in bioinspired design but how bionics—as a multidisciplinary approach—is acting as the driving force behind discoveries in plant anatomy and morphology. First, some state of the art, 3D technologies for analyses, and simulations are presented, with a focus on X-ray microtomography (μCT) and Finite Element Analysis (FEA), including examples of recent research studies that are benefiting from both multidisciplinary teams and technological advances. Secondly, the role of biomedical imaging in botany investigation is discussed, showing the importance of 3D technologies for new insights, and how they are improving biodiversity as a whole. Finally, the last topic gives an overview of upcoming technologies and the panorama of future studies in plant sciences and bionics.

2 3D Technologies: Analyses and Simulations

As the awareness of the inherent complexity of biological materials increases—particularly regarding plants—the need for 3D technologies for observing and analyzing samples became more evident. During the pioneer studies in plant anatomy of Hooke and Grew the need for a higher optical magnitude was necessary to allow visualization of small structures like tissues and cells. However, in the past decades, this demand has shifted to also understanding the 3D hierarchy and organization of those fine details, leading to a new comprehension of plant anatomy and morphology. In addition, in-depth functional analyses and simulations have been leading to insights regarding plant physiology, as well as mechanical and physical responses to external stimuli.

Advances in technologies are making them more accessible to researchers from many fields, which contributes directly to findings in many areas. Generally, plant materials have the intrinsic characteristic of presenting a fine, detailed architecture, a complex arrangement of structures, multiple levels of hierarchy, as well as a combination of delicate and soft with stiff and hard cells, tissues, and organs. One of the main ways of documenting, analyzing, and transmitting knowledge in botany is by means of illustrations, which can be found in a major part of specializing literature [27]. In possession of instruments such as microtomes and microscopes, anatomists are able to observe thin sections, which reveal valuable information about the constituent tissues and cells. The registers of each individual section can be done manually or digitally, in order to understand the entire structure. However, traditionally, the volumetric reconstruction of the general aspect of the biological material is done based on the deduction of the researcher, through the interpretation of the observations of the sequential sections [115]. Therefore, having the ability to investigate the biological material with a high-resolution and non-destructive technique is key not just to ensure the integrity of the sample but to facilitate the comprehension of its true morphology [93]. In this topic, we cover two of the main 3D technologies for investigating plants, both for qualitative and quantitative analyses, along with numerical simulations.

2.1   X-Ray Microtomography (μCT)

With the first developments of X-ray-based imaging in 3D, with computed tomography (CT) equipment in the mid-1970s [55], the technique has become a well-established and regularly employed modality in current diagnostic radiology in hospitals worldwide [45]. Shortly after its popularization, CT imaging has evolved with a focus on improving the resolution of the scans, giving birth to the now-called X-ray microtomography (or microcomputed tomography), or μCT. Stock [126] highlights that μCT technology has developed at a slower rate compared to that of clinical CT, because of clear economic reasons—i.e., investments in more expensive technologies are far more acceptable in hospitals, with straightforward benefits to patients than in research centers, where μCT is most commonly employed. However, it was with the need for investigating small animals for the study of human diseases that biomedicals started to emphasize the application of higher resolution CT scans, leading to investments in commercial μCT equipment, in the mid-to-late-1990s [126]. The development of more affordable lab-scale μCT technology was also a crucial step for the democratization of the technology. Once restricted to a small number of large synchrotron-based research facilities, high-resolution X-ray scanners have gained attention to major manufacturers, such as Bruker™ (Kontich, Belgium), Zeiss™ (Pleasanton, CA, USA), North Star Imaging™ (Rogers, MN, USA), General Electric™ (Wunstorff, Germany), and others.

It is noteworthy the recently growing number of research papers that utilize X-ray microtomography as the primary source of investigation methods, in a variety of fields [91]. Assessing them is rather difficult, though, due to the numerous nomenclatures used to define the technique—e.g., X-ray microcomputed tomography, X-ray microtomography, high-resolution computed tomography, X-ray nanotomography, X-ray microfocus, just to name a few—and even several abbreviations of the method: μCT, micro-CT/micro-CT, nCT, HRCT, HRXCT, and others. Even between μCT manufacturers, the terminologies used are diverse: micro-CT (Bruker™), X-ray microscopy (Zeiss™), X-ray system (North Star Imaging™), microfocus CT (General Electric™), etc. Still, the advantages of employing the technique in the research methodology workflow in plant sciences are vast. Firstly, there is the capability of digitalizing a biological sample volumetrically, instead of just superficially, similar to what can be found in other observation techniques, such as scanning electron microscopy and atomic force microscopy. This "bulk", volumetric scanning is particularly important to avoid multiple passes of superficial digitalization in the same sample (like with SEM, o AFM), where each one is performed with the removal of some material. Second, differently from conventional clinical CT, μCT utilizes a smaller X-ray focal spot size and higher resolution detectors, among others, achieving a much higher level of details in the obtained virtual model. Evidently, to avoid working with huge bioimaging datasets, the higher the chosen spatial resolution, the smaller the field of view needs to be. Third, just like clinical CT, nevertheless, μCT also allows the digital sectioning of the 3D model in virtually any plane, allowing the researcher to investigate the sample without the need to destroy it. When working with fragile or delicate specimens, like many plant-based materials, this is utterly important to ensure the integrity of cells, tissues, and structures. Forth, besides the realization of morphological analysis, binarization and segmentation procedures can be followed for the division of the 3D model into regions of interests (ROIs), which allows quantitative data to be collected and assessed, such as counting (number of cells), sizing (volumes, lengths, surface areas), and others, utilizing a number of proprietary software suites or open-source alternatives like Fiji, a distribution of ImageJ [113]. Many μCT types of equipment also include in situ capabilities of micro- and nanomechanical testing—including compression, tension, and indentation—while scanning, for the sample to be digitalized throughout the experiment. Fifth, unlike many lab-based techniques used in plant sciences, sample preparation often requires little to no additional steps prior to scanning—e.g., fixation, inclusion, staining, and metallization. The usage of contrast agents (e.g., phosphotungstate, osmium tetroxide, bismuth tartrate, and others) is sometimes followed, however with up-to-date μCT features, such as phase contrast and dark field, soft tissues can be distinguished more easily in untreated samples.

The functioning of μCT scanning relies on the effect electromagnetic waves in the X-ray spectrum has on the matter, and the digitalization procedure is often confused by researchers. Fundamentally, image acquisition is performed by the irradiation of the sample by the X-ray source, which is relatively attenuated by different local densities and captured by a detector (scintillator coupled to a CCD sensor), forming a 2D projection of the sample, perpendicular to the X-ray beam. The sample is then rotated, and a new set of 2D projections is captured. After a full—or half—rotation, all projections are combined with a particular CT algorithm that verifies regions with similar density patterns, merging them into a 3D representation. Finally, this 3D volume can be exported into a stack of grayscale slices—where the darker the value of each pixel the lower its density—which can be worked in the post-processing procedure, reconstructed in 3D—where each slice pixel would be treated as a voxel—and used for qualitative and quantitative analyses. In the X-ray tube, the higher the potential difference (voltage) between the cathode and anode, the higher the kinetic energy of the electrons and, consequently, the emitted photons. When striking the target in the X-ray tube (typically made out of tungsten due to its high atomic number and melting point), the electrons decelerate, producing Bremsstrahlung (continuous spectrum) and characteristic radiation (peak). In addition to tube voltage, the current—established by the electron fluency in the tube—and the exposure—defined by the time at each projection—are the main μCT parameters to considerer. In the short term, the higher the current-exposure time, the higher the number of photons focusing on the sample, which may reduce noise, but can also decrease visual resolution due to scattering, and the higher the tube voltage, the higher the X-ray energy and the broader its spectrum, which can be suited for larger samples, but may decrease the contrast levels due to saturation of the detector. As a rule of thumb, for plant samples, it may be interesting to define relatively small values for voltage (~70–80 kV) and current (~90–130 μA), based on the size of the specimen, due to the general low density of the material; besides, working with smaller samples is preferable for obtaining a higher resolution during digitalization, with less noise and more contrast ratio.

Further μCT physical principles are beyond the scope of this chapter and can be consulted in the works of [8, 45, 87, 126].

Many recent papers in plant sciences have been applying μCT as an innovative method for discovering or better comprehending features in a variety of samples. It is noteworthy still that, just like in bionic-related works, most research groups count on the presence of professionals with multiple backgrounds, like in engineering, physics, and design, to assist in the research workflow—ever since the set of acquisition parameters in the digitalization, to the post-processing and analyses steps. For instance [84, 86], studied the 3D architecture of congested inflorescences in Bromeliaceae that accumulate different amounts of water, in multiple Nidularioid genera (subfamily Bromelioideae). Authors segmented those regions based on high-resolution μCT images, and described their morphology, volume, and orientation in individuals, allowing the interpretation using a typology-based comparative approach of inflorescence development and branch patterns in this group that presents an obscured morphology to interpret with traditional methods. Due to the congested morphology of inflorescences, investigating their anatomical features and performing precise measurements would be rather difficult to realize via manually sectioning. Pandoli et al. [97, 98] explored ways of preventing microbial and fungal proliferation in bamboo, utilizing silver nanoparticles. Authors used μCT to analyze the distribution of antimicrobial particles in the vascular system of the plant, which contributed not only to increasing the protection of the natural material but to better understand the distribution of particles in its vascular anatomy. Brodersen et al. [11, 12] investigated the 3D xylem network of grapevine stems (*Vitis vinifera*) and found that these connectivity features could contribute to disease and embolism resistance in some species. The authors mentioned the benefits of applying μCT: "selected vessels or the entire network are easily visualized by freely rotating the volume renderings in 3D space or viewing serial slices in any plane or orientation" and that the automated method generated "orders of magnitude more data in a fraction of the time" compared to the manual one. Teixeira-Costa and Ceccantini [130] analyzed the parasite–host interface of the mistletoe *Phoradendron perrottetii* (Santalaceae) growing on branches of the host tree *Tapirira guianensis* (Anacardiaceae through high-resolution μCT imaging. Authors also highlighted the difficulties in "imbedding and cutting lignified and large materials", therefore manually performing such complex 3D investigation would require a "huge series of sequential anatomical sections"; the authors even mentioned the previous work of [17], where the anatomy of the endophytic system of *Phoradendron flavescens* required hundreds of anatomical sections. Investigating the intricate distribution of fine arrangements in 3D is one of the main advantages of employing μCT for the study of plants. More recently, for example, Palombini et al. [96] studied the complex morphology of the vascular system of the nodal region of bamboo with the technique. When comparing their findings with previous serial sections based on illustrations (Fig. 2) it is clear that the scattered arrangement of the atactostele distribution of monocots is hard to be spatially understood in 3D by using just anatomical sections (Fig. 2a). Even if the segmentation procedure after μCT imaging acquisition can also be labor intensive—authors had to manually select each one of the hundreds of vascular bundles, separating them into the corresponding

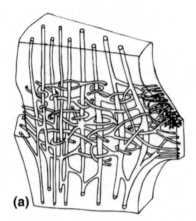 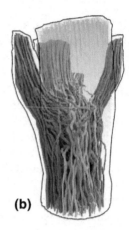

Fig. 2 Technological improvements for detailing the complex anatomy of vascular bundles of the nodal region of bamboo: **a** hand illustration of nodal reconstruction based on serial cuts by Ding and Liese [21], **b** actual morphology based on X-ray microtomography of the nodal region, where the bundles of each secondary axis are shown with a different color, and the ones of the primary axis are faded

axis (Fig. 2b)—due to the difficulties in automatizing the process, at the very least the obtained virtual morphology remained intact and undamaged, allowing it to be anatomically accurate represented.

2.2 Finite Element Analysis (FEA)

As seen before, the segmentation process of a μCT-based investigation permits the 3D visualization of a specific ROI of a sample along with morphologically accurate qualitative and quantitative analyses. Once the region of interest has a segmented mask, it is considered binarized, i.e., it is no longer represented by means of a grayscale image, at which the contour of a certain region gradually fades as it approaches its physical end—as we can originally find in an unprocessed μCT stack—but as a set of black and white voxels. What it means is that a region in a binarized stack can be spatially defined with the precision of a voxel: if the binarized, segmentally masked region is set, it can be distinguished by white voxels, otherwise, it is viewed as a black region, just like the background in μCT. And it is this absolute division between black and white voxels that make a region discrete, allowing further calculations and analyses to take place. Furthermore, since the masked region is spatially—i.e., mathematically—defined, it can be exported to be used for other purposes. One of the main file extensions used in μCT post-processing is STL, a superficial mesh with universal compatibility. For instance, a μCT-based STL mesh can be 3D printed for educational uses [93], allowing a microscopic, morphologically accurate region to be significantly magnified and physically handled by students. Like most biological materials, plants exhibit an intrinsic and complex hierarchical level of elements, arrangements, shapes, and materials that extend from the macro- to the nanoscopic world. Despite being fundamentally constructed by a small number of basic components, it is their build-up that adds natural materials to their remarkable properties and makes them desirable as both a straightforward building material and a source for bioinspiration projects [3, 123]. However, their heterogeneity and intricacy also make biological materials difficult to be accurately investigated, for instance, in terms of predicting their mechanical performance. This is the main reason why they present a natural variability in properties, and thus often require a sufficient number of individuals for the data sampling to be meaningful [46]. Therefore, another major possibility of employing 3D biological models obtained via μCT is their use in numerical analyses.

An important characteristic of scientific reasoning is the process of dividing a larger, complex problem into smaller, simplified components, which can then be separately resolved more effortlessly and, once combined, be used as an estimated solution for the original issue. If a problem—or a numerical model—is obtained with a finite number of components, it is called "discrete". On the other hand, if a model is considered so complex that it could be indefinitely subdivided, it would require it to be defined by assessing its infinitesimal parts, leading to the need for differential equations, and thus being called "continuous" [147]. This is the essence of the Finite Element Method (FEM) at which in the 1940s mathematicians, physicists, and particularly engineers worked on techniques for the analysis of problems in structural mechanics, and later applied them to the solution of many different types of problems [5]. Simplifying, the procedure is based on the division of a continuous problem into a finite number of elements, in the process called discretization, leading to the procedure known as Finite Element Analysis (FEA). It is worth noting that despite the assistance of desk calculators in the 1930s, FEA only had its most significant improvements with the upsurge of digital computing, at which a set of governing algebraic equations could be established and solved more effectively, leading to real applicability [5]. The origins of the subdivision of continuum problems into finite elements set are arguable, though. With the multiple solutions for the brachistochrone problem—i.e., finding the fastest descent curve between two non-superposed points with different elevations in the same plane—initially proposed by Johann Bernoulli (1667–1748), his brother Jacob Bernoulli (1655–1705) along with other known personalities like Galileo Galilei (1564–1642), Gottfried Leibniz (1646–1716), and later Leonhard Euler (1707–1783), it was observed that it would require reducing the variational problem into a finite number of equidistant problems [125]. Gadala [35] comments on even older possible roots of the discretization process, dating back to early civilizations, like Archimedes' (ca. 287–212 BC) attempts to determine measures of geometrical objects by dividing them into simpler ones, or the Babylonians and Egyptians search for the ratio of a circle's circumference to its diameter, i.e., π. As for the present day, numerous commercial software solutions are available, including Abaqus™ (Dassault Systèmes™, Vélizy-Villacoublay, France), ANSYS™ (Canonsburg, PA, USA), COMSOL™ (Stockholm, Sweden),

Nastran™, and others, for areas varying from structural mechanics to fluid dynamics, electromagnetics, and heat transfer.

Finite element analyses can be applied into a number of fields, from facilitating the resolution of well-defined CAD-based geometries to the simulation of specific, more abstract scenarios which otherwise could be considered much more difficult to achieve. The basic process of conducting an FEA is based on the definition of three essential sets of variables: geometry, constitutive properties, and boundary conditions. The first one is related to the shape, dimensions, and relative organization of the 3D model. As mentioned, the geometry can be obtained employing CAD software or, in the case of biological samples, via digitalization. Recently, μCT-based models are being increasingly used in FEA in many fields of biomedical research, allowing organic and complex geometries to be investigated and simulated in silico. In the process of discretization (Fig. 3), at which the original μCT image stack (Fig. 3a) is converted into a mesh suitable for FEA, two basic methods can be followed, voxel based and geometry based [9, 94]. The first one is the most straightforward due to the direct conversion of the image stack voxels into elements (Fig. 3b). As for the geometry based, the selected regions must first be segmented and then transformed into an FEA mesh (Fig. 3c). The main differences are that in the voxel-based meshing there is no need for segmentation processes, but the resulting mesh tends to be much

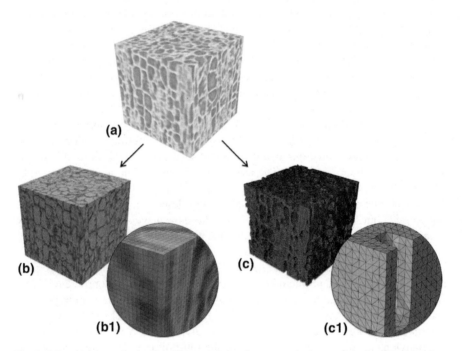

Fig. 3 Discretizing methods of μCT-based finite element analysis: **a** original μCT of bamboo parenchyma; **b** voxel-based discretization, with detail **c**1 of voxel mesh; **c** geometry-based discretization, with detail **c**1 of tetrahedral mesh

larger depending on the number of μCT slices in the stack and the resolution of each slice, i.e., the greater the stack size and resolution, the greater the number of voxels and elements in the mesh. As for the geometry-based method, once the mesh is defined automatically, users must also ponder the refinement-performance balance, whereas on the one hand a finer mesh would represent more accurately the original geometry, yet it would consume more computational resources. Another point worth mentioning is that the geometry-based mesh tends to be much more universally accepted in a variety of FEA solvers, consequently allowing a wider range of several types of analyses to be executed.

The second basic set of parameters required in any FEA is the constitutive properties. They are related to the physical characteristics of the geometry regarding its mechanical (elastic modulus, yield strength, dynamic viscosity...), thermal (thermal conductivity, specific heat capacity...), and electromagnetic (electrical conductivity, magnetic permeability...) properties, among others. Since the resulting μCT images are defined in the grayscale, at which the whiter the color the denser that region (Fig. 3a), the constitutive properties of a voxel-based analysis can be extracted directly from the images, like in Fig. 3b, where the warmer colors represent a denser material, i.e., higher values of mechanical properties were attributed. As for the geometry-based method, once the region is homogenized, so must be its properties for the analysis to be accurate [94]. The last main parameter to be configured is the boundary conditions, i.e., all the external parameters that affect the analyzed geometry, e.g., loads (including pressures, forces, temperatures, electric potentials...), displacements, and constraints, among others. Once defined all parameters, the analysis can be solved. FEA is a complex and comprehensive field of research and many other fundamentals and parameters should be considered, like non-linearities (geometric, constitutive, and boundary conditions), types of analyses (implicit, explicit, thermal, modal, buckling, vibration...), types of mesh elements (tetrahedra, hexahedra...), and more, which are also beyond the scope of the chapter, and in-depth information can be accessed in the works of [5, 65, 107, 147].

Despite being less commonly found in the plant sciences fields, the numerical analysis still appears in many important research studies, particularly when combined with biomedical imaging. As discussed before, the main advantages of a μCT-based FEA in plants are the ability to carry out investigations where (i) an accurate morphology is needed; (ii) the potential of conducting experimental analysis (i.e., with real samples) is somewhat difficult or impossible to accomplish due to limiting factors like diminished sample size, lack of control of boundary conditions, restrictions to perform experiments with quick modification of variables, etc. Still, similar to the process of applying the upcoming obtained knowledge in bionics, the investigation of μCT-based FEA itself also benefits from a team with multiple backgrounds, as may be seen by some examples. For instance [92], assessed the structural role of bamboo parenchyma as an important tissue in the plant using μCT and FEA. Due to its matrix-like behavior, in a composite analogy, the ground tissue distributes local stress into a larger region so the sclerenchyma bundles can act like reinforcements for greater stiffness and strength. The authors also observed, using the same set of techniques [96], that bamboo tends to spread compression stresses into secondary branches in order to preserve the main axis. Due to the resolution of workable fibers as well as the complexity of the vascular arrangement in the plant, only μCT-based FEA could be used for this type of analysis. Nogueira et al. [85] presented the first known literature application of a heat-transfer FEA based on μCT of biological material. In this research, the authors utilized high-resolution scanned images of a Bromeliad inflorescence to investigate the role of the accumulated water in the inflorescence tank of the plant, identifying it as acting as a thermal mass, by absorbing the external temperature and preventing internal structures to overheat. This feature results in the protection of the inflorescence and flower components, preserving it without the appearance of injuries and necrosis in bracts and leaves. Even though the authors also utilized an experimental procedure to complement the findings of the numerical analysis, the in silico tests were crucial for a fine assessment of the inflorescence internal structure's temperatures as well as the replication of the tests with the modification of multiple variables. Forell et al. [30] explored ways to prevent lodging at maize stalks for bioenergy, by means of altering morphological characteristics. Using μCT-based FEA, authors could digitally modify individual stalk variables like diameter and rind thickness to assess its performance modifications. By highlighting the benefits of using non-traditional, engineering-related techniques like FEA in plant analyses, authors state that "collaborations between plant scientists and biomechanical engineers promise to provide many new insights into plant form and function".

3   Bionics: From Design to Biodiversity

As seen before, bionics or biomimetics is fundamentally based on the relationship of knowledge interchanging between natural and applied sciences, and how it can be achieved is via technology. Not only a better understanding of organisms and biological materials can be obtained with unusual techniques of observation and analysis, but interdisciplinary teams in the context of bionics or biomimetics are also ensuring a more sustainable development [79], which also reflects into gains for biodiversity as a whole. From unraveling signals used by flowers to attract pollinators [102] to assessing the role of parenchyma wall thickness for the mechanical properties of bamboo [22], techniques considered almost exclusive to engineering, design, and architecture are unveiling new applications in the biological world. Likewise, technologies originally intended for the development of new bioinspired products are also a way to promote unexpected insights into the natural world [121], thus paving the street two ways: from nature-to-project and project-to-nature. This concept is also known as "reverse biomimetics" which was defined in a guideline for biomimetics from the Association of German Engineers (VDI or *Verein Deutscher Ingenieure*):

> Over the last few years, it has become apparent that knowledge gained through the implementation of biologically inspired principles of operation in innovative biomimetic products and technologies can contribute to a better understanding of the biological systems. This

relatively recently discovered transfer process from biomimetics to biology can be referred
to as "reverse biomimetics". [134]

By employing a "two-way bionics" like philosophy, research groups are able
to discover new tools, techniques, equipment, and methodologies, that otherwise
may never be considered for solving scientific inquires in a variety of disciplines.
Different from interdisciplinary, the transdisciplinary approach lies not only in the
interconnection among different disciplines, but in the total removal of the boundaries
between them [81], i.e., treating it as a uniform, homogeneous macro-science, where
no method is unique and immutable. And by considering the applications of digital
technologies, enabling the exchange in knowledge between fields of sciences is
becoming more accessible than ever, particularly regarding bionics and biomimetics
[62]. 3D technologies and numerical simulations are becoming more "integrated" and
more "integrative" in the scientific problem-solving workflow, iteratively questioning
the object of study and the object at which the natural characteristic is supposed to
be applied. "Within this so-called process of reverse biomimetics, the functional
principles and abstracted models (typically including finite element modeling) are
repeatedly evaluated in iterative feedback loops and compared with the biological
models leading to a considerable gain in knowledge" [122].

An organism, an animal, or a plant is a package of its adaptative history, written
in its genes, and seen as its phenotype. Each one of them brings itself its ecological history, strategies of reproduction, responses to the environment, and characteristics that make this unique organism well succeed today. In the same way that
our knowledge about the relationship in the life tree has grown with advances in
the techniques and analysis of molecular systematics, the understanding of biological morphology was also improved using new technologies, like confocal laser
scanning microscopy, nuclear magnetic resonance imaging, microcomputed tomography, and others. Nevertheless, knowledge about biodiversity is not a one-way street,
and related or even distant areas end up benefiting from technological advances in
different fields. For example, several well-succeed lineages distributed today have
some morphological feature associated with their evolutionary success, this morphological adaption or phenotype is also called "key innovation" [58]. This "new" feature
could occur in a specific lineage, like the evolution of the tank habit in Bromeliaceae
[118], or could be recovered to a large group, as the diversification of different flower
morphology among the Angiosperm lineages [26, 142]. Recovering these traits'
origins is one of the main challenges to reconstructing the life history in different
groups, and the fossil record is an essential step to our understanding of evolution
and diversification of these features, especially in groups that have low preservation
or do not have totally preserved specimens, like the monocotyledons [74, 119]. In
this sense, the use of μCT have been changing the fossil analysis, not only in plants
but in different fields of biodiversity, since using this tool is possible to visualize
internal delicate features without the need for a destructive or invasive preparation
[51, 127]. Using this approach, a new fossil botanical genus was proposed, *Tanispermum*, related to the extant Austrobaileyales and Nymphaeales, based on some

particular features of the seeds [32]. In the same way, [36], analyzing an inflorescence fossil of Fagales, present several distinct features of flower components, the inflorescence arrangement, and characteristics of the stem useful to understand the diversification of the stem, the evolution of pollination modes, and other character evolution in Fagales. On the other side of biodiversity research [14], studied the semicircular canals of the endosseous labyrinth of living and fossil Archosauria, represented today by birds and crocodylians, in order to understand several aspects of locomotor system diversification, and if shapes variation of this structure could be related of locomotor changes, like flight, semi-aquatic locomotion, and bipedalism. The authors bring to light several aspects of this structure evolution, showing high divergence in this trait and that the differences in the semicircular canal are related to spatial constraints among the analyzed lineages, besides the higher degree of divergence in this structure appeared early in the divergence of bird-crocodylian. The most interesting aspect of these discoveries in biodiversity research using 3D technologies, like μCT, is how significant three-dimensional analysis is for understanding each group. This means that the individual or the structure is three dimensional and the understanding of its form, function, physiological process, size, strain limitation, and growth form is difficult to be inferred by only using 2D images.

Indeed, images are everything, especially nowadays, and all of the images generated by these new technologies are in digital format. This information brings to us another important topic to three-dimensional approaches, which is the "Digitalization of the Biodiversity Collections" and how 3D technologies could be associated with the storage, shared, and popularization of these collections. The natural collection worldwide holds billions of specimens of biodiversity, including information like taxonomic position, geographic location, ecological, and temporal data [90]. In recent years, many institutes and initiatives concentrate efforts to digitalize specimens of natural collections. The reasons are the most diverse, like understanding the taxa distribution through time, climate change, the dynamic of invasive species, and others [75, 90]. One of these efforts is coordinated by the Rio de Janeiro Botanical Garden, in partnership with the CNPq—Brazil (National Council for Scientific and Technological Development), with the so-called "REFLORA" program. Its goal is the historical rescue or the repatriation of the specimens of the Brazilian flora deposited in foreign herbaria, through the digitalization with high-resolution images used to construct the "Virtual Herbaria Reflora". Besides, the program also includes the digitalization of the Brazilian Herbaria, allowing that researchers worldwide work in this platform, as they do in the physical collections [108]. This step of digitalization, which includes the digital image of the specimen and associated metadata, like taxonomic position and geographic location, is the first step to digitalizing a natural collection. However, with the improvement of technologies, like computed tomography, those specimens digitalized could provide much more details complementing this digital imaging, so the data associated with the specimen using, for example, μCT can be virtually analyzed, dissected, and manipulated by researchers worldwide [47]. Furthermore, the μCT digitalized models could be used for 3D printing, which can be employed as a didactical tool or for sharing these specimens. Recently [7], proposed a 3D model equivalent of a real and unique object, called VERO, which stands for "Virtual Equivalent of a Real Object", being proposed as a non-fungible token (NFT). In this model, the VERO and the NFT are interlinked, i.e., through this system, the VERO is transformed into a type of NFT. In accordance with the authors, "VEROs can be produced by museums to enable recreational collectors to own NFTs that represent virtual versions of the objects housed within museum collections". In this way, the idea of having the natural collections digitalize through high-resolution morphology using 3D technologies of the specimens also presents some opportunities for the popularization of the specimens held in these collections, especially that ones that present risk of degradation, such as fossils and unique and delicate specimens. Another major contribution of setting VERO is the possibility of raising funds for maintaining vulnerable or under-resourced collections, museums, and institutions. By defining it as NFTs, with the use of blockchain technology, buyers can both (i) ensure it is an original artwork, by preventing forgeries, which ownership can be traced back to its institution, and (ii) being a digital object, its trade avoids complicated copyright issues that can be quite common with physical objects [7].

4 Conclusions and Future Directions

Bionics and biomimetics should be considered as a means to obtain interchangeable pieces of knowledge between the "project" and the "object" of study. By blurring the lines dividing the work activities of scientists of classical fields and design engineers, we can gain access to new ways to promote more sustainable development, either by (i) better comprehending the functioning and requirements of the natural world as a way to reduce the human impact in nature and furthering healthier biodiversity, and by (ii) creating more environmentally sound projects, which includes more efficient materials and structures, and more ecologically integrated organizational systems. Increasing access to state-of-the-art technologies is a way to ensure a more transdisciplinary approach between sciences, favoring unified teaching and research [91]. μCT has been shown a versatile tool in plant research, due to its non-invasive character and, in most cases, due to the lack of intensive preparation requirements for samples. Some of the advantages of using this approach include high morphological resolution of complex and delicate internal structures, like visualizing the complex vascular system in the nodal region of bamboo [96], obtaining details of the 3D architecture, and the modeling of gas exchange in leaves [24, 52, 73], in addition, some opportunities as in vivo measures and experiments are possible [25]. However, the technique usually requires a long image acquisition time, and the ionizing radiation dosage difficult to repeat the imaging acquisition of some delicate specimens. Besides, some plant tissues like meristems present low density which reduces X-ray attenuation-based contrast, thus requiring a sample preparation to increase the contrast [124].

Available methodologies for imaging samples with different compositions that follow three-dimensional approaches include confocal imaging, Magnetic Resonance Imaging (MRI), Optical Projection Tomography (OPT), Macro-Projection Tomography (M-OPT) [66, 67, 104, 132], and other examples in [103]. The selection of the imaging method in biodiversity research is a matter of scale and resolution. The researcher needs to previously know the limitation of each technique and the trade-offs between the benefits and limitations of the methodology employed. For example, confocal microscopy and OPT present the opportunity for imaging live plants with restrictions of sample size [66, 104]. M-OPT allows imaging of large sample size, presenting 3D digitalization of the morphology of the samples, including visualizing of gene activity and internal details of structures histochemically marked [67]; however, it is limited by the field of view and the processing artifacts that affect morphology due to the partial cleaning of tissues [43]. MRI allows non-invasive and non-destructive analysis of intact and living plants [132], despite limitations in resolution. The technique is used to measure different physiological processes, like measuring and quantifying fluid movement in the xylem and phloem [140], and the development of embolism in vessels [34, 76]. Besides it also allows repeated access to plant organs development, revealing details of its morphology and anatomy [49, 78, 80, 133]. The main drawback of conventional MRI can be overcome by newer high-resolution magnetic resonance imaging techniques (or μMRI). However, the higher the intended resolution in the analysis, the more powerful (and expensive) the magnet of the equipment has to be, which increases the costs of the procedure, in addition to not achieving the same spatial resolutions [43] as in X-ray nanotomography (nCT) systems. Some recent works utilize combined approaches in order to compare size, resolution, and the procedures, showing that the advantages and limitations of each technique can be complemented when used in a combined way [16, 44, 77]. Overall, there is no one perfect technique that can be employed in every investigation, sample, or analysis, and therefore it is up to the scientist to choose the most suitable method or combination of methods to reach the research goals. Furthermore, a broader diversity in the members' training backgrounds in a research group must be seen as a possibility to accomplish more in-depth analyses, benefiting the whole knowledge transferring process between nature and project and contributing to a better relationship of humans with the environment.

References

1. Alberts B, Johnson A, Lewis J, Morgan D, Raff M, Roberts K, Walter P (2017) Molecular biology of the cell. W.W. Norton & Company
2. Anderson TF (1951) Techniques for the preservaation of three-dimensional structure in preparing specimens for the electron microscope*. Trans N Y Acad Sci 13:130–134. https://doi.org/10.1111/j.2164-0947.1951.tb01007.x
3. Ashby MF, Ferreira PJSG, Schodek DL (2009) Nanomaterials, nanotechnologies and design: an introduction for engineers and architects. Butterworth-Heinemann, Burlington, EUA

4. Batchelor PG, Edwards PJ, King AP (2012) 3D medical imaging. In: 3D imaging, analysis and applications. Springer, London, pp 445–495
5. Bathe K-J (1996) Finite element procedures. Prentice Hall, Englewood Cliffs, N.J.
6. Bhushan B (2009) Biomimetics: lessons from nature—An overview. Philos Trans R Soc A Math Phys Eng Sci 367:1445–1486. https://doi.org/10.1098/rsta.2009.0011
7. Bolton SJ, Cora JR (2021) Virtual equivalents of real objects (VEROs): a type of non-fungible token (NFT) that can help fund the 3D digitization of natural history collections. Megataxa 6. https://doi.org/10.11646/megataxa.6.2.2
8. Boyd SK (2009) Micro-computed tomography. Advanced imaging in biology and medicine. Springer, Berlin, Heidelberg, pp 3–25
9. Boyd SK (2009) Image-based finite element analysis. Advanced Imaging in biology and medicine. Springer, Berlin, Heidelberg, pp 301–318
10. Brazier LG (1927) On the flexure of thin cylindrical shells and other "thin" sections. Proc R Soc London Ser A, Contain Pap Math Phys Character 116:104–114. https://doi.org/10.1098/rspa.1927.0125
11. Brodersen CR, Choat B, Chatelet DS, Shackel KA, Matthews MA, McElrone AJ (2013) Xylem vessel relays contribute to radial connectivity in grapevine stems (Vitis vinifera and V. arizonica; Vitaceae). Am J Bot 100:314–321. https://doi.org/10.3732/ajb.1100606
12. Brodersen CR, Lee EF, Choat B, Jansen S, Phillips RJ, Shackel KA, McElrone AJ, Matthews MA (2011) Automated analysis of three-dimensional xylem networks using high-resolution computed tomography. New Phytol 191:1168–1179. https://doi.org/10.1111/j.1469-8137.2011.03754.x
13. Brodersen CR, Roddy AB (2016) New frontiers in the three-dimensional visualization of plant structure and function. Am J Bot 103:184–188. https://doi.org/10.3732/ajb.1500532
14. Bronzati M, Benson RBJ, Evers SW, Ezcurra MD, Cabreira SF, Choiniere J, Dollman KN, Paulina-Carabajal A, Radermacher VJ, Roberto-da-Silva L, Sobral G, Stocker MR, Witmer LM, Langer MC, Nesbitt SJ (2021) Deep evolutionary diversification of semicircular canals in archosaurs. Curr Biol 31:2520-2529.e6. https://doi.org/10.1016/j.cub.2021.03.086
15. Callister WD, Rethwisch DG (2012) Fundamentals of materials science and engineering: an integrated approach, 4th edn. Wiley, New York
16. Calo CM, Rizzutto MA, Carmello-Guerreiro SM, Dias CSB, Watling J, Shock MP, Zimpel CA, Furquim LP, Pugliese F, Neves EG (2020) A correlation analysis of light microscopy and X-ray MicroCT imaging methods applied to archaeological plant remains' morphological attributes visualization. Sci Rep 10:15105. https://doi.org/10.1038/s41598-020-71726-z
17. Calvin CL (1967) Anatomy of the endophytic system of the mistletoe, phoradendron flavescens. Bot Gaz 128:117–137. https://doi.org/10.1086/336388
18. Chen BC, Zou M, Liu GM, Song JF, Wang HX (2018) Experimental study on energy absorption of bionic tubes inspired by bamboo structures under axial crushing. Int J Impact Eng 115:48–57. https://doi.org/10.1016/J.IJIMPENG.2018.01.005
19. Demanet CM, Sankar KV (1996) Atomic force microscopy images of a pollen grain: a preliminary study. South African J Bot 62:221–223. https://doi.org/10.1016/S0254-6299(15)30640-2
20. Dickinson MH (1999) Bionics: biological insight into mechanical design. Proc Natl Acad Sci 96:14208–14209. https://doi.org/10.1073/pnas.96.25.14208
21. Ding Y, Liese W (1997) Anatomical investigations on the nodes of bamboos. In: Soc L, Chapman G (eds) The bamboos. Academic Press, London, pp 265–279
22. Dixon PG, Muth JT, Xiao X, Skylar-Scott MA, Lewis JA, Gibson LJ (2018) 3D printed structures for modeling the Young's modulus of bamboo parenchyma. Acta Biomater 68:90–98. https://doi.org/10.1016/j.actbio.2017.12.036
23. Dubos T, Poulet A, Gonthier-Gueret C, Mougeot G, Vanrobays E, Li Y, Tutois S, Pery E, Chausse F, Probst A V., Tatout C, Desset S (2020) Automated 3D bio-imaging analysis of nuclear organization by NucleusJ 2.0. Nucleus 11:315–329. https://doi.org/10.1080/19491034.2020.1845012

24. Earles JM, Buckley TN, Brodersen CR, Busch FA, Cano FJ, Choat B, Evans JR, Farquhar GD, Harwood R, Huynh M, John GP, Miller ML, Rockwell FE, Sack L, Scoffoni C, Struik PC, Wu A, Yin X, Barbour MM (2019) Embracing 3D complexity in leaf carbon-water exchange. Trends Plant Sci 24:15–24. https://doi.org/10.1016/j.tplants.2018.09.005

25. Earles JM, Knipfer T, Tixier A, Orozco J, Reyes C, Zwieniecki MA, Brodersen CR, McElrone AJ (2018) In vivo quantification of plant starch reserves at micrometer resolution using X-ray microCT imaging and machine learning. New Phytol 218:1260–1269. https://doi.org/10.1111/nph.15068

26. Endress PK (2001) Origins of flower morphology. J Exp Zool 291:105–115. https://doi.org/10.1002/jez.1063

27. Evert RF, Eichhorn SE (2006) Esau's plant anatomy: meristems, cells, and tissues of the plant body: their structure, function, and development. Wiley, New Jersey

28. Evert RF, Eichhorn SE (2013) Raven biology of plants, 8th edn. W. H. Freeman, New York

29. Fahn A (1990) Plant anatomy. Pergamon Press, Oxford, Fourth

30. Von FG, Robertson D, Lee SY, Cook DD (2015) Preventing lodging in bioenergy crops: a biomechanical analysis of maize stalks suggests a new approach. J Exp Bot 66:4367–4371. https://doi.org/10.1093/jxb/erv108

31. Fratzl P (2007) Biomimetic materials research: what can we really learn from nature's structural materials? J R Soc Interface 4:637–642. https://doi.org/10.1098/rsif.2007.0218

32. Friis EM, Crane PR, Pedersen KR (2018) Tanispermum, a new genus of hemi-orthotropous to hemi-anatropous angiosperm seeds from the early cretaceous of eastern North America. Am J Bot 105:1369–1388. https://doi.org/10.1002/ajb2.1124

33. Fu J, Liu Q, Liufu K, Deng Y, Fang J, Li Q (2019) Design of bionic-bamboo thin-walled structures for energy absorption. Thin-Walled Struct 135:400–413. https://doi.org/10.1016/j.tws.2018.10.003

34. Fukuda K, Kawaguchi D, Aihara T, Ogasa MY, Miki NH, Haishi T, Umebayashi T (2015) Vulnerability to cavitation differs between current-year and older xylem: non-destructive observation with a compact magnetic resonance imaging system of two deciduous diffuse-porous species. Plant Cell Environ 38:2508–2518. https://doi.org/10.1111/pce.12510

35. Gadala M (2020) Finite elements for engineers with ANSYS applications. Cambridge University Press

36. Gandolfo MA, Nixon KC, Crepet WL, Grimaldi DA (2018) A late cretaceous fagalean inflorescence preserved in amber from New Jersey. Am J Bot 105:1424–1435. https://doi.org/10.1002/ajb2.1103

37. Gibson LJ, Ashby MF, Harley BA (2010) Cellular materials in nature and medicine. Cambridge University Press, Cambridge, UK

38. Gierke HE von (1970) Bionics and Bioengineering in Aerospace Research. In: Oestreicher HL, von Gierke HE, Keidel WD (eds) Principles and practice of bionics. TechnivisionServices, Slough, England, pp 19–42

39. Goldstein J, Newbury DE, Joy DC, Lyman CE, Echlin P, Lifshin E, Sawyer L, Michael JR (2003) Scanning electron microscopy and X-ray microanalysis, 3rd edn. Springer Science & Business Media, New York

40. Grew N (1682) The anatomy of plants. W. Rawlins, London

41. Habibi MK, Lu Y (2015) Crack propagation in Bamboo's hierarchical cellular structure. Sci Rep 4:5598. https://doi.org/10.1038/srep05598

42. Halacy DS (1965) Bionics: the science of "living" machines. Holiday House, New York

43. Hallgrimsson B, Percival CJ, Green R, Young NM, Mio W, Marcucio R (2015) Morphometrics, 3D imaging, and Craniofacial development. In: Chai Y (ed) Current topics in developmental biology. Academic Press, pp 561–597

44. Handschuh S, Baeumler N, Schwaha T, Ruthensteiner B (2013) A correlative approach for combining microCT, light and transmission electron microscopy in a single 3D scenario. Front Zool 10:44. https://doi.org/10.1186/1742-9994-10-44

45. Hanke R, Fuchs T, Salamon M, Zabler S (2016) X-ray microtomography for materials characterization. In: Hübschen G, Altpeter I, Tschuncky R, Herrmann H-G (eds) Materials characterization using nondestructive evaluation (NDE) methods. Woodhead, Duxford, UK, pp 45–79

46. Harwood J, Harwood R (2012) Testing of natural textile fibres. In: Kozłowski RM (ed) Handbook of natural fibres. Elsevier, pp 345–390

47. Hedrick BP, Heberling JM, Meineke EK, Turner KG, Grassa CJ, Park DS, Kennedy J, Clarke JA, Cook JA, Blackburn DC, Edwards SV, Davis CC (2020) Digitization and the future of natural history collections. Bioscience 70:243–251. https://doi.org/10.1093/biosci/biz163

48. Hesse L, Bunk K, Leupold J, Speck T, Masselter T (2019) Structural and functional imaging of large and opaque plant specimens. J Exp Bot 70:3659–3678. https://doi.org/10.1093/jxb/erz186

49. Hesse L, Leupold J, Speck T, Masselter T (2018) A qualitative analysis of the bud ontogeny of Dracaena marginata using high-resolution magnetic resonance imaging. Sci Rep 8:9881. https://doi.org/10.1038/s41598-018-27823-1

50. Heywood VH (1969) Scanning electron microscopy in the study of plant materials. Micron 1:1–14. https://doi.org/10.1016/0047-7206(69)90002-8

51. Hipsley CA, Aguilar R, Black JR, Hocknull SA (2020) High-throughput microCT scanning of small specimens: preparation, packing, parameters and post-processing. Sci Rep 10:13863. https://doi.org/10.1038/s41598-020-70970-7

52. Ho QT, Berghuijs HNC, Watté R, Verboven P, Herremans E, Yin X, Retta MA, Aernouts B, Saeys W, Helfen L, Farquhar GD, Struik PC, Nicolaï BM (2016) Three-dimensional microscale modelling of CO_2 transport and light propagation in tomato leaves enlightens photosynthesis. Plant Cell Environ 39:50–61. https://doi.org/10.1111/pce.12590

53. Holzwarth JM, Ma PX (2011) Biomimetic nanofibrous scaffolds for bone tissue engineering. Biomaterials 32:9622–9629. https://doi.org/10.1016/j.biomaterials.2011.09.009

54. Hooke R (1665) Micrographia: or some physiological descriptions of minute bodies made by magnifying glasses with observations and inquiries thereupon. Jo. Martyn and Ja. Allestry, printers to the Royal Society, London

55. Hounsfield GN (1973) Computerized transverse axial scanning (tomography): part 1. Description of system. Br J Radiol 46:1016–1022. https://doi.org/10.1259/0007-1285-46-552-1016

56. Hsieh J (2014) History of x-ray computed tomography. In: Shaw CC (ed) Cone beam computed tomography. CRC Press, Boca Raton, Florida, USA, pp 3–7

57. Hu D, Wang Y, Song B, Dang L, Zhang Z (2019) Energy-absorption characteristics of a bionic honeycomb tubular nested structure inspired by bamboo under axial crushing. Compos Part B Eng 162:21–32. https://doi.org/10.1016/j.compositesb.2018.10.095

58. Hunter JP (1998) Key innovations and the ecology of macroevolution. Trends Ecol Evol 13:31–36. https://doi.org/10.1016/S0169-5347(97)01273-1

59. Kasprowicz A, Smolarkiewicz M, Wierzchowiecka M, Michalak M, Wojtaszek P (2011) Introduction: tensegral world of plants. In: Wojtaszek P (ed) Mechanical integration of plant cells and plants. Springer, Berlin, Heidelberg, pp 1–25

60. Kindlein Júnior W, Guanabara AS (2005) Methodology for product design based on the study of bionics. Mater Des 26:149–155. https://doi.org/10.1016/j.matdes.2004.05.009

61. Knippers J, Schmid U, Speck T (eds) (2019) Biomimetics for architecture. De Gruyter

62. Knippers J, Speck T, Nickel KG (2016) Biomimetic research: a dialogue between the disciplines. 1–5. https://doi.org/10.1007/978-3-319-46374-2_1

63. Knoll M, Ruska E (1932) Das Elektronenmikroskop. Zeitschrift für Phys 78:318–339. https://doi.org/10.1007/BF01342199

64. Kraus JE, Louro RP, Estelita MEM, Arduin M (2006) A Célula vegetal. In: Appezzato-da-Glória B, Carmello-Guerreiro SM (eds) Anatomia vegetal, 2nd edn. UFV, Viçosa, pp 31–86

65. Kurowski PM (2004) Finite element analysis for design engineers. SAE International, Warrendale, PA
66. Lee K, Avondo J, Morrison H, Blot L, Stark M, Sharpe J, Bangham A, Coen E (2006) Visualizing plant development and gene expression in three dimensions using optical projection tomography. Plant Cell 18:2145–2156. https://doi.org/10.1105/tpc.106.043042
67. Lee KJI, Calder GM, Hindle CR, Newman JL, Robinson SN, Avondo JJHY, Coen ES (2016) Macro optical projection tomography for large scale 3D imaging of plant structures and gene activity. J Exp Bot erw452. https://doi.org/10.1093/jxb/erw452
68. Li H, Shen S (2011) The mechanical properties of bamboo and vascular bundles. J Mater Res 26:2749–2756. https://doi.org/10.1557/jmr.2011.314
69. Liese W (1998) The Anatomy of Bamboo Culms. BRILL, Beijing
70. López M, Rubio R, Martín S, Croxford B (2017) How plants inspire façades. From plants to architecture: biomimetic principles for the development of adaptive architectural envelopes. Renew Sustain Energy Rev 67:692–703. https://doi.org/10.1016/j.rser.2016.09.018
71. Ma J, Chen W, Zhao L, Zhao D (2008) Elastic buckling of bionic cylindrical shells based on bamboo. J Bionic Eng 5:231–238. https://doi.org/10.1016/S1672-6529(08)60029-3
72. Masters BR (2009) History of the electron microscope in cell biology. In: Encyclopedia of life sciences. Wiley, Chichester, UK
73. Mathers AW, Hepworth C, Baillie AL, Sloan J, Jones H, Lundgren M, Fleming AJ, Mooney SJ, Sturrock CJ (2018) Investigating the microstructure of plant leaves in 3D with lab-based X-ray computed tomography. Plant Methods 14:99. https://doi.org/10.1186/s13007-018-0367-7
74. Matsunaga KKS, Manchester SR, Srivastava R, Kapgate DK, Smith SY (2019) Fossil palm fruits from India indicate a Cretaceous origin of Arecaceae tribe Borasseae. Bot J Linn Soc 190:260–280. https://doi.org/10.1093/botlinnean/boz019
75. Meineke EK, Davies TJ, Daru BH, Davis CC (2019) Biological collections for understanding biodiversity in the Anthropocene. Philos Trans R Soc B Biol Sci 374:20170386. https://doi.org/10.1098/rstb.2017.0386
76. Meixner M, Foerst P, Windt CW (2021) Reduced spatial resolution MRI suffices to image and quantify drought induced embolism formation in trees. Plant Methods 17:38. https://doi.org/10.1186/s13007-021-00732-7
77. Metzner R, Eggert A, van Dusschoten D, Pflugfelder D, Gerth S, Schurr U, Uhlmann N, Jahnke S (2015) Direct comparison of MRI and X-ray CT technologies for 3D imaging of root systems in soil: potential and challenges for root trait quantification. Plant Methods 11:17. https://doi.org/10.1186/s13007-015-0060-z
78. Metzner R, van Dusschoten D, Bühler J, Schurr U, Jahnke S (2014) Belowground plant development measured with magnetic resonance imaging (MRI): exploiting the potential for non-invasive trait quantification using sugar beet as a proxy. Front Plant Sci 5. https://doi.org/10.3389/fpls.2014.00469
79. Möller M, Höfele P, Kiesel A, Speck O (2021) Reactions of sciences to the anthropocene. Elem Sci Anthr 9. https://doi.org/10.1525/elementa.2021.035
80. Morozov D, Tal I, Pisanty O, Shani E, Cohen Y (2017) Studying microstructure and microstructural changes in plant tissues by advanced diffusion magnetic resonance imaging techniques. J Exp Bot 68:2245–2257. https://doi.org/10.1093/jxb/erx106
81. Nicolescu B (2010) Methodology of transdisciplinarity—Levels of reality, logic of the included middle and complexity. Transdiscip J Eng Sci 1:19–38
82. Niklas KJ (1992) Plant biomechanics : an engineering approach to plant form and function. University of Chicago Press, Chicago, EUA
83. Niklas KJ, Spatz H-C (2012) Plant physics. University of Chicago Press, Chicago
84. Nogueira FM, Kuhn SA, Palombini FL, Rua GH, Andrello AC, Appoloni CR, Mariath JEA (2017) Tank-inflorescence in Nidularium innocentii (Bromeliaceae): three-dimensional model and development. Bot J Linn Soc 185:413–424. https://doi.org/10.1093/botlinnean/box059
85. Nogueira FM, Palombini FL, Kuhn SA, Oliveira BF, Mariath JEA (2019) Heat transfer in the tank-inflorescence of Nidularium innocentii (Bromeliaceae): experimental and finite element analysis based on X-ray microtomography. Micron 124:102714. https://doi.org/10.1016/j.micron.2019.102714

86. Nogueira FM, Palombini FL, Kuhn SA, Rua GH, Mariath JEA (2021) The inflorescence architecture in Nidularioid genera: understanding the structure of congested inflorescences in Bromeliaceae. Flora 284:151934. https://doi.org/10.1016/j.flora.2021.151934
87. Orhan K (ed) (2020) Micro-computed tomography (micro-CT) in medicine and engineering. Springer International Publishing, Cham
88. Overney N, Overney G (2011) The history of photomicrography
89. Paddock SW (2000) Principles and practices of laser scanning confocal microscopy. Mol Biotechnol 16:127–150. https://doi.org/10.1385/MB:16:2:127
90. Page LM, MacFadden BJ, Fortes JA, Soltis PS, Riccardi G (2015) Digitization of biodiversity collections reveals biggest data on biodiversity. Bioscience 65:841–842. https://doi.org/10.1093/biosci/biv104
91. Palombini FL, Cidade MK, Oliveira BF de, Mariath JE de A (2021) From light microscopy to X-ray microtomography: observation technologies in transdisciplinary approaches for bionic design and botany. Cuad del Cent Estud en Diseño y Comun 149:61–74
92. Palombini FL, Kindlein Junior W, Oliveira BF de, Mariath JE de A (2016) Bionics and design: 3D microstructural characterization and numerical analysis of bamboo based on X-ray microtomography. Mater Charact 120:357–368. https://doi.org/10.1016/j.matchar.2016.09.022
93. Palombini FL, Kindlein Júnior W, Silva FP da, Mariath JE de A (2017) Design, biônica e novos paradigmas: uso de tecnologias 3D para análise e caracterização aplicadas em anatomia vegetal. Des e Tecnol 7:46. https://doi.org/10.23972/det2017iss13pp46-56
94. Palombini FL, Lautert EL, Mariath JE de A, de Oliveira BF (2020a) Combining numerical models and discretizing methods in the analysis of bamboo parenchyma using finite element analysis based on X-ray microtomography. Wood Sci Technol 54:161–186. https://doi.org/10.1007/s00226-019-01146-4
95. Palombini FL, Mariath JE de A, Oliveira BF de (2020b) Bionic design of thin-walled structure based on the geometry of the vascular bundles of bamboo. Thin-Walled Struct 155:106936. https://doi.org/10.1016/j.tws.2020.106936
96. Palombini FL, Nogueira FM, Kindlein Junior W, Paciornik S, Mariath JE de A, Oliveira BF de (2020c) Biomimetic systems and design in the 3D characterization of the complex vascular system of bamboo node based on X-ray microtomography and finite element analysis. J Mater Res 35:842–854. https://doi.org/10.1557/jmr.2019.117
97. Pandoli O, Martins RDS, Romani EC, Paciornik S, Maurício MHDP, Alves HDL, Pereira-Meirelles FV, Luz EL, Koller SML, Valiente H, Ghavami K (2016) Colloidal silver nanoparticles: an effective nano-filler material to prevent fungal proliferation in bamboo. RSC Adv 6:98325–98336. https://doi.org/10.1039/C6RA12516F
98. Pandoli OG, Martins RS, De Toni KLG, Paciornik S, Maurício MHP, Lima RMC, Padilha NB, Letichevsky S, Avillez RR, Rodrigues EJR, Ghavami K (2019) A regioselective coating onto microarray channels of bamboo with chitosan-based silver nanoparticles. J Coatings Technol Res 16:999–1011. https://doi.org/10.1007/S11998-018-00175-1
99. Peng G, Jiang Z, Liu X, Fei B, Yang S, Qin D, Ren H, Yu Y, Xie H (2014) Detection of complex vascular system in bamboo node by X-ray μCT imaging technique. Holzforschung 68:223–227. https://doi.org/10.1515/hf-2013-0080
100. Pfeifer R, Lungarella M, Iida F (2007) Self-organization, Embodiment, and biologically inspired robotics. Science 318(80):1088–1093. https://doi.org/10.1126/science.1145803
101. Pohl G, Nachtigall W (2015) Biomimetics for architecture & design. Springer International Publishing, Cham
102. Policha T, Davis A, Barnadas M, Dentinger BTM, Raguso RA, Roy BA (2016) Disentangling visual and olfactory signals in mushroom-mimicking Dracula orchids using realistic three-dimensional printed flowers. New Phytol 210:1058–1071. https://doi.org/10.1111/nph.13855
103. Prunet N, Duncan K (2020) Imaging flowers: a guide to current microscopy and tomography techniques to study flower development. J Exp Bot 71:2898–2909. https://doi.org/10.1093/jxb/eraa094

104. Prunet N, Jack TP, Meyerowitz EM (2016) Live confocal imaging of Arabidopsis flower buds. Dev Biol 419:114–120. https://doi.org/10.1016/j.ydbio.2016.03.018
105. Quinn S, Gaughran W (2010) Bionics—An inspiration for intelligent manufacturing and engineering. Robot Comput Integr Manuf 26:616–621. https://doi.org/10.1016/j.rcim.2010. 06.021
106. Radon J (1917) Über die Bestimmung von Funktionen durch ihre Integralwerte längs gewisser Mannigfaltigkeiten. Berichte der Sächsischen Akad der Wiss 69:262–277
107. Reddy JN (2004) An introduction to nonlinear finite element analysis. Cambridge University Press, New York
108. REFLORA (2021) Reflora—Herbário virtual [Virtual Herbarium]. http://reflora.jbrj.gov.br/ reflora/herbarioVirtual/. Accessed 28 Oct 2021
109. Röntgen WC (1896) On a new kind of rays. Science 3(80):227–231. https://doi.org/10.1126/ SCIENCE.3.59.227
110. Roosa SA (2010) Sustainable development handbook, 2nd edn. The Fairmont Press Inc., Lilburn
111. Rowley JR, Flynn JJ, Takahashi M (1995) Atomic force microscope information on Pollen Exine substructure in Nuphar. Bot Acta 108:300–308. https://doi.org/10.1111/j.1438-8677. 1995.tb00498.x
112. Salmén L (2018) Wood cell wall structure and organisation in relation to mechanics. In: Geitmann A, Gril J (eds) Plant biomechanics. Springer International Publishing, Cham, pp 3–19
113. Schindelin J, Arganda-Carreras I, Frise E, Kaynig V, Longair M, Pietzsch T, Preibisch S, Rueden C, Saalfeld S, Schmid B, Tinevez J-Y, White DJ, Hartenstein V, Eliceiri K, Tomancak P, Cardona A (2012) Fiji: an open-source platform for biological-image analysis. Nat Methods 9:676–682. https://doi.org/10.1038/nmeth.2019
114. Schmitt OH (1963) Signals assimilable by living organisms and by machines. IEEE Trans Mil Electron MIL 7:90–93. https://doi.org/10.1109/TME.1963.4323055
115. Schultz AR (1972) Estudo prático da botânica geral, 4th edn. Globo, Porto Alegre
116. Şener LT, Albeniz G, Külüşlü G, Albeniz I (2020) Micro-CT in artificial tissues. Micro-computed Tomogr Med Eng 125–137. https://doi.org/10.1007/978-3-030-16641-0_9
117. Shao Z, Wang F (2018) The fracture mechanics of plant materials. Springer, Singapore
118. Silvestro D, Zizka G, Schulte K (2014) Disentangling the effects of key innovations on the diversification of Bromelioideae (Bromeliaceae). Evolution (N Y) 68:163–175. https://doi. org/10.1111/evo.12236
119. Smith SY, Iles WJD, Benedict JC, Specht CD (2018) Building the monocot tree of death: progress and challenges emerging from the macrofossil-rich Zingiberales. Am J Bot 105:1389–1400. https://doi.org/10.1002/ajb2.1123
120. Speck O, Langer M, Mylo MD (2021) Plant-inspired damage control—An inspiration for sustainable solutions in the Anthropocene. Anthr Rev 205301962110184. https://doi.org/10. 1177/20530196211018489
121. Speck O, Speck D, Horn R, Gantner J, Sedlbauer KP (2017) Biomimetic bio-inspired biomorph sustainable? An attempt to classify and clarify biology-derived technical devel-opments. Bioinspir Biomim 12:011004. https://doi.org/10.1088/1748-3190/12/1/011004
122. Speck O, Speck T (2021) Biomimetics and education in Europe: challenges, opportunities, and variety. Biomimetics 6:49. https://doi.org/10.3390/biomimetics6030049
123. Speck T, Bold G, Masselter T, Poppinga S, Schmier S, Thielen M, Speck O (2018) Biome-chanics and functional morphology of plants—Inspiration for biomimetic materials and structures. Plant biomechanics. Springer International Publishing, Cham, pp 399–433
124. Staedler YM, Masson D, Schönenberger J (2013) Plant tissues in 3D via X-ray tomography: simple contrasting methods allow high resolution imaging. PLoS One 8:e75295. https://doi. org/10.1371/journal.pone.0075295

125. Stein E (2014) The origins of mechanical conservation principles and variational calculus in the 17th century. Lecture notes in applied mathematics and mechanics. Springer, Berlin, Heidelberg, pp 3–22
126. Stock SR (2009) MicroComputed tomography: methodology and applications. CRC Press, Boca Raton, Florida, USA
127. Tafforeau P, Boistel R, Boller E, Bravin A, Brunet M, Chaimanee Y, Cloetens P, Feist M, Hoszowska J, Jaeger J-J, Kay RF, Lazzari V, Marivaux L, Nel A, Nemoz C, Thibault X, Vignaud P, Zabler S (2006) Applications of X-ray synchrotron microtomography for non-destructive 3D studies of paleontological specimens. Appl Phys A 83:195–202. https://doi.org/10.1007/s00339-006-3507-2
128. Tang Z, Wang Y, Podsiadlo P, Kotov NA (2006) Biomedical applications of layer-by-layer assembly: from biomimetics to tissue engineering. Adv Mater 18:3203–3224. https://doi.org/10.1002/adma.200600113
129. Taylor D, Dirks J-H (2012) Shape optimization in exoskeletons and endoskeletons: a biomechanics analysis. J R Soc Interface 9:3480–3489. https://doi.org/10.1098/rsif.2012.0567
130. Teixeira-Costa L, Ceccantini GCT (2016) Aligning microtomography analysis with traditional anatomy for a 3D understanding of the host-parasite interface—phoradendron spp. Case study. Front Plant Sci 7. https://doi.org/10.3389/fpls.2016.01340
131. Thomas S, Durand D, Chassenieux C, Jyotishkumar P (eds) (2013) Handbook of biopolymer-based materials. Wiley-VCH Verlag GmbH & Co, KGaA, Weinheim, Germany
132. Van As H (2006) Intact plant MRI for the study of cell water relations, membrane permeability, cell-to-cell and long distance water transport. J Exp Bot 58:743–756. https://doi.org/10.1093/jxb/erl157
133. van Dusschoten D, Metzner R, Kochs J, Postma JA, Pflugfelder D, Bühler J, Schurr U, Jahnke S (2016) Quantitative 3D analysis of plant roots growing in soil using magnetic resonance imaging. Plant Physiol 170:1176–1188. https://doi.org/10.1104/pp.15.01388
134. VDI 6220 (2012) VDI 6220 Blatt 1:2012-12—Biomimetics—Conception and strategy—Differences between biomimetic and conventional methods/products. VDI-Richtlinien 36
135. Vincent JF, Bogatyreva OA, Bogatyrev NR, Bowyer A, Pahl A-K (2006) Biomimetics: its practice and theory. J R Soc Interface 3:471–482. https://doi.org/10.1098/rsif.2006.0127
136. von Gierke HE, Lauschner EA (1970) Foreword. In: Oestreicher HL, von Gierke HE, Keidel WD (eds) Principles and practice of bionics. TechnivisionServices, Slough, England, pp 13–14
137. Wayne R (2010) Plant cell biology. Elsevier
138. Wegst UGK, Ashby MF (2007) The structural efficiency of orthotropic stalks, stems and tubes. J Mater Sci 42:9005–9014. https://doi.org/10.1007/s10853-007-1936-8
139. Williams DB, Carter CB (1996) The transmission electron microscope. Transmission electron microscopy. Springer, US, Boston, MA, pp 3–17
140. Windt CW, Vergeldt FJ, De Jager PA, Van As H (2006) MRI of long-distance water transport: a comparison of the phloem and xylem flow characteristics and dynamics in poplar, castor bean, tomato and tobacco. Plant Cell Environ 29:1715–1729. https://doi.org/10.1111/j.1365-3040.2006.01544.x
141. Wollman AJM, Nudd R, Hedlund EG, Leake MC (2015) From Animaculum to single molecules: 300 years of the light microscope. Open Biol 5:150019. https://doi.org/10.1098/rsob.150019
142. Woźniak NJ, Sicard A (2018) Evolvability of flower geometry: convergence in pollinator-driven morphological evolution of flowers. Semin Cell Dev Biol 79:3–15. https://doi.org/10.1016/j.semcdb.2017.09.028
143. Xiang E, Yang S, Cao C, Liu X, Peng G, Shang L, Tian G, Ma Q, Ma J (2021) Visualizing complex anatomical structure in Bamboo nodes based on X-ray microtomography. J Renew Mater 9:1531–1540. https://doi.org/10.32604/jrm.2021.015346
144. Xu Y, Sheng K, Li C, Shi G (2010) Self-assembled graphene hydrogel via a one-step hydrothermal process. ACS Nano 4:4324–4330. https://doi.org/10.1021/nn101187z

145. Zhang S (2003) Fabrication of novel biomaterials through molecular self-assembly. Nat Biotechnol 21:1171–1178. https://doi.org/10.1038/nbt874
146. Zhang T, Wang A, Wang Q, Guan F (2019) Bending characteristics analysis and lightweight design of a bionic beam inspired by bamboo structures. Thin-Walled Struct 142:476–498. https://doi.org/10.1016/j.tws.2019.04.043
147. Zienkiewicz OC, Taylor RL, Zhu JZ (2013) The finite element method: its basis and fundamentals, 7th edn. Butterworth-Heinemann, Oxford

The *Maniola, Lycaenidae,* and Other *Lepidoptera* Eggs as an Inspiration Source for Food Storage and Packaging Design Solutions

Massimo Lumini

Abstract Eggs are nature's successful evolutionary design tricks, well designed to deliver multi-task biofunctional strategies for life's challenges. They appear in the vital *scenario* in the form of original and surprising bio-tech design solutions affected by the genetic and environmental constraints they are called to interact with. For these basic survival needs, the eggs must work very well: capturing the sperm of the male for a correct optimization of the fertilization processes, protection from physical and mechanical trauma, climatic mediation, and fine aeration of the internal larvae. These surprising embryo packagings are a sort of lifeboat laid down and often left alone by females in front of the intricate, complex, and highly wild food interweaving the planet's ecosystems. We found eggs in the reproductive cycles of many living species: fish, cephalopods, birds, and above all, individual insects. Butterfly eggs constitute a class of exciting and still little studied solutions, considered for possible bionic and biomimetic inspirations. Many *Lepidoptera* eggs generally have an external textured shell, the *chorion*, made up of waxed surface *keratin*, which maintains the correct humidity of the egg throughout the growth cycle. *Keratin* is a fibrous protein rich in sulfur amino acids, cysteine, and self-assemble into fiber bundles. It has the characteristic of a very tenacious mineralized fabric and is remarkably impermeable to water and atmospheric gases. Each egg is glued by the mother's butterfly to the support of branches or leaves of the *nourishing plants* by a gluey substance of chemical still largely unknown constitution, so adhesive that it is impossible to detach the eggs if not breaking them. In some butterfly species, like the *Maniola* and *Lycaenidae* family, the shell's structure has a spatial organization in the form of complex geodesic ribbed micro domes that resemble Buckminster Fuller's geodesic structures. Another exciting aspect of butterfly's eggs design concerns the *micropyle* and *aeropyles* layers system, which ensure the proper introduction of the male sperm, air, and oxygen needed to larva's growth. This study, conducted by the *BionikonLab&FABNAT14* laboratory of Iglesias-SU Italy, considers the structural, morphological, and geometric aspects of some types of butterfly eggs

---

M. Lumini (✉)
BionikonLab&FABNAT14, IIS "G.Asproni,", Iglesias, SU, Italy
e-mail: m.lumini57@gmail.com

© The Author(s), under exclusive license to Springer Nature Singapore Pte Ltd. 2022   45
F. L. Palombini and S. S. Muthu (eds.), *Bionics and Sustainable Design*, Environmental Footprints and Eco-design of Products and Processes,
https://doi.org/10.1007/978-981-19-1812-4_3

that await internal ventilation. The purpose is to define a list of essential design problem-solving concepts that apply to creating food packaging, considering the crucial aspects of preserving freshness and commercial and nutritional qualities, reducing food waste, and the additional use of chemicals, antioxidants, and plastics packs.

Keywords Bionics · Biomimetics · Chemical food preservatives reduction · Chorion · Jun mitani · Origami · Food packaging · Food home preservation · Lepidoptera · Plastic waste reduction · Tsutsumu

1 Biomimesis: Hints of History and Methodology

Searching technological solutions inspired by observing and studying organisms' forms and living processes has its oldest roots in the *mimesis* process of Greek aesthetic thought. For the ancient classical culture, the terms *art* and *technique*, which they translate as *téchne*, were an ordinary meaning of *knowing how to do* and *knowing how to work*, referring to a set of rules applied and followed in the performance of creative and craft activities. With the development of technical and technological knowledge, this complex and refined set of procedures that guided thought became more complex and sophisticated. Its development has concretely contributed to the emergence of a methodological scientific doctrine. Historically, and without going into the complexities of specific epistemology, we find the result of two procedural processes to arrive at rational solutions and theories: the inductive and deductive methods. These two methodological procedures are somehow translated into the two research systems: *top-down* and *bottom-up* as briefly represented in Fig. 1.

In general, design research on a contemporary bionic and biomimetic basis makes use of these two particular thought processes. These different approaches develop opposing research streams, but in R&D practice, they very often become complementary to each other. In a typical bionic problem-solving process, they often interact with each other, complementing and integrating through different feedback steps. Let's consider a typical *top-down* industrial design process. The creative path focuses on the definition of one or more specific technical and technological design problems and esthetic-functional aspects of form and function that create a plan of potentially acceptable solutions. This initial level may consist of the need to completely rethink and redesign a specific aspect of an object or existing technical or technological performance or develop ideas that are essentially linked to the innovation of strategic and functional needs in a human context. Implementing top-down methodological process cycles is expected to lead to the conception of entire classes of new and competitive useful objects and performances, prototypes, patents, and other types of R&D solutions. In summary, after the narrowing down of the problem domain in question, we move through *cascade steps* to the realms of natural technology in searching for possible biological morphological-functional analogs. This information, appropriately contextualized and transferred to the design context, can inspire potential

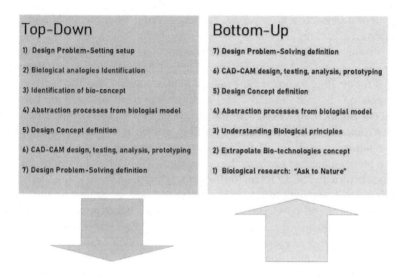

Fig. 1 Step by step of *Problem-Setting* and *Problem-Solving* in bionics' and biomimetics' research

technical and functional solutions. The designer must carefully manage the possible analogies between the two technological levels, natural and artificial. The inopportune transference of biological principles into inappropriate design contexts can lead to forced, scale-less design solutions of considerable aestheticizing bio-formalism. In this sense, through successive processes of abstraction and sublimation of biological models, they overcome the slavishly imitative tout court and formalistic aspect of natural models. Interesting in this context is the concept of lightweight structures developed in the 1960s by the German architect Frei Otto, co-founder of the *Biologie und Bauen* design group [1].

In the conception of this pioneering group of designers and biologists, the mimetic research does not place the natural organism on a primary or antithetic level to the human, technical and technological product. BuB's investigation leads to the discovery of unprecedented principles of observation and analysis and laboratory experiments that reveal surprising functional and structural similarities between the constructs of nature and technology. The bottom line is that if you compare a blade of grass and a high voltage tower and properly contextualize them in terms of their specific dimensions and structural levels, they must function in terms of parameters and operating conditions in which the physical forces, resistance, applied loads, and optimization structure of the materials behave according to similar static and structural models and schemes. The famous models of minimal membranes made of soap bubbles and films used by biology and construction clarified and guided in architectural and civil engineering problem-solving both the static and performance behavior of the morphology of a spider web or a biological membrane and that of a circus tent or a tensile structure of a textile roof of an airport terminal or a stadium. From the 1970s onwards, these concepts are stigmatized by Frei Otto in his conception of innovative architectural and engineering intuitions such as gridshells and strain architectures [2].

Proper analysis of the allowable stresses imposed by a shape or configuration of a given material and the structural design of an optimized static response is essential for creating natural organisms and human engineering. The German group had understood how to develop a correct biodesign methodology long before the current development of a design culture based on bionics and biomimetics, which explores transversal solutions between biology and design. This bionic estheticism trend of *digital decoration* [3] appears in some design approaches that, through a creative process that can be defined *as digital rococò* [4], risk becoming a kind of aesthetic rhetoric that itself refers to Biomimicry [4].

A significant tendency in the design culture of design, engineering, and architecture was born and developed thanks to the viral diffusion on the Internet, of studies and creations inspired by the imitation of SEM images of extraordinary microorganisms such as radiolarians and diatoms and their brilliant siliceous ultrastructure. This contamination of biology domains into the design Imaginarium not always correct sustainable and functional inspiration has been unleashed, leading to a slavish and detached transposition and morphological translation of natural microstructures, lifted from the realms of the invisible until they reach the physical and tectonic consistency of a human scale of function. Biologist and innovator Janine Benjus define Biomimicry as:

> a technology-driven approach that focuses on putting nature's lessons into practice, seeing nature as model, benchmark, and mentor," then proper design aimed at truly sustainable architecture and design must move beyond evocative esthetic proposals to mimic nature to produce a world of artificial objects that know how to respond in the most natural, and thus environmentally friendly, way to the challenges that human and terrestrial evolution will soon require. [5]

It is worth remembering as possible anticipation of these contemporary bioinspired hypotheses that in the theory and practice of ancient design, the three cardinal principles of human *technè* (Latin tr.*ars-artis*), closely linked to the imitative source of nature, were established by the architect and historian of architecture Marco Vitruvio Pollone in his treatise *De Achitectura*:

> Haec autem ita fieri debent, ut habeatur ratio firmitas, utilitas, venustatis. [6]

The iconography, later taken up by Renaissance treatises, describes this concept with the vertices of an equilateral triangle to illustrate the equivalence of each of them in the complexity of their dynamic relationship. The balance and harmony of classical art and some of Renaissance's cultural aspects are partly due to this proxemic conception of the design approach's variables. The equidistance of the vertices of the equilateral triangle requires that the practical and utilitarian aspects necessary for the execution of an artificial artifact (*utilitas*, from *utilis*-utility) are necessarily balanced and balanced with the aesthetic ones (*venustas*, from *Venus-Veneris*-Vinus like the goddess of beauty) and technical-engineering (*firmitas,* from *firmus*-solid). None of the three fields of action should surpass the others in importance: the triangle that

Fig. 2 The Vitruvian *equilibrium* of design activity

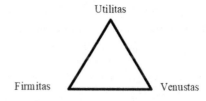

connects them should never turn into an irregular, scalene triangle in which one side arrogates to itself the right to enlarge itself to the detriment of the size of the other (Fig. 2).

It is necessary to propose a three-dimensional update of this Vitruvian methodological model in light of the current global environmental imbalance affecting planet Earth and the heavy responsibilities and complexity of designing. The new figure, which could rework the contemporary methodological vision of a unique harmony between Man and Nature, is a *tetrahedron,* the first of the Platonic polyhedra. Starting from the Vitruvian triangular base, let us imagine adding a fourth vertex in space equidistant from the other three: a *tetrahedron* has been created. In this new 3D point, we place the influence of the domain of *propinquitas,* a term translated from Latin-*propinquus* as *kinship, affinity, consanguinity,* and, by inference, *closeness.* This latest acquisition ensures that in each design methodology, all the specific aspects of the project (technical, functional, and aesthetic) must necessarily evaluate the network of causes and effects that trigger each other, taking into account the consequences of artificial impacts in terms of proximity ecological and fraternity between human beings and natural creatures (Fig. 3).

An authentic natural, bionic, biomimetic, or bioinspired future design approach must go beyond bio-formalist proposals' risky aestheticizing and fashionista infatuations. The culture of the project, combined with the needs of the economy, mass production, and consumption, must know how to experiment and accept the great challenge of sustainability. The future urgently requires a vision of global *propinquitas;* a deep, renewed, and healthy balance between the effects of doing and the utilitarian interests of technoscience and economic profit, mediated with the vital and aesthetic needs of human and earthly communities, in a renewed, more harmonious and balanced relationship. Let us now return to the broken thread of the analysis

Fig. 3 Updating the Vitruvian triadic model

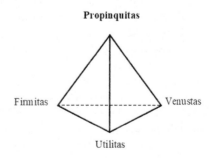

of the top-down working method. After defining the field of interest of the techno-logical problem, a process of research of possible biological analogies develops. A series of suitable design principles emerge from this phase of specialist research. Through successive processes of abstraction, geometrizations, and formal and func-tional passages that strictly distance design from the biological model, we arrive at a possible hypothesis of contextualization in artificial domains and the definition of a sustainable concept of bio-inspiration. The following engineering, modeling, CAD/CAM prototyping, testing, and design corrections through feedback and the checklist perfect the product. Finally, the project is communicated through graphic models, mockups, or virtual digital or augmented reality renderings. The bottom-up complementary process rides in a diametrically opposite way, intercepting in a biological database one or more organisms that represent exciting solutions that may concern the various bionic and biomimetic application domains: morphological, techno-technological, chemical-physical, and functional [3].

An iconic case is the invention of $Velcro^{©}$ adhesive strips by G. De Mestral, at the forefront since the post-war period in developing a methodology for scouting potential patents inspired by natural solutions. Biomechanical, functional, morpho-logical, and anatomical aspects are defined based on research and biological obser-vations. In the context of industrial design, the molecular and nanotechnological elements, which refer to mimetic analysis at the level of physical chemistry and material science processes, are neglected. These forays through biological analyzes define a set of fundamental design principles that, when appropriately subjected to processes of abstraction and geometrization, lead to a progressive detachment from the original natural model. The concepts developed are implemented technically and technologically to bring the R&D process to defining precise design solutions. The *bottom-up* approach requires long lead times, as typically, three to seven years may elapse between identifying a biological function or structure of interest for technical development and the actual production of an innovative biomimetic product inspired by it. The *top-down* process can lead to the development of efficient solutions in a relatively shorter time frame, estimated at six to eighteen months. In this regard, the research presented in this paper developed within the didactic and design activities of *BionikonLab&FABNAT14*. It is an experimental design laboratory located inside a public scientific and artistic high school in Sardinia-Iglesias SU (Italy). We are skilled in teaching the methodologies of the bionic and biomimetic basic design approach. The work team is formed by teachers, architects, and designers, supported by expert biologists and zoologists who work with students aged between 14 and 18. Early *BionikonLab* was founded in 1996; since 2014, it has been supported and integrated with *FABNAT14* a maker-space for 3D-Printing.

We use rapid prototyping technologies available in the laboratory to develop prototypes and mockups (Figs. 4 and 5). BionikonLab & FABNAT14 is a unique educational project of its kind in Italy. The students can develop a set of design skills more typical of the world of university and specialist research (such as in high schools of architecture, engineering, and design). The open space of the laboratory allows working groups, coordinated by teachers and designers, to carry out various activities

Fig. 4 The *BionikonLab&FABNAT14's* working space layout

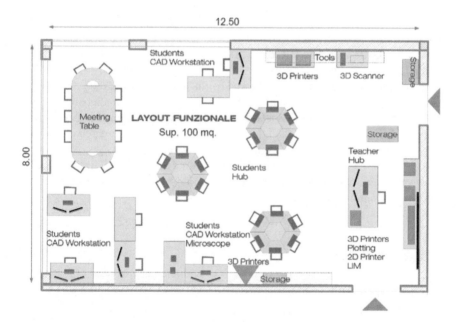

Fig. 5 The *BionikonLab&FABNAT14's* working space layout

WORKFLOW DIAGRAM

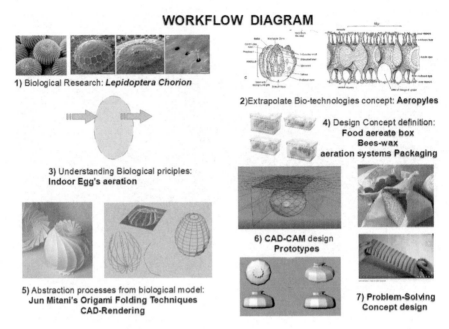

1) Biological Research: *Lepidoptera Chorion*

2)Extrapolate Bio-technologies concept: **Aeropyles**

4) Design Concept definition:
Food aereate box
Bees-wax
aeration systems Packaging

3) Understanding Biological priciples:
Indoor Egg's aeration

6) CAD-CAM design
Prototypes

5) Abstraction processes from biological model:
Jun Mitani's Origami Folding Techniques
CAD-Rendering

7) Problem-Solving
Concept design

Fig. 6 This infographic illustrates the main research and design steps, based on the application of a *bottom-up* design workflow

simultaneously: web research, meetings, brainstorming activities, lectures, workstations for CAD drawing, and different CAD-CAM fixtures for rapid prototyping and 3D digital microscopic observation.

The methodological approach followed by the teamwork to develop this specific research, based on the observation of the morpho-functional characteristics of Lepidoptera eggs, can be described as a *bottom-up* method. The inspiration that triggered the flow of analysis, research, and design solutions start from research on butterfly eggs. The designers defined a blueprint of different analog concepts related to the food packaging sector through creative brainstorming sessions. Follow the workflow below in Fig. 6, which exemplifies the application of the steps suggested by the *bottom-up* design methodology.

2 Eggs in Nature

Eggs are a successful evolutionary nature's trick. They are designed and built to provide multifunctional strategies for overcoming life's challenges. More than 99% of all animals that have ever lived on planet Earth hatched from an egg at the first moment of life. An egg is an organic vessel that contains the zygote in which an embryo develops until it can survive on its own. The fertilization of an ovum forms an egg. The evolutionary design concept underlying the idea of the egg is to package their reproductive genetic material in the most efficient and optimized way possible, beyond the individual specific needs that have led living things to create millions of different solutions. There are interesting similarities between problem-solving in embryo packaging and natural protection and the need for safety in food packaging: preservation of hygrometric data, protection from microbial attack, mold, fungi, and contaminants. For this reason, eggs must perform very well: capturing the male's sperm for a proper fertilization process, protecting the zygote from physical and mechanical environmental trauma, climatic mediation, and satisfactory ventilation of the interior laid by females and often left alone, from the intricate, complex, and highly wild food interrelationships of planet Earth's ecosystems. They emerge in the vital *scenario* related to the birth and evolution of insects, the oldest fossil of which is currently *Rhyniognayha hirsti,* a creature that dates back to the *Devonian* era, some 396–407 million years ago. The evolution of eggs continued 350 million years ago, laid by the *Captorhinidae,* the most primitive reptilian group. One of the essential evolutionary inventions occurs in the *Carboniferous*: giant reptiles evolve a solution that allows them to perform the delicate phase of reproduction regardless of the presence of water. The embryo in the egg was initially enclosed only by a membrane. Then a rigid, thin, and porous shell is formed, which protects the genetic content from moisture loss and proper oxygenation while protecting it from predators. Inside, a store of nutrients such as fats, proteins, and sugars is created. Thanks to this innovative solution, the embryo can reach a higher level of development than the larvae of amphibians. Isolated and protected in the healthy habitat guaranteed by this safe spaceship, the creature has a good chance of survival even in a dry environment. Birds are one of the animal groups that have adopted and maintained the solution of the egg during their evolution, which began about 200 million years ago, to ensure the continuity of species. Bird eggs have a rigid, multilayered, biomineralized eggshell (calcium carbonate) with three layers: an outer cuticle, a biomineralized layer, and the outer and inner shell membranes of collagen fibers. The calcareous shell has a resistance that in some respects surpasses that of ceramics and concrete: it is a highly resistant material to compression, tension, and bending loads. Although a fragment of the shell of a bird, such as a chicken, is extremely thin, it obtains data of excellent resistance when subjected to a compression test in the longitudinal direction. On the other hand, when stressed perpendicular to its structure, it is fragile and can easily be cracked by the delicate beak of a small unborn child. The bird's egg has an overall asymmetrical shell structure with a sharper upper pole than the base, which is perfectly designed to withstand the compressive loads exerted on the surface by the muscles of the anal oviducts as they are expelled from the female body. At the same time, as being highly resistant to impact, it must ensure easy breakage by the chick that bursts the egg to hatch. You can quickly try out this extraordinary resistance in terms of geometry and shape that an egg has by taking a chicken egg, placing it in the hollow of one palm, and at the same time applying firm vertical pressure with the other palm. It is almost impossible to break it. The structure combined with the chemical-physical and morphological properties of the building material is very resistant to compressive forces. When you apply a load to the poles, the shell is mainly subjected to compression because the bending stress is proportional to the distance between the axis on which the load is applied and the egg's surface furthest from it. If you compress the egg from its equator, the distance to the poles is much greater, and the bending stress, which is proportional to that distance, is many times greater, making the egg much less resistant. When we compress an egg in this way, and the tensile stresses that are discharged radially in the shell reach a critical level equal to the tensile strength of the calcium carbonate, the egg breaks. Thus, it is not the vertical compressive stresses that break the structure but the radially generated stresses. Some tests on chicken eggs have shown that their compressive strength is more than 100 pounds, while ostrich eggs have values of more than 1000 pounds. Understanding the biomechanics of eggs could lead to exploring more effective forms of packaging and industrial handling. Some sources indicate that 83.1 billion consumer eggs were produced in America in 2015, of which about 4/6% were broken or damaged between production and reaching the consumer. Using the unit cost of an egg of $2.47 equates to a loss of over a billion dollars per year in the United States alone [7].

On the contrary, if we exclude the species that give birth to their young, almost all reptiles also produce eggs. Still, unlike those of the hard-shelled birds, they are soft and leathery, created by a protein cast coated with calcium carbonate crystals, and only occasionally do the minerals present in the shell make it similar in biomechanical characteristics to those of birds. Reptile eggs are *teleolecitic* (eggs with large yolk in the vegetative hemisphere and scarce in the animal pole, therefore with marked polar differentiation), with many nutritional reserves, making them hatch at very advanced stages of embryo development compared to amphibians, for example. Unlike the bird's egg, it lacks an air chamber. It has a shell that is permeable to oxygen but such as to avoid internal dehydration. It generally consists of a membrane (*chorion*) covered with a limestone coating of aragonite or calcite units. An underlying membrane called amnios, a true evolutionary innovation of reptiles for land conquest, protects the embryo from dehydration and mechanical shocks. Gelatinous albumen provides an additional nutritional reserve of proteins, isolates the embryo mechanically, hydrates it, and protects it from microbial infections. There is a large egg cell (yolk or yolk sac) composed of lipids and proteins and an extensive membrane called allantoid, which has the dual function of collecting the semi-solid urine of the embryo and coming into contact with the shell to ensure respiratory exchanges: carbon dioxide comes out, and oxygen enters. From a bionic and biomimetic point of view, the chemical-physical nature of the egg itself generates a whole series of possible inspirations for engineering and design studies and applications. Insect eggs, of which approximately 900,000 species represent the largest *phylum* on Earth, constitute another massive class of differentiated solutions both in morphology and in the functional specificities of adaptation and environmental competition. From various research sources in the zoological field, it was possible to obtain a series of data, information, and images, which guided the subsequent development phases of the various design concepts. In particular, the essay "*Evolution of Oviposition Techniques in Stick and Leaf Insects*" (*Phasmayodea*) was of great interest and usefulness [8]. *Phasmayodea* are an order of insects whose members are variously known as stick insects, stick-bugs, walking sticks, stick animals, or bug sticks. The accurate images of the eggs produced by these families of tropical insects by the French entomologist François Tetaert describe in detail the incredible and imaginative variety of types and morphologies [9]. The creativity of natural design is expressed here with such characteristics, which translated into human imagination, have led us to define this incredible world of biodesign and eco-styling as *Atelier Embryo* (Fig. 7). In insects' crowded and competitive universe, laying, fertilizing, and hatching eggs represent the crucial node within the individual life cycles. The efficiency of these phases marks the success or failure of the evolutionary process of each individual, inserted in the articulated and complex dynamics of the hierarchies of the victim-predator ecosystemic food cycles. From this need for significant biological differentiation, the world of insects offers us an impressive variety of formal and functional solutions. The study and classification of their morphological matrices are highly interested in analyzing the factors that guide natural design [10].

Within the boundless universe of insects, this research has focused on *Lepidoptera.* This is a vast order of insects to which more than 158.000 species belong, known as butterflies and moths, and their name, from the Greek *lepis*-scale and *pteron*-wing, literally *means wings with scales,* concerning the presence of minute shaped structures that cover their wings. These microscopic elements are arranged one on top of the other similar to the roof tiles. Seen with the naked eye, they appear as a sort of dust. Thanks to the current development of electron scanning microscope-SEM

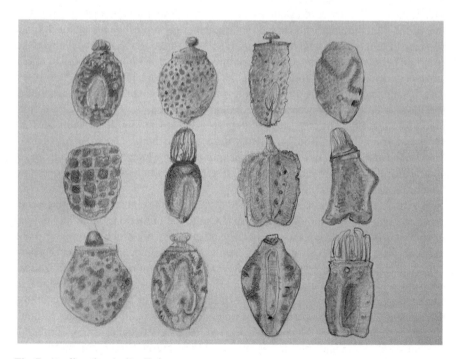

Fig. 7 An alienating *Atelier Embryo*

imaging techniques and other refined digital microscopy techniques, this ultra-thin *powder*, which in the popular imagination is linked to the butterflies' flying power, has shown extraordinary morphological and ultrastructural complex hierarchies and unveiled a microworld of exceptional beauty and efficiency. Butterflies have become the subject of in-depth bionic and biomimetic research. In their fascinating high-tech performance, there are still many unexplored aspects that present potential for transferring ideas in industrial production processes. The present research, for its part, has focused on the analysis of the particular metamorphic life cycle that the *Lepidoptera* have created in the millions of years of their evolution, focusing on the structural and functional nature of their eggs, the interest in the bionic specific.

3   The Butterfly Egg Effect

One of the first studies concerning the description of the world of insects and *Lepidoptera* was published in 1669 in Italy. The essay, written by the doctor and father of microscopic observation in medicine, embryology, histology, and physiology, Marcello Malpighi, is dedicated to the study of *Bombix mori*, and which contains the first scientific documentation of ovarioles and mature eggs of a *Lepidopteran* species. But another astonishing text is undoubtedly that due to the talent of Maria Sybilla Merian. Maria Sybilla, daughter of the German publisher and engraver Matthäus Merian, was a researcher of excellence, a reference point for naturalists and illustrators of the old continent who carved out a space of great prestige and authority in a universe of purely male scientific research. Self-taught, Maria Sybilla challenged the common belief that she wanted insects, defined as *beasts of Satan*, to be born by spontaneous generation originating from the putrefaction of organic matter. She devoted herself with enormous care and scientific value to observing the life of caterpillars; she discovered that they were born from eggs and then described their metamorphic cycle up to the butterfly stage with graces and splendidly illustrated plates. In 1679 and 1683, she published the two volumes of his book: *Der Raupen wunderbare Verwandelung und sonderbare Blumennahrung*, in which, through detailed texts and beautiful watercolor engravings, she described the metamorphosis of over one hundred species of butterflies [11].

A growing interest in insects developed throughout the eighteenth and nineteenth centuries, and real entomology was born. Despite the limitations due to the optical instruments of microscopy, the descriptive and artistic work of the entomologists of the time gave us images of great interest and scientific precision. In recent years, thanks to the advanced technologies of SEM magnification, digital visualization, and sophisticated video shooting techniques, the study of insects and butterflies, in particular, is offering research ideas of extreme interest for biomimicry. Generally, the life cycle of insects is particularly fascinating and complex as it passes through complex metamorphoses. Metamorphosis is a Greek word that means transformation or change in shape and into an insect's world are present two common types of this. Some individuals like grasshoppers, crickets, dragonflies, and cockroaches have incomplete metamorphosis. The young nymph usually looks like small adults but without wings. The complete cycle of metamorphosis is observable in butterflies, moths, beetles, flies, and bees. The young larva has a very different morphology from the adult insect and usually eats different food types. The complete cycle is organized into four stages: *egg*, *larva*, *pupa*, and *adult*. The life cycle can take over a year in some species, while in most specimens, it ends in a much shorter period, ranging from two weeks to a month. The relatively few earliest known *Prodryas persophone* butterfly fossils are from the *mid-Eocene* epoch, between 40 and 50 million years ago, but their development is closely linked to the evolution of flowering plants since both adult butterflies and caterpillars feed on flowering plants (Fig. 8) [12].

Many scientists suggest this specialized association between *Lepidopteran* groups and flowering plants, developed during the *Cretaceous* Period, often called the *Age of Flowering Plants*, 65–135 million years ago, dinosaurs also lived on the earth. Entomologists do not sufficiently understand this fact, and recent theory suggests that the origin of butterflies started 200 million years ago, in the *Triassic* period, from an obscure moth family, the *Hedylidae*. Scientists in 2018 found fossilized butterfly scales the size of a speck of dust inside ancient rock from Germany. This fact can reveal that butterflies came first, even before flowering plants, but a scientist says that modern colored butterflies evolved only after dinosaurs' mass extinction (Schootbrugge 2018) [13, 14].

Butterfly eggs vary in size from about 1 to 3 mm. in diameter, they are oval or spheroidal in shape, and their external structure of the *chorion* eggshell can be smooth or textured. The colors vary from yellow, white, green, and others typical

Fig. 8 *Prodryas persophone*. *Credit* Illustration by author's pencil drawing from fossil photography

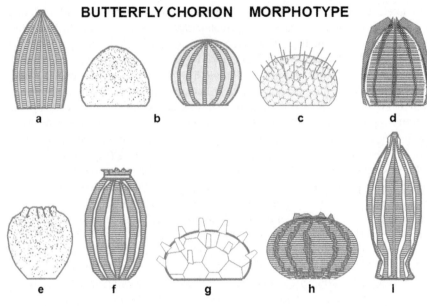

BUTTERFLY CHORION   MORPHOTYPE

a   b   c   d

e   f   g   h   i

(a) *Danainae* **(b)** *Satyridae* **(c)** *Acraeinae* **(d)** *Nymphalidae* **(e)** *Charaxinae*

(f) *Libytheinae* **(g)** *Lycaenidae* **(h)** *Hesperidae* **(i)** *Pieridae*

Fig. 9 Schematic representation of the main *chorion sculpturing* types

for each species. As we will see below, our research interest for a specific species of moths (*Lycanidae, Libytheinae, Hesperidae,* and *Pieridae*)) arose from the particular morphological conformations that their eggs have (Fig. 9).

Eggs appear in the life cycle of a butterfly, usually in springtime, when the female and the male coming out of the larval cocoon, attract each other through their *pheromones* in delicate and ephemeral dances that are a prelude to mating that can take place in the air or on the ground. Mating is achieved by joining the extremities of the abdomen for a few minutes or hours, depending on the species. After mating, the female secretes the eggs made by keratin through the ovarian follicular cells, which can vary from 100 to 300, and lays them. Some butterflies lay their eggs directly on particular plants that will house the voracious larvae. Females choose the most suitable ones thanks to very sensitive taste sensors present in their legs, laying a few eggs on the underside of a leaf of a specific plant of the species, to increase the chances that the larvae have sufficient food resources, thus guaranteeing their survival. The butterfly's eggs are typically attached to the plant, leaves, stems, or branches, with a special bio-glue. This natural adhesive is powerful; it works so well that it is impossible to remove an egg from a leaf surface without destroying the egg itself. The female of *Opodiphtera*, an Australian endemic moth belonging to the *Saturnidae* family, has been the subject of careful studies to understand the superpower of eggs' gluing [15].

The researchers analyzed the characteristics of the adhesives produced by more than 32 insects and found that the strongest was the egg attachment glue produced by saturniid gum moths of the genus Opodiphtera. This adhesive is made by a protein bio-complex hydrogel and is similar to the consistency of molasses, and once released by the female collateral glands, it undergoes an irreversible orange-brown gelling, which shows a very remarkable adhesive power. According to the research team, the recombinant mimics of this material could be functional as adhesives for biodegradable raw materials or as specialty biomedical products. This trend direction of bionic and biomimetic research of naturally inspired super glues can be found in many scientific publications that take into consideration both the characteristics of the eggs of butterflies, moths but also insects with exciting morphological characteristics, as in the case of *Phyllium Philippinicum* (*Phasmatodea: Phyllidae*) generally known as *leaf insects* [16].

The adhesive capacity of the bio-glue used by this insect has been subjected to very sophisticated investigations and measurements that promise open scenarios for developing bioinspired materials. It is fascinating to analyze the chemical relationships between the laid eggs and the host leaves. The butterfly lays tiny eggs that stick to the leaf margins. Consequently, the plant tries to adopt biochemical defenses to eliminate the intruder, such as, for example, activating rotting areas corresponding to the areas infested by egg-laying to try to damage the future intruder. Instead of laying their eggs on leaves and branches, some butterflies spread them in flight, leaving the larvae to choose their host plant. The interest in possible inspirations of bionic or biomimetic basic designs that consider the eggs of moths appears to be a relatively new sector. There are few detailed descriptions of eggs' surface and structure morphology of a closely related butterflies group based on scanning electron microscopy. Our first interest in addressing research on the morphological and functional characteristics of butterfly eggs was inspired by the presentation in a TED on Instagram of a documentary shot by director Louie Scwartzberg: Hidden Miracles of the Natural World. Louie Schwartzberg is a cinematographer, director, and producer who captures breathtaking images that celebrate life, revealing connections, universal rhythms, patterns, and beauty [17].

Through the use of sophisticated shooting directing and digital rendering technologies, Louie's work shows the incredible beauty and complexity of the *natural technologies* at work, such as the production of filaments by the silk glands of spiders. In this video, some stunningly beautiful sequences show the structure and surface pattern of an enlarged, high-resolution butterfly egg. This visual material has triggered the desire to deepen the interest in this still underdeveloped naturalistic aspect and prompted us to search for possible technological and design transfers of a bionic and biomimetic matrix. One of the first specific studies we have found and consulted, devoted to the surface morphology of *Lepidoptera* eggs, is the research work of Thomson [18] of the *Department of Biological Science University of Stirling, Scotland* [19]. This essay considers the external morphologies of the eggs of 14 butterflies belonging to the *Satyridae* species *Manioline*, examined through the use of SEM-scanning electron microscope technology. In addition to the considerable amount of specific zoological and biological information that it was possible to obtain from

Fig. 10 *Manióla cypricola*
chorion's egg

its reading, the aspect that we found most interesting for the biomimetic-inspired process of a particular problem-solving and basic design is linked to the images and the concept re-edited by Thomson of egg's surface *chorionic sculpturing*, used by Abrogast [18] (Fig. 10).

Every species of Maniolina butterfly, M. jurtina, M. megala, M. chia, and M. telmessia and all those subjected to photomicrographic analysis reveal morphological variants of the basic design structure). This structural model, which we could define as the egg dome, is described and classified variously as spherical, truncated conical, subspherical to subcylindrical, barrel, and subcylindrical to truncated conical shaped.

This first classification of the spatial geometric typologies assumed by the various *chorions* allowed us, in the design phase, to create a typological matrix that identifies the standard variants present in the production of lepidopteran eggs (Fig. 11).

Exciting, in the bottom-up process, was the evaluation, present in all the analyzed chorions, of the presence of a series of *micropyle* canals, primary cell petals, polygonal surface prominences, and curved ribs. Subsequently, we have deepened the knowledge related to the functional conformation of a moth egg to understand in more depth the design concept that underlies this natural artifact. Butterflies entrust the possibility of their evolutionary success to the careful construction and engineering of their eggs. To rationally and aesthetically fulfill this fundamental ecosystem function, over the evolutionary arc of millions of years, a dome morphology has been maintained and tested that originates from a spheroidal morphing, which is necessary as a genetic packaging for energy optimization of the volume/surface. The geometric

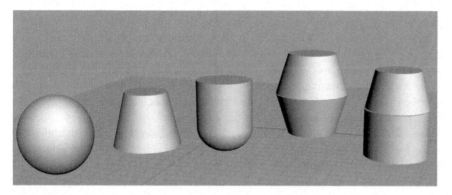

Fig. 11 *Chorions'* basic 3D-shapes on CAD rendering

and structural researches of the first observations of Thompshon (1917) and, in particular, the full development of the *synergetic geometry* and geodetic architecture of Buckminester Fuller and Frei Otto have established how the optimization of a dome shape satisfies more than others possible potential spatial schemes, to the fulfillment of a checklist of energy and functional performances [20].

However, the whole dome architecture is referable to the energetic-spatial behavior of soap bubbles, foams, the surface tension of liquids such as water, and minimal soapy membranes. Inside and outside the genetic capsule of the butterfly egg dome, these are the best functional responses.

3.1   Ensure a Resistant Bond to the Lower Surface of the Sheets or Stems from Counteracting Gravitational Forces

The gluing aspect was previously introduced in the course of the discussion. Since these data await a technical and exquisitely biochemical analysis, they are beyond the scope of our research, and therefore we leave to the bibliographic notes the possible insights for different bionic and biomimetic concepts.

3.2   Optimization of the Closet Package of the Spatial Distribution of the Aggregation of the Brood of Numerous Eggs on the Target Support

On the topic of spatial optimization of spherical or spheroidal cells or domes, there are numerous analyses and research papers, starting with the theoretical schemes proposed by Thompson of the hexagonal arrangements of cells that are like the closest geometric packaging of 2D discs. By putting a series of equal circumferences on a surface in tangential contact, a specific geometric structure of spatial optimization with a hexagonal matrix is generated. The female butterfly that lays a series of eggs, as occurs in an infinite variety of oviparous larval deposition morphologies (see mosquitoes, frogs, and amphibians, spiders, etc.), sequentially expels from the ovarian duct a continuous jet of hundreds of elements that initially they appear to be in a gelatinous, semi-dense state. This sort of *3D-bioprint* produces modules that are all hypothetically identical, which, in the space time intended for the deposition function, must adhere tenaciously to the leaf support. In a short time in contact with the air, they begin to dry and harden, taking on the typical shape of the species. The deposition takes place through groupings of eggs in small clusters that are arranged spatially in such a way as to guarantee the efficiency of future vital functions. In particular, the correct interstitial space must be provided for energy optimization of the hot/humid exchange, aeration, and oxygen/carbon dioxide gas exchange (Figs. 12 and 13). Starting from the construction of the base of every single egg, the shape that we find in all the models analyzed starts from a system of foundations that are practically, ideally, perfectly circular. From this planimetric system, which is subject to the law expressed by Thompson, extruding into space, the eggs assume, as already mentioned, infinite spatial variants of spheroidal, truncated cone, dome, which in most of the morphologies chosen, in principle resemble the shape adopted in the paper construction of the lampshade and flying Chinese lanterns.

Fig. 12 Eggs' arrangements on a leaf's surface

Fig. 13 Eggs' arrangements
on a leaf's surface

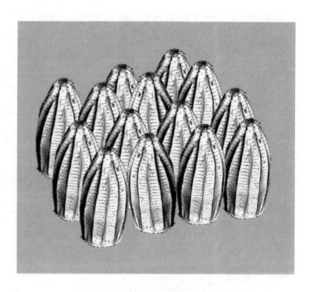

3.3   Offer Mechanical Protection from External Physical Agents (Wind, Compressions, Impacts, Etc.)

Making a thin material shell a resistant spatial structure involves interesting problems both at the micro-scale of insects and at the macro scale of human architectural designs. Considering the due differences in dimensional scale about the action of gravitational and static and mechanical forces on the structures and materials adopted in the two construction domains, the problem of optimizing lightweight frames appears to be a similar case between insects and humans. Biological engineering tends to work by adopting a series of structural solutions which, in compliance with the limitation of the previous point, provide for the creation of specific design strategies such as the creation of surface embossing, conveying, texturing, or radial ribs that shape and the fundamental spatial matrices identified to vary. The trend is to develop spatial structural models that offer effective resistance to form. We have already encountered the efficacy of resistant shell structures in terms of shape, talking about the static behavior of the body of birds' eggs. This structural aspect can be easily verified by experimenting with the resistance to the compressive forces acting on a simple sheet of paper resting on the ends of two vertical supports as if it were a horizontal beam or slab. In *BionikonLab* the student-researchers work to design and make simple structures by folding A4-size cardboard sheets in such a way as to resist the stresses of weight strength. Structural analogies are then observed in some natural living models, such as the leaves of cabbage or other broad-leaved plants, in which the veins help to make fragile leaf surfaces resistant (Fig. 14).

In this extremely thin horizontal structure, the distribution of the construction material through the dimensional ratio between the width, the depth is such that it flexes under its weight, assuming a configuration of reaching static equilibrium,

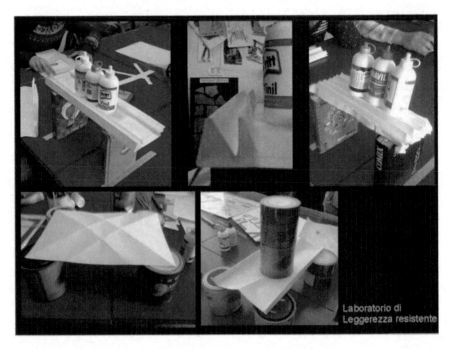

Fig. 14 Simple and intuitive resistance tests by bending cardboard surfaces

typically catenary. In mathematics, the catenary is a particular hyperbolic plane curve, the trend of which is the characteristic one of a homogeneous, flexible and non-extensible rope, bound to two vertical supports and subject only to its weight, as happens for example to an electric cable between two pylons. If we want to allow the sheet of paper to resist at least its weight, we are led to introduce a fold along its entire length. Thanks to the fold, the sheet will change its static behavior and will be able to maintain a perfectly horizontal spatial configuration. The introduction of a series of parallel folds creates a surface treatment such that the sheet, from a fragile structure, unable to support its weight, without adding reinforcement material, is transformed into a horizontal layout as well resistant to be able to carry additional loads. By experimenting with various configurations of folds, it is possible to study the efficiency of resistance to compressive forces. Folding structures, and *Origami* patterns, help apply thin and resistant industrial materials such as variously folded and corrugated sheets (Fig. 15).

The analysis of the ribs and embossing present in a *chorion*, gives the idea of how the moths make the structures of the tiny eggs extremely resistant by acting on the surface structure [21].

In nature, this resistance behavior due to shape is observed in the folds and veins present in the leaves, especially in the extensive and thin ones such as in the banana plant, in all palms like *Saw palmetto* genre.

Fig. 15 Some basic design models of paper folding structures (*shape memory surfaces*)

In structural engineering, a shell construction model is a structural body generated starting from a circular or curvilinear surface, whose thickness is minimal concerning its dimensions, just precisely like butterfly eggs. Due to their structural character-istics, the shell structures are not very suitable for supporting concentrated loads but only distributed loads. The loads applied to the shell surfaces are transferred to the ground by developing compressive tensile. Tangential stresses are acting in the tangent plane locally to the surface, given the complexity of a geometric design capable of opposing an appropriate resistance by shape to the compressive forces of the smooth spheroidal model, which for example, in the case of the hen's eggshell, requires sophisticated engineering of super materials, in the eggs of Lepidoptera it is rarely encountered. Furthermore, butterflies' eggs present the urgent need for a support base that is stubbornly glued to flat support. The forms that present appropriate *chorion sculpturing* is much more widespread through the introduction of concave/convex models, ribs, embossings, and various surface corrugations that guarantee a better structural stiffening and the absorption and structural discharge of the tension forces along the lines of the meridian and the transversal ones along the circumferential or parallel lines. In a hemispherical shell, the direct efforts according to the meridians are always compressive, while the direct efforts according to the parallels can be both traction and compression. The meridians tend to open in the lower area, and the circling forces acting at the base are traction (Fig. 16).

This fact explains how in the structure of the *Lepidoptera* eggs, a sort of foundation ring is adopted, glued to the base plane and to the leaf margin, necessary to absorb the horizontal thrusts of the shell. The chorion's surface shows a system of concave-convex ribs and corrugations, both along the meridians and the parallels. Nature adopted shell structures to design and construct exoskeletons of insects and arthropods, in the abdomen of various insects, spiders, claws and carapaces of crabs, and arthropods in general. As we have seen, in the eggshells of birds, the design of the shape and engineering of the materials probably reaches the heights of efficiency and the most performing formal purity. Shell structures are also found in the bones of the skull or shoulder blade, in the carapaces of turtles, in mushroom hats, in the pneumatic forms of jellyfish, to name a few. In human technique and technology, we ascertain the use of shell structures in the design and construction: of domes, silos, deposits for hydrocarbons and other materials, in inflatable marquees for ephemeral and temporary coverings of sports fields, concerts, stands, etc., in the creation of handicrafts such as pots, bowls, tableware and more and the rounded bodies of automotive and industrial bodies in general. In practice, it can conclude that the need to build a dome shell with relatively thin walls in elevation by the light structure code

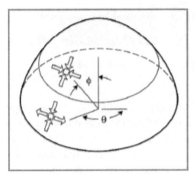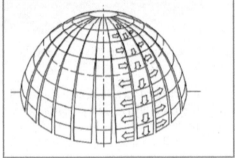

Fig. 16 Static behavior of the compressive forces acting on a dome

Fig. 17 Morphological affinities between a *Lycaenidae* egg and a *michetta* bread, due to the need to organize the surface of a *dome-shaped* structure through a polygonal mesh

leads both in natural and human design, with due differences in scale, to a series of similar structural solutions [22].

It is amazing to note some morphological similarities between a typical form of Italian bread, the Milanese *michetta* bread (a crunchy shell of flour and yeast, light, empty, and thin walls), with those of some eggs studied, *Lycaenidae* particularly (Fig. 17).

As will be seen later, in the course of our research, an interesting exploratory aspect led us, in the basic design experimentation phase, to deepen, starting from the use of a sheet of paper, the possibilities of spatial construction of shell structures using the refined folding techniques of modern *Origami*. The shell structure generates a series of complicated structural requirements. In the history of architecture, we find many examples of dome and shell structures, starting with the Mycenaean *tholos*, passing through the structural masterpieces of Roman engineering that see the highest peak in the Pantheon. All Gothic architecture led to considerable development in the construction of ribbed lightweight membranes structures. The reinterpretation of Gothic structural engineering by the Catalan architect Antonì Gaudì was also made possible by adopting *funicular models* based on the properties of the catenaries to respond to traction loads which, by inverting the arrow, could be assumed as a response curve to compression loads. The domes with more excellent light were divided mathematically and structurally into beams and ribs as in the extraordinary construction of the Centennial Hall in Wroclaw (1911–1913), up to the emblematic cases of structures such as the Zeiss dome-shaped planetarium built in Jena in 1924. With a span of 25 m, it has a shell thickness of only 6 cm. One of the most sophisticated fields of development that offers many natural mimetic implications is Buckminster Fuller's structural geodesic research and Pier Luigi Nervi's extraordinary reinforced concrete architecture.

3.4   Maintain the Correct Initial Permeability to the Male Sperm Flow to Ensure the Fertilization of the Genetic Material

The SEM analysis of the *chorion* reveals the presence on its surface of numerous *micropyle* (from the Greek *mikros*, small and *pulé*, gate), a sort of micro openings that allow male gametes to enter and fertilize the ovum found in a wide diversity of *taxa*, including insects, fishes, cephalopods and plants. Among insects and *Lepidoptera* orders, *micropyles* exhibit considerable variation in position, arrangement, and number. In some species, such as *Drosophila*, *micropyles* protrude from the egg chorion on *stalks* in *Lepidoptera* are located in *micropylar* pits while others are superficial [23]. Some research shows that in some cases, an attractive chemical agent appears to act by the *micropyle* to better direct and channel the male sperm inside the egg [24]. The reproductive biology of egg formation in *Lepidoptera* and generally of insects had an intense development in the Twentieth Century. Even if the in-depth study of these complex and fascinating subjects of zoology is beyond the scope of this study, we consider it interesting to cite an essay that shows and clarifies the fundamental aspects of morphology and physiology related to the production of eggs in *Lepidoptera* [25].

3.5   Guarantee the Conservation of the Correct Internal Valuable Humidity for the Vital Processes of the Larva and Allow the Metabolic Gaseous Oxygen/Carbon Dioxide Exchanges

Generally, every single egg represents in the butterfly's life, and all oviparous beings, the guarantee policy for the survival of their species. For this crucial evolutionary reason, its functioning has been optimized and tested over millions of years to allow the best reproductive success. The tiny nurse-capsules of the dome *chorion* are survival craft launched into its ecosystem's wild and unknown space. Each butterfly abandons its offspring to its destiny, and for this fundamental survival need of the same species, it tries to offer the best conditions of defense and protection to the growing small embryo. The control of the microclimate and indoor gas exchanges represents one of the most delicate problems of environmental engineering that the *chorion* has to face. The surface of the *chorion*, often within the various ribs, is dotted with a layer of other inputs called *aeropyle*. Insects use an *aeropyle* design system to allow the larva to breathe and aerate the internal egg space to optimize gas exchange with the environment, whether the egg is dry or wet. The formation of this layer acts as an efficient distribution system of gases for the developing embryo, as is found in some insects, including *Drosophila* [26].

Very accurate SEM scans show, through splendid photographic images, the variety of solutions adopted in the yards of animal life [27].

Egg *chorion* of castor butterfly *Ariadne merione* exhibits well-organized regional and radial complexity [28].

The *chorion* surface bears longitudinal ridges, grooves, protuberances, and several respiratory filaments bearing *aeropyles* to make correct gas exchange egg's outdoor-indoor. A cross-section of tiny keratin *chorion's* membrane through SEM analysis, shows an ultrastructural bio-materials sandwich created by four distinct layers:

- Internal layer
- Trabecular layer having trabecular air spaces
- Principal layer
- External layer.

The particularly *chorion's* air spaces system offers analogies with the *aerenchyma,* or *aerial parenchyma,* a spongy plant tissue, like mangroves aerial roots, mainly composed of air spaces enabling gas exchange to take place by diffusion underground. This functional aspect has been considered in the design abstractions to stimulate new ideas on the possible engineering of foods-box and particular food preservation films, replacing the PVC-made products widely used by the mass market and the agri-food industry. In these products, the addition of chemicals such as *phthalates* causes contact with foods and food products containing lipids, their solubility, and the consequent release of toxic contaminants and harmful to human health. Therefore, in addition to the architectural and structural suggestions of the overall morphology of the eggshell domes as a source of design inspiration, this extraordinary functional aspect present in the ultrastructures of the eggshell designed to guarantee perfect control of the internal microenvironment of the tiny butterfly eggs, has focused in particular the transfer design interests of our bottom-up process. *BionikonLab's* work teams, through creative passages of brainstorming sessions, decontextualized a series of functional and formal analogies from *chorion's* eggshell technologies. They then elaborated a series of practical and morphological metaphors useful for transferring *chorion* eggshell technologies to potential design solutions for the proper storage of food, fruit, and vegetables through their micropyled-aerial food packaging.

3.6 Offer the First Food to the Caterpillar Immediately upon Hatching

The fertilized eggs hatch after an incubation which varies according to the species and sometimes also by the temperature from the environmental hygrometric conditions and is generally between a few days and 2–3 weeks. When the cleft is approaching, the eggs change color and become darker and reveal the small caterpillar that lives and moves inside them. The eggs do not tear apart violently as the caterpillar carefully carves a sort of circular porthole in the upper part of the chorion cap, pushing it outwards, making an opening, and coming out with a little contortion. This birth phase is very delicate as the caterpillar is very vulnerable and needs food to strengthen itself. Nature provides its first food with the shell of the egg itself that the caterpillar eats greedily to have the first energy to face the difficulties of the world it will have to enter. Once the caterpillar has consumed its first meal, it immediately searches for new food, which is usually provided by the tender leaves that have housed the colony of laid eggs.

4   The Lycanidae's Construction Eggs Design

The *Lycanidae* is the second-largest family of *Lepidoptera*, comprising several 6.000 species worldwide. They are brilliantly colored, and the adult individuals are generally under 5 cm. in size with a rapid flight. This family comprises seven subfamilies, including the blues (*Polyommatinae*), the coppers (*Lycaeninae*), the hairstreaks (*Theclinae*), and the harvesters (*Miletinae*). The whole life cycle of *Lycanidae* offers fascinating behavior and functional aspects that demonstrate the complex evolution of terrestrial ecosystem relations between individuals of different species. The caterpillar born from the eggs has a cylindrical shape with often mimetic colors and, in general, numerous pearls and bristles along the whole body. It moves whit three pairs of short thoracic legs, and to move quickly on the vegetation, it uses particular false abdominal legs provided with a structure similar to a sucker. Its buccal chewing apparatus is equipped with powerful jaws capable of finely chopping the leaves of the host plant. The caterpillar grows very quickly, and during its life cycle, it performs a series of 4–5 molts until it turns into a pupa or chrysalis. The *Lycanidae's* complex metamorphic process exposes them to a thousand dangers due to the attack of parasites and predators. But nature has provided these extraordinary creatures with many vital tricks. Both the caterpillars and the pupae, for example, possess glands that secrete secretions capable of attracting and conditioning the aggressive attitude of the ants toward them and being able to receive mutual protection against parasites and predators from them. Individuals of some species harassed by parasites or predators, thanks to particular structures placed on the eighth abdominal tergite, can emit specific chemical signals that alert the ants that rush to their defense. Some larvae can produce vibrations and a low range of sounds that are transmitted through the substrates they inhabit to communicate with the sentinel ants. The type of social behavior that *Lycaenidae* establish with ant communities is particularly complex and varies according to the species: in some, the relationship (*myrmecohily*) can be mutualistic, parasitic, or predatory. The larvae that establish a positive symbiosis with the ants while feeding on the leaves of the host plant receive protection as if they engage a sort of private army of mercenaries in their pay. The currency with which they repay these services by the army of ants is a particular sugary substance that they sell and of which the ants are very fond. Sometimes some larvae, after a first leaf phase, manage to illegally settle inside the precious nest of the anthill, parasitizing the community that will continue to look after the pupa until the butterfly is born, which takes place inside the anthill. Upon exiting the chrysalis cocoon, the adult butterfly remains with its wings folded and dangling, drying in the sun and pumping *hemolymph* inside the wing ribs, like an inflatable, to take off the nuptial flight. Beyond all this interesting zoological information found on this important family of butterflies, the most significant aspect of the bottom-up research within this study was that related to the analysis of the particular morphology that their eggs present. To accompany us in the knowledge of the structural details of the design of the construction site of the *Lycanid's* domes was, on the one hand, consult a series of specialist entomological researches and, on the other hand, learn about an interesting biomimetic work by the designer Tia Kharrat [29].

In 2016 Kharrat developed the design concept *Metamorphosis: Inception, an architectural structure based on mimicry of the shape of Lycanidae butterflies egg* as a University of Westminster graduate. As seen previously, each species of butterfly produces its particular interpretation of the fundamental design matrix of the egg-shaped dome. The *Hesperidae* explore a variant of a smooth chapel-like dome-shell chorion with little sculpturing, the *Nymphalidae* concentrate around a sort of barrel shape with ribs and various ribs, *Satyridae* produce eggs with forms reminiscent of vases with elaborate lids and finally the *Pieridae* that create tapered eggs, rich in ribs and sculpturing surface. But is the family of *Lycanidae* that have the most geometrical and intricate egg shape like: *L. tityrus, L. virgaurea, L. alciphron, Surendra vivarna amisena, Arhopala abseus, Miletus biggsii,* or *Megisba malaya sikkima.* Tia's research starts from considering that the metamorphosis of butterflies is a well-known and studied entomological aspect, but few know about the start of this journey: the caterpillar egg. The designer applies several methodologies of geometrization and basic design to describe the morphological *chorion sculpture* belonging to basic bionic and biomimetic design methods. The tiny keratin constructions of *Pratapa deva relata,* the Singaporean White Royal Butterfly, presents complex patterns and geometric progressions of the Fibonacci Sequences and the domes geodesics Bucky Ball. Through spatial subtraction of negative spherical tiling and rendering in 3D printing, Tia gets closer to the ratio that structures the geodesy of the egg. Analyzing the patterns that govern these chiseled eggs' structural and spatial organization as jewels continue with identifying subdivisions that can be described and modeled using fractal logics and Lloyd's algorithm to create more uniform Voronoi polygons iterations. The Kharrat project, after various steps of successive geometric abstractions that provide, for example, the CAD graphic rendering of a series of Voronoi meshes and offsets in iterations, produces a whole series of CAD-CAM models of mapping patterns, extruded patterns, and Lloyd's patterns that come to an interesting final consideration. They created subdivisions of the concave-convex surfaces of the chorion through a generative model in a self-referential mode that activates a series of infinite subtraction schemes of spheres at an increasingly reduced dimensional scale. Using the same rules of increasing the density toward the edges, as presented with Lloyd's Algorithm, a pattern can be produced that resembles that of the original White Royal Egg *chorion's* structure. Through further iterations of the fractal process, the abstraction of the model continues up to the final result of a design concept of a sculptural installation in which people can interact with live butterflies. This idea stands as a work of denunciation regarding the ever-declining numbers of butterflies worldwide and in the UK to raise awareness of more significant conservation and education on eco-sustainability. The author imagines an ethereal space filled with dappled light where people can come for contemplation and perhaps their metamorphosis. This concept results from our research to be the only bioinspired bionic and biomimetic design linked to the study of butterfly eggs' morphological and functional characteristics, excluding some still pioneering research works, which try to reveal the biochemistry of super-adhesiveness that allows gluing of their cylindrical bases to the leaf surfaces of the host plants. Our work has moved in a completely different direction, which concerns the theme of food design packaging and conservation, as mentioned several times in the course of the discussion.

5   Food Storage and Conservation in Human and animal's Eating Behaviors

The need for food's preservation, storage, packaging, and transportation has afflicted human civilization since the dawn of time, like hunger and thirst. In every step, country, and society of history, human inventiveness has devised many efforts, technical solutions, and design inventions to set aside food resources safely, often linked to the seasonal or periodic availability of water, fruit, vegetables, etc. cereals or animal's bodies parts. The development of the various preservation technologies has made it necessary to experiment and test solutions capable of guaranteeing the conservation of biochemical characteristics, hygienic safety, and the pleasure of the taste of daily consumption. Over time, countless gimmicks, technological cycles, and food processing methods have been tested. These multiple food technologies inventions they depended on and have been heavily conditioned by several aspects like:

- **climate regional conditions**;
- **food typology disponibility**;
- **energy and water resources availability**;
- **hygienic, cultural, and religious conditioning**;
- **level of one's technoscientific evolution**;
- **economical and marketing global scales strategies**.

Dehydration through solar energy in tropical and warm climates, artificial heating and smoking by fire in areas with wood availability, cooking in modified atmosphere, salting among the coastal peoples, glaciation with ice and snow in cold countries, and many other culinary arts tricks like fermenting, pickling, curing or canning have been experimented to ensure food's organoleptic characteristics. Food's edible quality and flavor preservation were designed not only for actual human survival. Economic and market strategies consider very carefully the aesthetic treatment, custom differentiation of social and financial classes, and pure eating pleasure behavior related to the consumption of foods. Food preservation and backlog are aspects that also affect many other living beings on the planet. Surviving in the wild is a challenge for all animals, even for those that live at the top of their ecosystem food chain, and one strategy that animals use is storing food for the future time. Storage food behavior is a result of the survival instincts of all living creatures, and they hold resources to consume later in possible seasonal changes or to evade the presence of competing for other animal species in the same territory. Can find many examples: tayra (*Eira barbara*) is a mustelid that lives in Central America; it is an omnivore but is particularly greedy for fruit and picks unripe bananas and keeps them until they ripen. The exciting thing is that tayra seems to demonstrate a temporal knowledge linked to the passage of time, as they collect banana food that they would like to eat in the future when it is ripe. Ants store food in their *gaster* (drop or social stomach) and can share this food with other ants, and will do so when requested; *Myrmecocystus mexicanus* lives in arid regions where water and food are in short supply. However, the spreading areas of these ants are affluent in resources during the wet seasons, at which time all the food surplus produced by the colony is stored inside the *honey ants.* Honey ants are specialized individuals that act as living pantries; they accumulate water and honey resources in their abdomen, swilling like a wineskin and hanging from the ceiling of the burrows of the colony nest. When the workers-ants need food, they have to feed the food regurgitated from this living container. Dogs, especially terriers, dachshunds, beagles, are more predisposed than other dog breeds to hide food reserves like bones in burrows dug into the ground. This custom could improve the taste of food, like our raw ham preservation technologies, for example. In this way, they can build food reserves for the low season, keeping it cool and out of sunlight. Squirrels hide seeds and other food in pantry hiding places for the winter season. Many birds also behave similarly, creating food stacks as a food supply. Some birds show extreme behavior, as in the case of having it (*Lanius*), which impales its prey on thorny stems, leaving them to dry to obtain a convenient pantry of food to nourish its chicks. If the caiman (*Melanosuchus*) captures a large prey, they keep it underwater for a certain period to dry it to make it softer.

5.1 The Fragile Ecosystem of the Domestic Refrigerator

The creation of preserves and iceboxes to refrigerate food is lost in the mists of time. The Romans, for example, kept ice, snow, and salt in underground tanks made of masonry to preserve perishable foods such as fish or meat. The first domestic iceboxes were called *neviere (snow)*, which appeared in early 1600, there were sellers of ice who took it from mountain areas, and especially in the summer season, they sold it on the street, trying to preserve it with salt and woolen pieces. The first domestic refrigerators are cabinets with a zinc-coated interior in which ice blocks refresh food and drinks. In 1875 the inventor James Harrison, applying the expansion pump designed by Jacob Perkins fifty years earlier, conceived a prototype of a refrigeration cabinet. Later, Carl von Linde patented the liquefied gas exchanger. The fluid to be compressed and expanded was methyl ether, which had the defect of being easily explosive. Linde replaced it with ammonia, others with sulfur dioxide and metichloride. These gases did not explode but being poisonous if they escaped from the pump, they caused massacres. The first electric refrigerator was invented around 1913 by Fred W. Wolf, but the product was not initially very successful. Later, in 1915, the American inventor Alfred Mellowes was the first to build refrigerators similar to modern ones, equipped with a compressor to produce cold locally and independently. More than a million units of this appliance were sold, thus constituting the first mass refrigerating appliance. In 1926 Alber Einstein and his pupil Leò Szilárd conceived the so-called *Einstein-Szilárd cooler*, subsequently patented in the United States in 1930. It was a design of an absorption cooling device, operating at constant pressure and which required exclusively a source of heat and as an alternative to the device invented in 1922 by the Swedes Baltazar von Platen and Carl Munters. He was also inspired by a news case that told of the death of a Berlin family killed by the toxic fumes of ammonia leaking from their refrigerator due to a broken gasket. The two physicists thought of an apparatus without moving parts capable of significantly reducing the deterioration of the seals. It also worked without using electricity. The system patented by Einstein-Szilárd was simple but was beaten by the American invention of the non-toxic Freon gas with the characteristics of greater thermodynamic efficiency, which became the gas used by all the compressors of modern refrigerators. Only a few decades ago, scientists discovered that Freon destroys the atmospheric ozone layer, and now its use has been banned, and by 2030 it will have to stop its production. All these green anti-time aspects that characterize Einstein's invention have interested a group of Oxford scholars led by researcher Malcolm McCulloch, who in 2008 took up this patent, perfecting some parts to make it usable for the appliance market. Indeed the rapid spread of domestic refrigeration technologies from the second post-war period to today has improved things. The refrigeration industry continuously develops technological innovations to obtain a product with increasingly functional and energy consumption characteristics evolved. Poor storage of food and domestic foods causes food waste that reaches 50/70% in the most developed countries. In contrast, in the rest of the world, the problem of food hygiene is linked to difficulties in accessing domestic refrigeration technologies. And still causes enormous damage to the health of billions of human beings. The correct technology of domestic and professional food refrigeration (canteens, restaurant kitchens, fast-food distributors, etc.) provides for strict protocols in almost all countries of the world (HCCP). Despite everything, the general problem is still far from a solution since, in addition to the technological aspects of the refrigerator appliance, it is necessary to evaluate the incorrect storage behavior by consumers. Recent studies show that many pathologies linked to allergies and food poisoning derive either from inefficiencies of household appliances or from errors in methods and storage cycles. These diseases generally increase during the summer and are linked to the consumption of cold dishes and cross-contamination between foods due to poor storage and hygiene of nutrition and food products. Diseases due to the consumption of contaminated food constitute one of the world's most widespread public health problems. In industrialized countries alone, 30% of the pool is affected by food poisoning.

One of the most common and harmful bacteria is *Listeria* (*Listeria monocytogenes*) which comes from the soil and animal feces and which proliferates in fresh foods such as cured meats, raw and undercooked meat, cheeses made from unpasteurized milk, soft paste, milk vegetables, and in frozen products as it tolerates salty environments and at temperatures between +2 and 4 °C. The internal environment of a domestic refrigerator, where we store food of various types, origins, and organic consistencies, is comparable for the complexity of the environmental variables and the relationships between the parts, like a proper natural-artificial ecosystem. Heat, cold, humidity, thermal exchanges due to the very characteristics of food preservation induce particular cyclical events and interrelationships between these. If we don't consider these complex elements with due attention, the food stored inside can deteriorate and create an altered, toxic, contaminated, and dangerous microenvironment for the consumer's health. The domestic refrigerator is suitable for storing perishable foods in general for a relatively short period as due to its construction type and frequent openings, it is unable to maintain a uniform temperature, with dangerous and continuous changes in degrees and contamination due to the external environment and the constant changes in time, food, and food products. A first general rule, straightforward, to guarantee correct preservation of the food inside is the *first in, first out* one: take out and consume first the foods that we placed in the fridge first. Should always put recently purchased foods behind those already purchased for the longest time. This mode helps to consume old food and reduce the amount of food you will have to throw away. The right temperature for a domestic refrigerator is generally +5 /7 °C on the central shelf and +8 °C on the upper ones; the middle area is the most suitable for hosting dairy products, cold cuts, and cakes. Above the vegetable crate, the lower shelf is the coldest (+2°/4°). Should store fresh meat and fish in this part of the fridge. The lower boxes have temperatures around +10 °C and are suitable for preserving fruits and vegetables that a colder temperature would damage. Compartments, shelves, and crates placed inside the door are the hottest points of the refrigerator (10°–15 °C) suitable for soft drinks, preserved sauces, and other liquids. The refrigerated compartments should not be crammed with goods as it is necessary to ensure the correct circulation of air and the homogeneous distribution of the various temperatures. Frost and ice, which generally form on the back wall in no frost systems, hinder internal refrigeration, causing condensation and poor storage with the formation of dangerous molds. Many vegetables and fruits, such as exotic ones, are best stored out of the fridge. The management of leftovers needs some rules: they must be placed in the fridge within two hours to avoid spores, bacteria contamination, and degenerative processes. It is good practice to wrap or cover foods to prevent them from losing moisture and flavor and store them in special low food containers with lids. Covering bowls and plates with food film also helps preserve them correctly, avoiding biological cross-contamination and odors. Boiling or too hot foods should never be stored in the fridge and keep the door open for long periods. Proper maintenance and periodic internal hygiene also guarantee the correct conservation of the refrigeration ecosystem. Finally, it is necessary to remember all the problems of the freezer compartment, which, through sub-zero cooling, allows food and food to be stored for long periods. A separate discussion concerns the correct conservation of cheeses, which in the specific case of this research constitutes the type of food that has mainly focused on the analysis and production of design concepts. Upstream of the problem of proper domestic storage, all the issues related to the needs of transport, storage, display, sale of food in large and small commercial networks and the complex world of industrial packaging act.

5.2 Sustainability in the Food Packaging and Preservation Sector

Packing, storing, and food delivery are actions and technologies accompanying daily and domestic humanity's gestures and rituals for millennia. One of the anthropological and cultural aspects that we found most interesting in our search concerns *Tsutsumu*, the art of Japanese packaging (Fig. 18). *Tsutsumu*, which means to wrap, to tuck in, is authentic traditional art, linked to the great tradition of *Origami*, *Ikebana* and other Japanese aesthetic and design forms, which attaches great importance to respect for a specific object through a rigorous and elegant packaging process with appropriate and harmonious wraps and materials [30].

The original ideogram shows two curved lines that wrap around each other and come from the logogram of the *fetus* enclosed by a radical covering it as if it were in the mother's womb; that is the image par excellence of care and love [31].

Oka Hideyuki lists three constant characteristics found in traditional packaging:

- Almost exclusive use of natural or derivative materials such as straw, leaves, bamboo cane, fabrics, and papers;
- Aesthetic awareness of decoration that has its roots in Japanese Shinto religious thought, which considers the act of wrapping as a ritual that delimits pure and impure areas that, separating the inside from the outside, purifies it through order and cleanliness;
- The extreme care with which each material is manipulated through a patient, skilled manual wisdom means giving great importance and respect through the gift to its recipient.

With the progress and the advent of synthetic materials of these three fundamental characteristics for standard packages, only consideration for the recipient survives in the modern culture of mass production. In fact, in modern Japanese packaging, the reference to natural materials remains persistent, and the attention to food's orderly, rational, and hygienic handling. However, packaging, especially food and disposable packaging, is considered one of the leading causes of environmental pollution and aesthetic degradation. Every day in the world, millions of packaging, primarily made of plastic materials, are poured into the soil and waters that constitute massive landfills that tragically impact the planet's social contexts and natural ecosystems. Food packaging in particular, however, plays a fundamental role in the protection of human health, as it prevents and limits the possible serious diseases linked to

Fig. 18 Flayer of Oka's traditional *Tsutsumu* collection held at Meguro Museum of Art, 2021

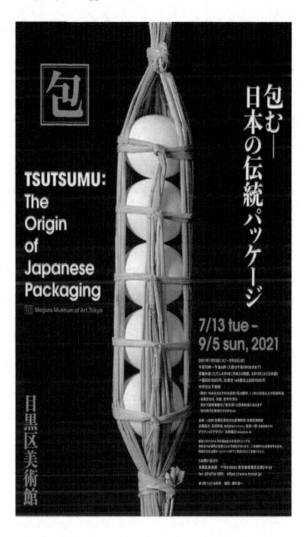

pathogenic contamination and incorrect forms of conservation of products related to the human nutrition of billions of individuals in the world. Due to the enormous socio-economic impact of the sector, new areas of research are developing worldwide that converge in redefining a possible overall redesign of the packaging itself and the consumption and use rites associated with it toward greater sustainability. Sustainable Development has been defined since 1987 as:

> development that meets the needs of the present without compromising the ability of future generations to meet their own needs. [32]

This definition is of particular interest if we consider the growing trend of the world population, which is estimated to reach 10 billion individuals by 2050, a process that will trigger enormous pressures on global production chains, while every year, a third of all the food produced for human consumption. Nowadays, the publication of the 2030 Agenda and its 17 SDG sustainable development goals aims:

- to end hunger, achieve food security, improve nutrition and promote sustainable agriculture (Obj 2);
- to guarantee a healthy life and promote the well-being of all at all ages (Obj 3);
- to ensure sustainable consumption and production models (Obj 12).

In this context of a future marked by the general development of sustainability, food packaging, thanks to its primary functions, such as the preservation, protection, and ease of distribution of food, can play fundamental importance in improving the sustainability of the food sector. A turning point that could slow down and reverse the current trend of severe environmental pollution is represented by a total rethinking of the entire supply chain even if many controversial and contradictory aspects remain. In fact, on the one hand, sector legislation is focused on supporting forms of prevention and reduction of the quantities of material plastic, incredibly disposable, given the problematic issue of its disposal and recycling. On the other hand, recent scientific studies show that food packaging can bring environmental benefits through the correct storage of food and, therefore, food waste. It is necessary to develop integrated and systemic intervention models that consider all these aspects in a sustainability model that evaluates the interrelationships between environmental, social, and economic factors: People, Planet, and Profits. Life Cycle Initiative, coordinated by UNEP (United Nations Environment Program), defines the Life Cycling Thinking approach in this regard. In the logic of this approach, a sustainable future for Food Packaging must always foresee and integrate the following aspects:

- **Strategic tools for sustainability**
- **Innovative packaging materials**
- **Packaging design features**
- **Shelf-life optimization**
- **Consumer's behaviors**.

The life cycle assessment or Life Cycle Assessment (LCA) methodology is currently one of the most used in determining the environmental impacts of products or services. Thanks to its results, it can guide the choices for stakeholders in preparing and managing new options of sustainable strategic decisions. In particular, it appears essential in the impact assessment of the food packaging sector. It is possible to evaluate, in an integrated way, those that are considered as *direct effects* with *indirect effects*. In food's packaging researches they're necessary five challenges to arrive at the evaluation of natural, sustainable solutions [33, 34]:

- Identify and obtain specific data on the functions of the packaging that influence food waste;
- Understand the total environmental impact of packaging-food systems by evaluating trade-offs between product protection needs and its environmental impact;
- Understand how should consider the functions of packaging in environmental impact studies;

- Improve the characteristics of the packaging solutions bearing in mind the possible reduction of food waste;
- Studying incentives for stakeholders to reduce food waste.

In particular, the correct conservation cycle and the package of food products of animal origin, such as meat and cheeses, generally have a high degree of environmental impact, can lead to an overall environmental benefit. The rethink in LCA's forms of usual aspects of packaging, such as materials, technologies, and design, could positively influence food waste. Several elements seem to be crucial in this process: intervening on the shelf life of the product, rethinking the technologies and functionality of food preservation systems, especially in the home, and influencing, through forms of food education, the behavior of consumers toward correct methods of food's conservation and consumption with a higher rate of organic deterioration. In this research, we have acquired them to configure the contours of the general problem-setting and the framework of possible alternatives and solutions through a series of bioinspired design problem-solving concepts. From a methodological and didactic point of view, after the bottom-up biological analysis phase, which made it possible to identify some central aspects of the problem, such as natural aeration systems and technologies, the meta-design phase began. The first step of this research step start from a creative brainstorming set of question to define our level of perception of the problem, of which the most significant were:

- How significant is the impact of packaging on the psychological perception, the choice, and sale of a food product?
- Does packaging influence the perception of healthy food?
- How much does the current food packaging system contribute to global pollution due to its packaging, transport, storage, sale, consumption, and disposal?
- What styles of use and conservation of food do we adopt every day?
- Do we think alternatives are possible?

Initially, all these questions were filtered by our experience as consumers who daily contact food, drinks, and derivatives and their package design and communication system. Once we analyzed our behaviors and styles of consumption, we made a checklist and then compared them with a series of sector studies. Most food products are packaged for sale, and their environmental footprint depends not only on the product itself but also on its package. To reduce the environmental footprint, you can modify both intrinsic attributes and extrinsic attributes of the product. Most products are packaged in an attractive casing, often to seduce consumers' attention, indispensable to store and protect a product and ensure its hygiene. At the same time, the container informs the consumer about the product's characteristics and prevents the content from coming into unwanted contact often, however, the packaging is only noticed at the end of life, when it is transformed into waste to dispose of once its contents are used. Companies of any merchandise consider packaging to be a fundamental element for consumers' perception of the brand, positively impacting its value and recognizability. Its function goes beyond the purely practical aspect of protecting and transporting the product. It has the task of differentiating it on the shelf of shopping and, above all, of significant distribution, where the shape, material, and overall graphic element must be beautiful. Guided by this series of considerations, we explored and identified the main aspects that characterize food packaging, focusing in particular on those that in some way recalled the characteristics of vital conservation that we had found in the functional design of eggs in general and of moths in particular. Like the tiny larva that inhabits the inside of the *chorion*, food, in particular fruit, vegetables, dry foods, grains, meats, and cheeses, can be considered as a living organic material, whose vital characteristics must be preserved at the highest level. Rice and *pasta* do not grow and develop in a vital metamorphosis inside their packaging, even if they undergo a series of molecular transformations and degradations over time. They often don't need to breathe; in fact, the packaging techniques try to avoid the harmful and destructive oxidation of contact with air, oxygen, and bacterial loads of the external environment for most food products. The recent vacuum technology, which involves the forced extraction of the air inside the plastic bags that contain the food, combined with the technology of the cold cycle in the case of fish, meat, and cheese, has solved many diseases. It is a food preservation technology that limits the use of preservatives that are harmful to human health and the pockets of manufacturing companies. Depending on the food considered, vacuum technology can lead to a third, double, or even three times the storage time compared to standard refrigeration. In considering the conservation needs of the organoleptic qualities of foods and their safety and hygiene standards, we have evaluated that the functionality of the chorion is a source of bionic and biomimetic inspiration that may initially seem a contradiction. The aerophilic layer system manages the gas fluxes, to breathe and expel carbon dioxide, internal–external gas exchanges of the egg as the larva needs oxygen. The Chorion *aeropyle* layer system and the vacuum refrigeration technology seemed to position themselves at the antipodes of the scale of values adapted for a *bottom-up* transfer. But in the research continuation, we discovered that in the correct conservation of some foods, the management of the aeration is a fundamental aspect of their organoleptic preservation, as in the particular case of cheeses. The *aeropyle* layer system manages the indoor-outdoor gaseous exchanges of the egg as the larva needs oxygen to breathe and expel carbon dioxide. *Chorion aeropyle* layer system and vacuum-refrigerated technology seemed to position themselves at the antipodes of the scale of values that can adapt for a *bottom-up* transfer. Some foods in which aeration management is a fundamental aspect of their conservation, as in the particular case of cheeses.

5.3 Cheese and Other "Living" Food

Cheese is a dairy product obtained from the acid or rennet coagulation of whole milk, partially or full skimmed by coagulation of caseins, with the addition of enzymes and table salt. Western European mythology traces the origin of cheese to Aristeo, son of Apollo and the nymph Cyrene. In reality, the origins of the cheese get lost in the notes of the times, and it has deep roots in the civilizations of the Mediterranean basin, in North Africa, and Asia Minor. Its discovery is probably due to the sense of seeing in the herding culture of curdled milk in the bowels of animals slaughtered during lactation. From the production of the curd, the following types of cheese are obtained:

- **Soft paste** (*stracchino, brie, quark*)
- **Semi-hard paste** (*fontina, provolone, pecorino*)
- **Hard cheese** (*Grana Padano, Parmigiano Reggiano*).

Due to their delicate organoleptic characteristics, cheeses are one of the most challenging types of food to preserve, as they are live foods, subject to continuous changes in their very chemical constitution. They oxidize quickly on the surface, hardening. Phenomena such as proteolysis and the increase of olfactory and gustatory notes can reach harmful and unpleasant levels for the taste and human nutrition. The sun's rays and UUVV rapidly degrade its fatty substances and vitamins in its organic mass. In the peasant and rural world, the ideal place for storing cheeses as well as cured meats and wines were and are the cellars and pantries, generally underground places or located in particular areas in the house, which have unique constructional characteristics of isolation and orientation in terms of temperature, humidity, and air circulation. The cellar generally has a temperature between 10° and 15° and a humidity rate between 80 and 90% and guarantees proper ventilation that does not have to foresee violent flows of air and sudden temperature changes. For some soft cheeses, temperatures drop between 5° and 10°, while for cooked cheeses, more temperate cellars are needed, between 12° and 20°. The basement for the maturing and professional preservation of cheeses is a natural ecosystem. Its delicate balance and a slight variation of its parameters are enough to alter the proportions between the various species of molds and bacteria, negatively impacting some cheese taste. In practice, the perfect conservation of the cheese, even for domestic use, should simulate as much as possible the characteristics of these unique places. The complex techniques of creation and aging of the hundreds of types of products available on world markets were born and developed and refined precisely based on this extreme variability in temperatures, acidity, contacts with air and oxygen, of the molecular chemical system of the milk and its dairy products. Most dairy products sold in large food distribution chains and cheeses are vacuum packed, in polystyrene containers, and wrapped in PVC films or bags. Once purchased, the conservation of the cheeses requires the use of the refrigerator, in particular refrigerated areas, and generally in closed containers for food, as the smell they emanate is powerful, characteristic, not always appreciated by the modern sense of smell and thus risks contaminating the characteristics of all the other foods in the refrigerator or pantry. The enemies of cheese conservation are:

- Excessive contact with the air and therefore excessive drying causes it to harden and rapidly lose those qualities of flavor and fragrance that are typical of it;
- The lack of air that suffocates and prevents its natural fermentations;
- The excess of cold that blocks its organoleptic qualities;
- Excess heat accelerates excessive fermentation.

As you can see, cheese is a type of food that requires constant isolation from the air, unlike many others, requires controlled ventilation. Neither too much oxygen nor too little, and for this reason, the home preservation of the purchased cheese, can therefore represent a problem on which to develop biomimetic problem-solving. There are, therefore, rules to ensure good home and catering conservation of particular foods such as cheeses, capable of maintaining the flavors and organoleptic qualities typical of their food status as unaltered over time as possible:

- Temperature: fresh cheeses must be placed in the coldest area of the fridge (2–4°); The aged cheeses with pasta cooked in the less cold one (10–12°) while the other types in the compartment at temperatures between 6°and 8°. The sudden changes in temperature are very harmful and often affect the organoleptic characteristics of the product and even in some cases lead to the formation of pathogenic microorganisms for human health;
- Freezing: the cheese must never be frozen, as the subsequent defrosting irremediably alters the molecular structure of the pasta, compromising its gustatory and olfactory quality characteristics;
- Counter cut cheeses: the slices must be stored wrapped in greaseproof paper to maintain the proper humidity and never in transparent PVC films, as the phthalates present in the plastic material to soften it, react with the fats, releasing toxic substances;
- The ideal of conservation, according to the most accredited gastronomic traditions, involves the use of moistened linen cloths to wrap the cheese that is stored under glass bells, suitably closed to allow the dispersion of condensation water (the cheese must breathe);
- To let the cheeses breathe, we recommend quick storage in perforated polyethylene sheets to ensure a certain air circulation or in food-grade paper bags, which are porous and gradually permeable to oxygen;
- Since the correct tasting of most cheeses is around 16°, the cheese must be removed from the refrigerator at least ½ h 1 h before being consumed;
- Limit the vacuum storage of hard cheeses to short periods and never for soft cheeses.

In the ways of preserving the past and in the old-style culinary traditions that remain, stored the cheeses in special containers, the so-called *cheese safe, cheese cage, moscaiola (fly cage), cave a fromage* (Fig. 19).

These objects, prevalent in kitchens and pantries worldwide before the advent and widespread diffusion of domestic refrigerators, were used and placed in a niche of the walls more exposed to the North, where even in summer, the temperature could be lower and higher humidity. Made of wood and with a thin metal mesh, it allowed the aeration of the cheeses and cured meats stored inside, preventing unsanitary contact with insects such as flies, midges, ants, cockroaches, moths, etc. From the construction point of view, most of the models analyzed correspond to small cabinets equipped with one or more shelves and with walls made of metal mesh framed by wooden frames. There are also other table solutions, such as the paramosche food covers that function as a kind of openable umbrellas placed on

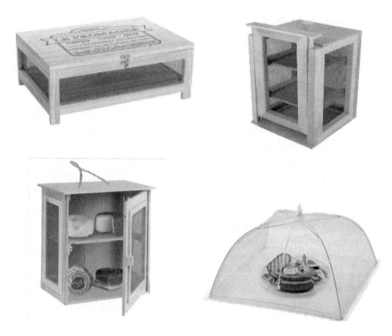

Fig. 19 Different vintage cheese safe models

a serving dish to defend it from the attacks of insects and dust, and other hostile environmental agents. Like the small butterfly eggs, these simple objects with an anonymous design solve some problems related to the correct conservation of *living foods* such as cheeses. The research phase on packaging and food containers also provided interesting images of objects conceived on an industrial level and created to address this type of problem within the space of the domestic refrigerator. In most of the models analyzed, the basic idea is a transparent box usually made of glass or PVC with a lid. Some models integrate a removable grid-like bottom to ensure ventilation of the parts in contact. In some models, there is a cutting board made of either wood or plastic materials, designed to present cheeses directly on the consumer's table. There is the possibility of ensuring ventilation through adjustable micro-perforated hatches in specific models, but without these solutions relating to a strategic and specifically designed distribution of the ventilation system. In general, we can conclude that the sector does not offer particular design innovations and shows inevitable poverty of ideas and alternative solutions with a more specific design. Our concept fits precisely into this production niche, experimenting with a series of hypotheses and solutions that take formal and functional cues from the results and bio inspirations of bottom-up research and the conservation solutions adopted in the past and current products used for the preservation of food, especially in the context of domestic refrigeration.

6 Design Inspirations

6.1 *Bee's Wrap*® *and Other Bees' Wax Food Package*

Specific market research on alternative products to PVC and aluminum food films has allowed us to collect many valuable suggestions for the definition of our problem-solving. For years, a whole series of food packaging and storage products has been on the market, made from natural and sustainable materials such as organic cotton and beeswax, which, when properly used, are proposing themselves as valid, totally recyclable, and sustainable alternatives. To products based on synthetic polymers. These products for household packaging are organic cotton cloths soaked in a mixture of beeswax, which, when cooled, becomes waterproof and flexible. It is used by heating them with the heat of the hands and modeling them around foods or containers for food storage, constituting an ecological and natural alternative to aluminum and plastic films. *Bee's Wrap*® proposes using reusable and compostable products is a small yet meaningful step toward reducing single-use plastic and food waste. It is not recommended for preserving meat or fish, incredibly if raw. Similar products are making their appearance in world markets by proposing a sustainable and, in a certain sense, *Tsutsumu*-oriented approach to the daily life of gestures linked to food consumption by billions of individuals in the world. In addition, several studies have shown that beeswax-based films help fight bacteria and microbes in food, and their breathable properties favor the preservation of food that remains fresh and fragrant for longer. *Bee's Wrap*©, *beeskin*®, and many other similar products, now available on the international market, are sustainable and 100% natural product and does not derive from industrial processes and hydrocarbon derivation. They have been washable and reusable for a long time, reducing domestic storage costs. Projects for the inclusion of this technology in large-scale distribution cycles are being studied.

6.2 Origami, Kusudama, Oribotics, Snapology, *and Other Paper Folding Techniques*

In addition to the aforementioned Japanese *Tsutsumu* Art of packaging, we have considered another series of sources of inspiration for our project, such as the in-depth study of particular artistic paper folding techniques such as *Origami* or *Kusudama*, especially in the more advanced contemporary versions.

6.2.1 Origami

The traditional Japanese art of Origami, from *ori*-folding and *kami*-paper (*kami* changes to *gami* due to <u>*rendaku*</u>), dates back to the Edo period, around 1600 AD. In modern usage, *Origami* is an inclusive term for all folding practices, regardless of their culture of origin. This technique of paper manipulations aims to transform a flat square sheet of paper into a finished 3D shape through folding and sculpting techniques. Traditional *Origami* start from a square sheet of paper, and modern origamists, generally discourage using cuts, glue, or markings on the forms. We can combine the small number of basic origami folds in various ways to make intricate designs. The principles of *Origami* help designers and engineers create a new approach in stents, packaging, robotics, and other engineering applications. The potentialities inherent in the various paper folding techniques and the fabric materials are currently mainly developed thanks to many studies and research from academic fields of structural engineering, mathematics, industrial design, and computer science [35].

The Japanese computer engineer Jun Mitani, Professor of Computer Engineering at the University of Tsukuba is one of the most interesting contemporary origamists. His work was of great help in the basic design and engineering phase of our projects. Mitani has also developed various open-source software, thanks to which it is possible to create quickly, particularly complex spherical and 3D curved origami forms [36].

Through a series of simple tracing of mountain and valley folds, Jun developed an unseen world of 3D shapes. The similarity of many of Mitani's models to the primary morphologies found in the analysis of the *Lepidoptera chorion* appears extraordinary. These two-dimensional media folding techniques support the creation of possible innovative solutions for industrial packaging products. Another valuable source of suggestions for constructing exciting and unpublished 3D models is undoubtedly the site origamitutorials.com, which proposes a series of models to realize complex geometries that have many affinities with the morphologies of the analyzed butterfly eggs.

6.2.2 Oribotics

Oribotics [laboratory] is a project of collaboration between Matthew Gardiner and Aphids, especially Artistic Director David Young as Composer. It was shown over two nights at the Asialink center at Melbourne University in 2005. Oribotics is a concept created through the fusion of *Origami* and technology, specifically "bot" technology, such as robots or intelligent computer agents known as bots. Ori-bots is a joining of two complex fields of study. Robotics is Origami that is controlled by robot technology; paper that will fold and unfold on command. An *Oribot*, by definition, is a folding bot. By this definition, anything that uses robotics/botics and folding together is a robot. This search includes some industrial applications already in existence, such as Miura's folds taken to NASA's space design high-tech tricks. Robotics is a field of research that thrives on the aesthetic, biomechanical and morphological connections between nature, *Origami*, and robotics. At the highest level, Gardiner's *Oribotics* tech evolves toward the future of self-folding materials, and the focus of the author is the actuation of fold-programmed materials like paper and synthetic fabrics. The design of the crease pattern, the precise arrangement of mountain and valley folds, and how they fold and unfold directly informs the mechanical design, so a critical area of current research is discovering patterns that have complex expressions that can be repeatedly actuated. Studying the potential inherent Oribotics technology has allowed us to deepen the *self-folding* and *shape memory* characteristics inherent in particular materials. It has been of great use in redesigning the home packaging and storage foods using PVC and aluminum rolls-films.

6.2.3   Kusudama

Kusudama derives from the Japanese words for medicine and ball and alludes to the fact that it was traditional to create spheres of herbs considered medicinal in ancient times. Introducing the popular culture of paper objects folded into spherical shapes, scented and enriched with incenses and balms slowly replaced this custom. Since the various modules are tied together with threads or glue, purists tend not to consider this simple origami technique.

6.2.4   Crumpling

This technique consists of crumpling and reopening a sheet of paper several times and, at the same time, creating a series of particular models with a solid natural appearance of the surface and shape to obtain "organic" figures and objects. Proposed for the first time by Paul Jackson, the French Vincent Floderer developed it.

6.2.5   Curve Folding

We can find the first examples of curved paper folding starting from some experiments in the Bauhaus design courses around the 1920s. Pioneers of this surface modeling approach were Ron Resch and David Huffman. A lot of interest has been created around this technique and its potential, especially from the academic world, and worthy of note are the works of Ekaterina Lukasheva and Philip Chapman-Bell.

6.3   Design Concept: ChorionPack©—3D-Textured and Ventilated Bees-Waxed food's Cardboard Emballage

6.3.1   Premise

The first food films were produced using cellophane, a transparent cellulosic material whose synthesis was discovered in the early twentieth century. In the decades following it, various other organic polymers such as polyethylene were joined to produce food films. Starting from the 1930s, PVC began to be made, which in the remainder of the twentieth century dominated the materials used for this purpose thanks to its cost-effectiveness and ease of use. In recent years, due to the recognized danger of the additives used in the production of PVC film, the share of food films produced with other plastic materials has increased considerably, including, in particular, polyethylene and EVA (ethylene vinyl acetate). For several years it has been known that the use of some types of plastic films can cause the contamination of stored food with substances harmful to health. In particular, phthalates, a class of substances added to PVC to improve its flexibility and mouldability, are under accusation. The risk of contamination is more excellent in foods that contain a significant amount of lipids, in which *phthalates* are more easily soluble. The percentage of these substances added in the films has been very limited by current legislation, and various companies market polyethylene-based films in which PVC is absent. Alongside PVC film, in the post-war period, the use of thin aluminum foil, also known as aluminum foil, was imposed to preserve food at home in many homes, kitchens, and markets around the world. Recent studies show that this food preservation system also hides a danger to the environment and human health. From a production point of view, aluminum produces a lot of pollution and consumption of primary energy. Furthermore, its contact with acidic foods causes chemical processes of molecular degradation and risks transferring aluminum microparticles into protected foods. Aluminum is a highly toxic metal for the human body. If taken in excess, it can affect the bones and central nervous system. As previously considered, for some years now, on the market for products intended for home preservation of food, there are a series of substitute products, more ecological, which use eco-sustainable materials such as whole cotton fabrics, essential oils, etc., and beeswax. The design concept that we propose through this article, is inspired on the one hand by the bottom-up analysis of the *chorion*, with particular reference to its internal aeration and ventilation by *aeropyle* layer and, on the other, by the paper folding techniques suggested by the research by Jun Mitani and contemporary origamists. The novelty is to imagine producing rolls of sheets for food, based on recycled cardboard, which is industrially punched according to the upstream fold patterns as in *Origami*. The sheet's surface has a series of micro-perforations and, employing the shape memory property impressed in the sheet through the series of folds, by manipulating them quickly, it is possible to obtain *Origami/Tsutsumu* type envelopes of various sizes. After analyzing multiple construction schemes proposed by Jun Mitani and tutorials of origami sites, we have developed a series of prototypes designed through specific software. The final idea is to replace PVC films and aluminum foil for preservation whit an ecological package system. We checked different forms of models, inspired by the morphology of the eggs, creating packages of various sizes. One aspect of our design research considers the properties of high environmental sustainability and potential alternatives to plastic films in these products. In the context of this research, we have developed a problem-solving concept for the industrial creation of a series of packages. The *ChorionPack©* is a way of packaging and home storage of solid foods that *need to breathe,* such as hard and semi-hard cheeses. This system can store food inside and outdoors, like in cheese-cages and the refrigerated environment. Custom packages are inspired by the *Tsutsumu* philosophy, which envisages giving particular attention to the ritual of food packaging. Greater attention and respect in the handling and conservation and presentation at the table of the food we eat daily is a guarantee of education and hygiene of the body and mind and regards the saving of food waste for the improvement of the state of the environment and the economy. In addition to the *Tsustumu* inspiration, the suggestions offered by the potential of Jun Mitani's innovative cylindrical-spherical origami were essential to creating personalized homemade packages with an original folding design in paper impregnated with beeswax and essential oils. The *ChorionPack*[©] system exploits the shape memory that can be imposed on ecological cardboard sheets by punching and creasing the surface according to the origamic scheme of mountain-valley folding. Inspired by *Chorion sculpturing* and aeropyle layers, the various sides of the package have a series of small holes to allow the aeration of the internal food content. Let's think about the *ChorionPack*[©] structure differently. The first hypothesis involves making this domestic package system through industrial processes or on a semi-artisanal scale, using sheets of full organic, 100% ecological cardboard with a weight of around 200gr./mq., which are industrially impregnated with beeswax and essential oils, compatible with contact food fats. Specifically, we propose evaluating the opportunity to experiment and transfer the wax impregnation technology from the organic fabric, and beeswax packaging system, to the cardboard material. This innovative domestic food packaging and storage system is particularly suitable for cheeses, fruits, vegetables, and foods that require a ventilated storage environment. This whole new design approach can be defined as *ShapeMemoryPackagingSystem*[©]. The concept proposed by the *BionikonLab&FABNAT14* teamwork opens a possible alternative direction to aluminum foil emballage and plastic boxes in creating and developing new self-assembling packaging methods. This new design area would be part of the current research in bionics and biomimetics solutions that, by combining origami and nature, create industrial solutions inspired by orobots. The predisposition of an original folding system in the packaging material, mainly cardboard and cardboard-waxed fabric couplings, could quickly guide the user's handling in creating customized food-preservative home containers. Different starting formats with various sizes could be created and offered to the consumer. Each of these is useful for realizing different final 3D shapes of packaging and storage.

6.3.2 Problem-Solving

The first problem-solving step was to generate an accurate morphological analysis of the geometry of a butterfly *chorion*. For the characteristics identified in the checklist of the problem-setting relating to the aerated domestic packaging and storage of particularly delicate foods, fruit, and vegetables, the choice fell on the *Maniola* genus eggs and others (Figs. 20, 21, 22, 23 and 24)

They have a cylindrical-barrel shape whit *aeropyle* layers, which lends itself to generating interesting types of cardboard-waxed drilled boxes. Their *chorion* appears, in SEM enlargements, scaffolded through a sequence of vertical ribs (we

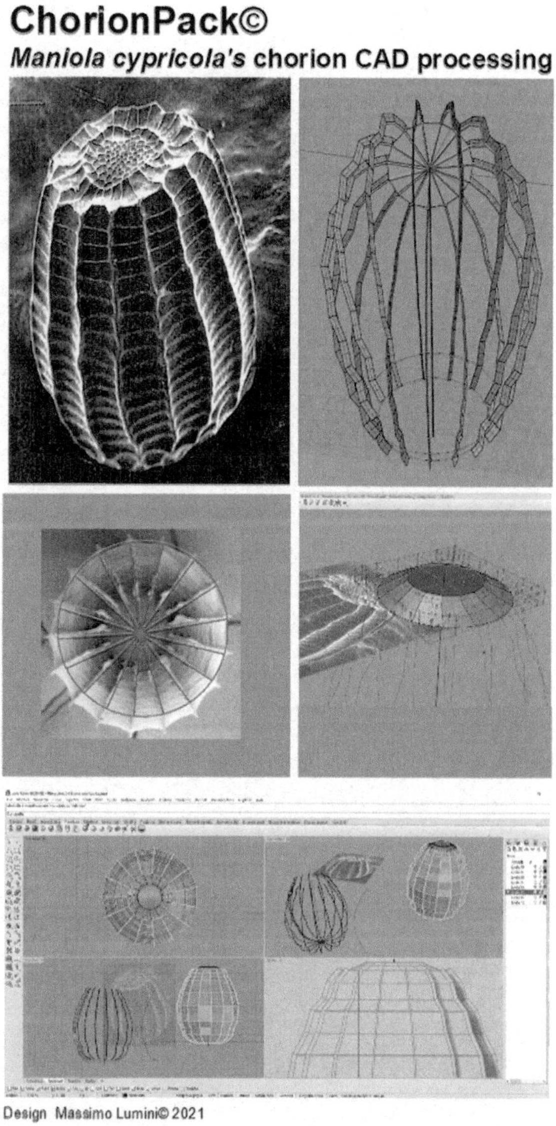

Fig. 20 Morphological analysis and CAD rendering of *Maniola's* and other butterfly species *chorion sculpturing*

ChorionPack©

Maniola cypricola's chorion CAD processing

Design Massimo Lumini© 2021

counted 15–17) which, like staves of a barrel, stiffen and shape the *chorion sculpturing*. This system of ribs, which initially appeared to us placed on an orthogonal plane at the base. Upon closer observation, they suggested a slightly curved trend to give a kind of rotation to the whole structure. A rib-depression etched system into the tiny keratin *chorion's* surface is comparable to an *Origami* mountain-valley folding paper. We have observed that the various ribs that run along the entire longitudinal surface of *chorion*, tend to give a twisted organization, as shown in Fig. 25.

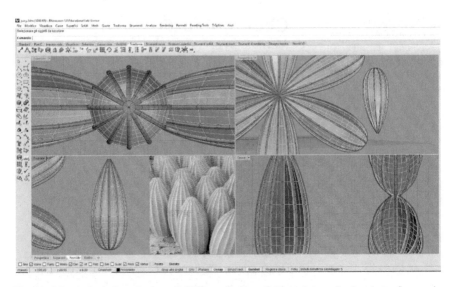

Fig. 21 Morphological analysis and CAD rendering of *Maniola's* and other butterfly species *chorion sculpturing*

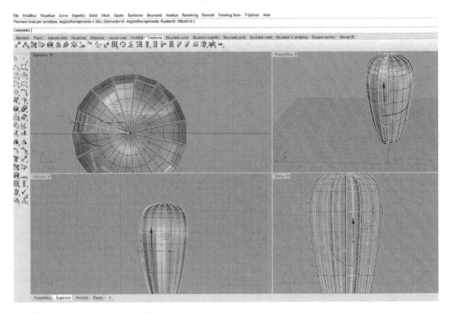

Fig. 22 Morphological analysis and CAD rendering of *Maniola's* and other butterfly species *chorion sculpturing*

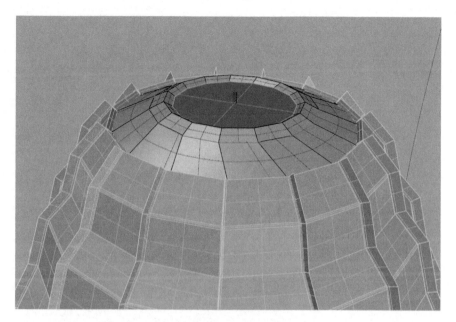

Fig. 23 Morphological analysis and CAD rendering of *Maniola's* and other butterfly species *chorion sculpturing*

This image has been associated, by formal analogy, to the characteristics of folded cardboard model, proposed by Japanese Prof. Jun Mitani. This particular spherical 3D pattern is generated by the twisted spiral motion of various impressed onefold-ribs on its surface (Fig. 26). We made a prototype of it, using Mitani's origamic software. The 3D shapes created by Mitani works based on what we define as *Shape Memory Packaging System*$^©$ (Fig. 27). These rotational spherical shapes are easily obtained by manipulating a structured cardboard cylinder with a series of folds. Combining it with a compression and rotation action of the two hands makes it easy to impart a rotation generator to the entire surface. From a tubular folded shape, a spheroidal corrugated shape is created. Subsequently, the design process was to generate, with the use of ORI-REVO$^©$ software, a series of morphological variants of the basic barrel model. The set of programs developed by Mitani (ORI-REVO$^©$, ORI-REVO-MORPH$^©$, and ORI-REF$^©$) are available on the web and can generate different folding plans, and their effectiveness verified. The ORI-REVO$^©$ program allows us to create a 3D shape generated by drawing more or less curved lines over a two-dimensional grid. A real-time rendering model shows the effect of tracing red and blue lines (fold mountain and fold valley) in a rectangular sheet and a render 3D model. By saving the file in the.obj extension, it is possible to view it in other CAD-CAM software. With ORI-REVO$^©$ we can generate a folding model of a sheet with cylindrical projection up to a maximum number of faces of 32 (Figs. 28, 29, 30, 31, 32, 33, 34, 35, 36 and 37).

Fig. 24 Morphological
analysis and CAD rendering
of *Maniola's* and other
butterfly species *chorion
sculpturing*

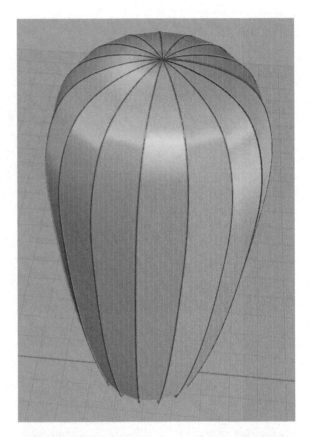

Fig. 25 In this *Maniola's*
egg, fifteen ribs adorn the
chorion sculpturing in a
twisted direction

Fig. 26 Example of Jun
Mitani's spherical Origami
model handling. The design
was generated through his
ORI-REVO© software

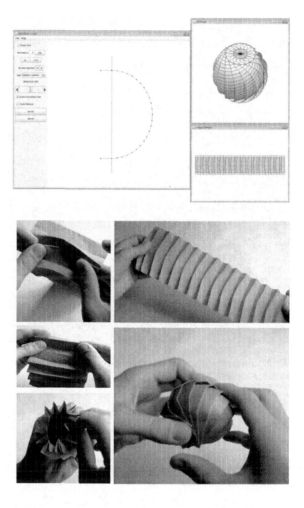

We subsequently made a series of cardboard prototypes using a cutting plotter to ascertain the idea's feasibility. Depending on the weight and volume characteristics of the food to be preserved, the consumer can choose the better *ChorionPack©* typology. By manipulating the S model, the smallest, a sheet of 18 × 29 cm., for example, we can obtain a spherical envelope of 10 cm. about in diameter, which can contain a portion of cheese around 150/200 gr. The following models M, medium, large, and XL, can accommodate a quantity of food up to 500 gr. (Figs. 38, 39, 40, 41, 42, 43, 44, 45, 46, 47, 48 and 49).

Alternatively, we envisioned creating a sort of coupled material, with a second heat-sealed inner layer, made with a sheet of ecological cotton fabric, also perforated, impregnated with beeswax and essential oils compatible with contact with fatty foods. Its contact surface can be easily washed and sanitized with a cloth moistened with warm water and well wrung out. In this way, following small maintenance attentions,

Fig. 27 Example of *shape memory* proprieties, applied to cardboard folding shape. *Credit* BionikonLab photo archive

Fig. 28 CAD rendering of morphological variants of different *chorion* eggs shapes

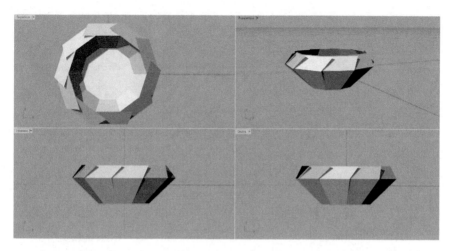

Fig. 29 CAD rendering of morphological variants of different *chorion* eggs shapes

Fig. 30 CAD rendering of morphological variants of different *chorion* eggs shapes

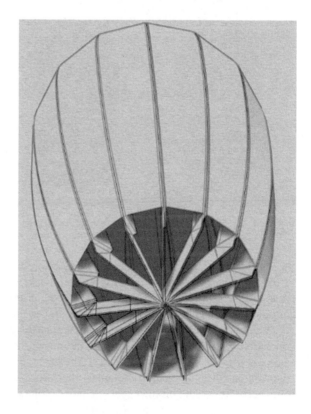

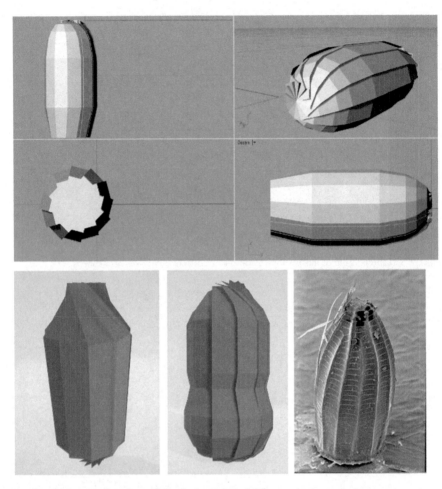

Fig. 31 CAD rendering of morphological variants of different *chorion* eggs shapes

the *ChorionPack©* can be reused several times according to the product characteristics of the bee's-wax system. Alternatively, we envisioned creating a sort of coupled material. The supporting structure is a punched non-impregnated cardboard folding chorion. A second heat-sealed *aeropyle*-inner layer, made with a sheet of ecological cotton fabric, also micro-perforated, is impregnated with beeswax and essential oils compatible with contact with fatty foods. Its contact surface can be easily washed and sanitized with a cloth moistened with warm water and well wrung out. In this way, following small maintenance attentions, the *ChorionPack©* can be reused several times according to the product characteristics of the beeswax products system. The last alternative can foresee that the supporting structure of the package is waxed to increase its duration and for better hygienic maintenance.

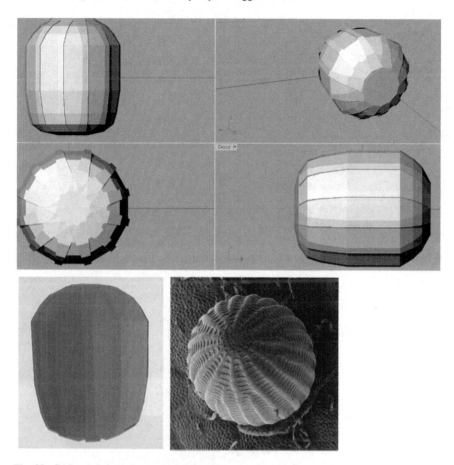

Fig. 32 CAD rendering of morphological variants of different *chorion* eggs shapes

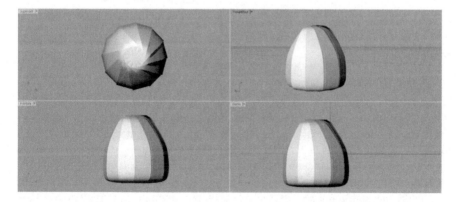

Fig. 33 CAD rendering of morphological variants of different *chorion* eggs shapes

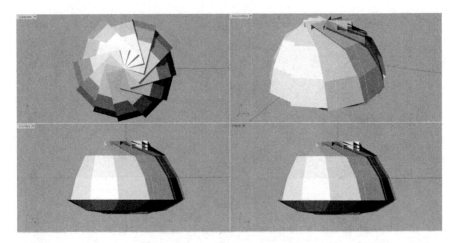

Fig. 34 CAD rendering of morphological variants of different *chorion* eggs shapes

Fig. 35 CAD rendering of morphological variants of different *chorion* eggs shapes

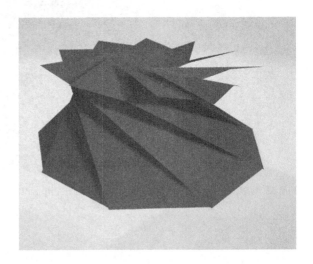

7 Conclusions

Since 2019, still coordinated by Prof Massimo Lumini, *BionikonLab&FABNAT14's* work team of designers, teachers, and students, analyzes the morphology and physiology of some particular *Lepidoptera's* eggs. They have produced over time, a series of observations that have been transformed into a *problem-solving* design concept, linked to the problems of packaging and preservation of particular foods, fruits, and vegetables through a *bottom-up* process of bionic research. In particular, we have focused on the problems of foods that must *breathe* as they require proper ventilation during their storage outdoors or in the fridge. Some original considerations have been developed and are defined as copyright, such as the *ShapeMemoryPackagingSystem*©

Fig. 36 CAD rendering of morphological variants of different *chorion* eggs shapes

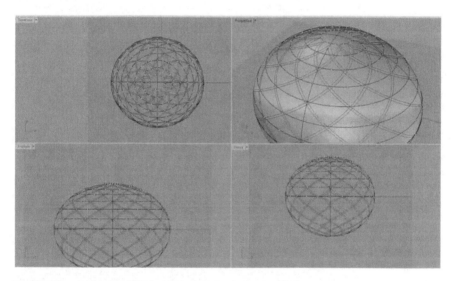

Fig. 37 Fibonacci spiral CAD rendering of the ribs' textures on a *chorion* surface (*Lycaenidae*). *Credit* E. Manconi

concept. The *ChorionPack*© system, as presented in the general guidelines in this publication, is also bestowal pending. The general idea tends to transfer some solutions developed in the folding tech sector, particularly the *orobotics* field. This storage system, which seeks to propose a possible replacement for traditional refrigerated boxing systems in plastic or aluminum foil packaging in a better sustainable way, was initially designed for domestic users. Nothing prevents us from imagining locations in large-scale distribution scenarios.

Fig. 38 The model is drawn in CAD and a cardboard mockup is cut out using a craft plotter. The sheet's waxed surface of a *ChorionPack.* © is perforated to ensure proper aeration of the inner food. The mountain/valley folds obtained mechanically, generate a memory of shape. To obtain the packaging, the two margins of templates are glued together (*ShapeMemoryPackagingSystem*©)

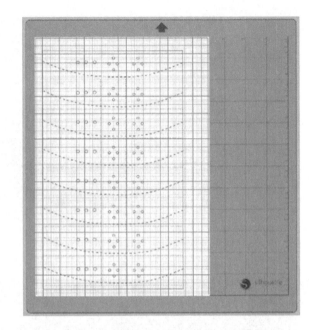

Fig. 39 The model is drawn in CAD and a cardboard mockup is cut out using a craft plotter. The sheet's waxed surface of a *ChorionPack.* © is perforated to ensure proper aeration of the inner food. The mountain/valley folds obtained mechanically, generate a memory of shape. To obtain the packaging, the two margins of templates are glued together (*ShapeMemoryPackagingSystem*©)

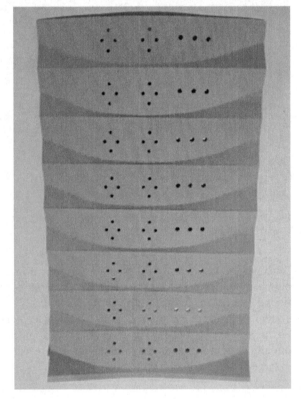

Fig. 40 The model is drawn in CAD and a cardboard mockup is cut out using a craft plotter. The sheet's waxed surface of a *ChorionPack.* © is perforated to ensure proper aeration of the inner food. The mountain/valley folds obtained mechanically, generate a memory of shape. To obtain the packaging, the two margins of templates are glued together (*ShapeMemoryPackagingSystem*©)

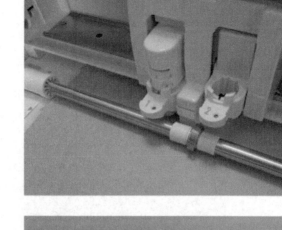

Fig. 41 The model is drawn in CAD and a cardboard mockup is cut out using a craft plotter. The sheet's waxed surface of a *ChorionPack.* © is perforated to ensure proper aeration of the inner food. The mountain/valley folds obtained mechanically, generate a memory of shape. To obtain the packaging, the two margins of templates are glued together (*ShapeMemoryPackagingSystem*©)

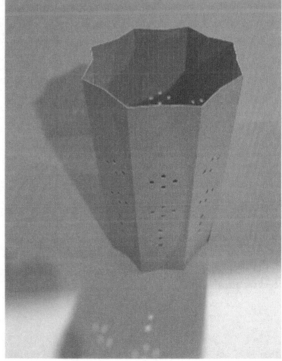

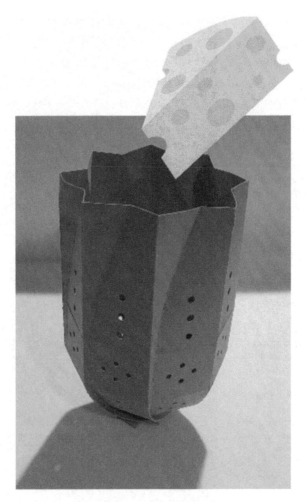

Fig. 42 The bottom of the *ChorionPack*. © closes quickly by moving the palm in the correct direction. Once the bottom is closed, and the food to be stored has been inserted, the upper parts of the package are rotated in the opposite direction to the bottom. The opening and closing of the container are reversible

Fig. 43 The bottom of the *ChorionPack.* © closes quickly by moving the palm in the correct direction. Once the bottom is closed, and the food to be stored has been inserted, the upper parts of the package are rotated in the opposite direction to the bottom. The opening and closing of the container are reversible

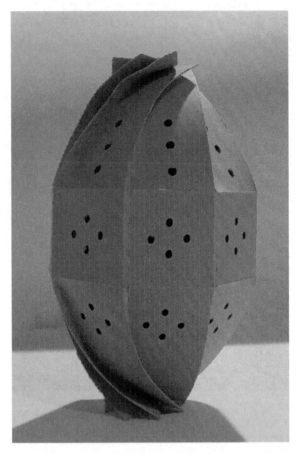

Fig. 44 The bottom of the *ChorionPack*. © closes quickly by moving the palm in the correct direction. Once the bottom is closed, and the food to be stored has been inserted, the upper parts of the package are rotated in the opposite direction to the bottom. The opening and closing of the container are reversible

Fig. 45 The bottom of the
ChorionPack. © closes
quickly by moving the palm
in the correct direction. Once
the bottom is closed, and the
food to be stored has been
inserted, the upper parts of
the package are rotated in the
opposite direction to the
bottom. The opening and
closing of the container are
reversible

Fig. 46 The
waxed-cardboard sheet of
the *ChorionPack*©, before
being glued can be boxed in
a cylindrical package for its
sale and home storage. A
variant of the *chorion,* with a
flat bottom; this and all the
models can be made in
various typologies and sizes

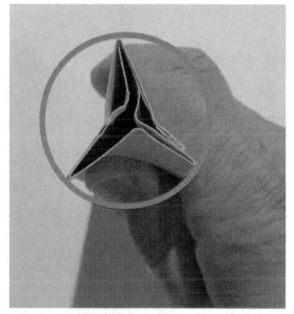

Fig. 47 The waxed-cardboard sheet of the *ChorionPack*©, before being glued can be boxed in a cylindrical package for its sale and home storage. A variant of the *chorion,* with a flat bottom; this and all the models can be made in various typologies and sizes

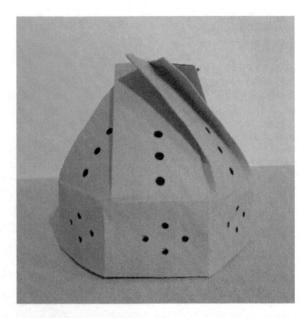

Fig. 48 A variant of the basic model

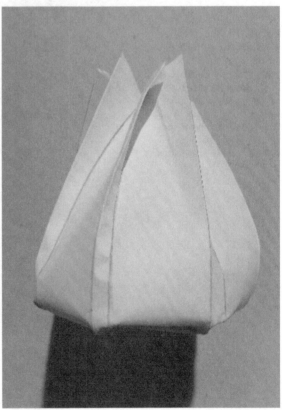

Fig. 49 A variant of the basic model

Acknowledgements Prof. Massimo Lumini is the only author and editor of the present chapter but some inspirations, CAD rendering, and design suggestions, come from the results of several lectures, workshops, and bionics and biomimetics training courses held during the schools-years 2019/2020 e 2020/2021 at BionikonLab&FABNAT14 in Iglesias-SU Italy. All the infographics, photos, images, and cardboard-origami prototypes accompanying the text, are original artwork of the author, except CAD in Fig. 39 (Emanuela Manconi). The author is thankful for the BionikonLab&FABNAT14 team, including the teachers Emanuela Manconi and Silvia Musa, and students Francesco Concas, Giovanni Concas, Enrico Congia, and Valerio Antonio Filippi.

References

1. Frei O (1984) L'Architettura della natura. Forme e costruzioni della natura e della tecnica e processi della loro formazione. Milano, Il Saggiatore, pp 7–9
2. Liddell I (2015) Frei Otto and the development of gridshells. Case Stud Struct Eng 4:34–39. https://doi.org/10.1016/j.csse.2015.08.001
3. Salvia G, Rognoli V, Levi M (2009) Il Progetto della natura. Gli strumenti della biomimesi per il design. Milano, Franco Angeli, p 46
4. Lumini M (2018) About biomorphic exuberance and digital Rococò in design and parametric contemporary architecture. In Following Forms, following functions. Pratcice and disciplines in dialogue. In: Pau F, Vargiu L (eds) Newcastel upon Tynes. Cambridge Scholars Publishing, p 141
5. Benyus JM (1998) Biomimicry. Innovation inspired by nature. HarperCollins Publishers Inc, NY
6. Marco Vitruvio Pollione, De architectura, liber I, 2
7. Nahan EN, Sherman VR, Pissarenko A, Rohrbach S, Fernandes DJ, Meyers MA (2017) Nature's technical ceramic: the avian eggshell. J R Soc Interface. https://doi.org/10.1098/rsif.2016.0804

8. Robertson JA, Bradler S, Whiting MF. Evolution of oviposition techniques in stick and leaf insects (Phasmatodea). National Identification Service, APHIS PPQ, USA Department of Agriculture, Betsville, MD, USA. Department of Biology and M. L. Bean Museum, Brigham Young University, Provo, UT, USA. Department of Morphology, Systematics and Evolutionary Biology, Johann-Friedrich-Blumenbach Institute of Zoology and Anthropology, University of Göttingen, Göttingen, Germany. https://doi.org/10.3389/fevo.2018.00216

9. Tetaert F (2016) Une base de références pour les oeufs des phasmes, in Insects. n° 183. http://www7.inra.fr. Ohasmid Studies, Volume 19, Edward Baker & Judith Marshall, January 2018

10. Law Y-H, Sediqi A (2010) Sticky substances on eggs improves predation success and substrate adhesion in newly hatched Zelus renardii (Hemiptera:Reduviidae) instars. In: Annals of the entomological society of America, vol 103, no 5. Oxford University Press, pp 771–774. https://doi.org/10.1603/AN09143

11. Merian MS (2015) La meravigliosa metamorfosi dei bruchi. Elliot Lit Edizioni Srl, Roma

12. Briggs H (2018) Meet the butterflies from 200 million years ago. http://www.bbc.com/news/science-environment-42636275

13. Osterath B (2018) Rethinking evolution: butterflies came first, flowers came second. http://p.dw.com/p/2qgma

14. van de SchootbruggeSchootbrugge B et al (2017) Lepidoptera (moths and butterflies) thrived in gymnosperm forests following the end-Triassic extinction. EGU General Assembly Conference Abstracts. Available at https://ui.adsabs.harvard.edu/abs/2017EGUGA..19.7184V/abstract

15. Li D, Huson MG, Graham DL, Proteinaceous adhesive secretions from insects, and in particular the egg attachment glue of Opodiphtera sp. moths. CSIRO Molecular and Health Technologies, Sydney Laboratory, NSW, Australia. https://doi.org/10.1002/arch.20267

16. Büscher TH, Quigley E, Gorb S.N (2020) Adhesion performance in the eggs of the Philippine leaf insect Phyllium Philippinicum (Phasmatodea: Phylliidae). Department of Functional Morphology and Biomechanics, Institute of Zoology, Kiel University, Am Botanischen Garten 9, 24118 Kiel, Germany. https://doi.org/10.3390/insects11070400

17. https://www.ted.com/speakers/louie_schwartzberg?language=it

18. Arbogast RT, Leonard Lecato G, Byrd RV (1980) External morphology of some eggs of stored-product moths (Lepidoptera pyralidae, gelechiidae, tineidae), Int J Insect Morphol Embryol 9(3):165–177. ISSN 0020-7322, https://doi.org/10.1016/0020-7322(80)90013-6

19. Thomson G (1992) Egg surface morphology of Manioline butterflies (Lepidoptera, Nymphalidae, Satyrinae). Atalanta 23(1/2):195–214. Würzburg. ISSN 0171-0079

20. Synergetics. https://www.bfi.org/about-fuller/big-ideas/synergetics

21. Nieves-Uribe S, Flores-Gallardo A, Llorente-Bousquets J, Luis-Martinez A, Carmen P, Use of exochorion characters for the systematics of Hamadryas Hübner and Ectima Doubleday (Nymphalidae: Biblidinae: Ageroniini). https://doi.org/10.11646/zootaxa.4619.1.3

22. Evolution of domes in architecture. https://www.re-thinkingthefuture.com/architectural-styles/a2615-evolution-of-domes-in-architecture/

23. Iossa G, Gage MJG, Eady PE (2016) Micropyle number is associated with elevated female promiscuity in Lepidoptera. https://doi.org/10.1098/rsbl.2016,0782

24. Yanagimachi R, Cherr G, Matsubara T, Andoh T, Harumi T, Vines C, Pillai M (2013) Griffin F, Matsubara H, Weatherby T, Kaneshiro K, Sperm attractan in the micropyle region of fish and insect eggs. https://doi.org/10.1095/biolrepro.112.105072

25. Telfer WH (2009) Egg formation in Lepidoptera. J Insect Sci 9(1):50. https://doi.org/10.1673/031.009.5001

26. Al-Dosary MM, Al-Bekairi AM, Moursay EB, Morphology of the egg shell and the developing embryo of the Red Palm Weevil, Rhynchophoforus ferrugineus (Oliver). Entomology College of Education for Girls, Scientific Departments, Al-Kharj Univ.,Saudi Arabia. https://doi.org/10.1016/j.sjbs.2010.2.012

27. Particularly striking in this regard are the images of Martin Oeggerli, a Swiss molecular biologist. He used his microscope to probe the tiniest enclave of nature, taking his sharp eye for beauty to the nanoscale. www.micronaut.ch

28. Srivastava AK, Kumar K (2016) Ultrastructure of egg chorion of castor butterfly Ariadne merione (Crammer) (Lepidoptera: Nymphalidae). Department of Zoology, University of Allahabad, Allahabad 211002, India. J Compar Zool 263:1–5. https://doi.org/10.2016/j.jcz.2016.03.015
29. Kharrat T. https://wewanttolearn.wordpress.com/2015/11/25/the-butterfly-egg/
30. See exhibition: Tsutsumu: the origin of Japanese Packaging. Meguro Museum of Art, Tokyo. https://mmat.jp/en/exhibition/archive/2021/20210713-364.html
31. Misciagna A (2013/2014) Il packaging tradizionale giapponese. Thesis on Corso di Laurea magistrale in Lingue e civiltà dell'Asia e dell'Africa mediterranea. Università Ca' Foscari-Venezia, AA. http://dspace.unive.it/handle/10579/5464
32. Brundtland GH (1987) Our common future. Report of the World Commission on Environment and Development, United Nations. https://www.are.admin.ch/are/it/home/media-epubbl icazioni/pubblicazioni/sviluppo-sostenibile/brundtland-report.html
33. Francesca L (2018/2019) Does packaging influence the perception of healty food?, Pag. 3, LUISS Department of Economics and Finance-Course of Marketing, A.Y. www.tesi.luiss.it
34. Wilkström F, Verghese K, Auras R, Olsson A, Williams H, Wever R, Grönman K, Kvalvåg Pettersen M, Møller H, Soukka R, Packaging strategies that save food: a research agenda for 2030. Accessed 21 Apr 2018. https://doi.org/10.1111/jiec.12769
35. Kenney KL (2021) Folding Tech. Using Origami and nature to revolutionize technology. Twenty-first Century Books, Minneapolis
36. https://mitani.cs.tsukuba.ac.jp/en/cp_download.html

Transport Package and Release of Ladybug Larvae with Biomimetic Concepts

Fernando José da Silva, Cynara Fiedler Bremer, Sofia Woyames Costa Leite, and Verônica Oliveira Souza

Abstract Agrotoxics, pesticides, and other agricultural chemical inputs have long been used for pest control in crops and plantations around the world. As people have become aware of this problem of food with pesticides, the demand for vegetables and greens produced in gardens, with natural pest control, without the use of pesticides, has increased. The Biofábrica de Joaninhas (Ladybug Factory), an agency linked to the city of Belo Horizonte/MG, produces and distributes ladybug and chrysopid larvae, to communities and vegetable gardens, with the aim of protecting the production of vegetables, respecting the environment and people's health in general. This project was inspired by a similar work in Caen, France, started in 1980. The first official distribution to the local population took place in 1984 and remains a public policy in that country to this day. The objective of the work presented here was the development of an alternative packaging, produced with biodegradable materials, avoiding environmental problems, instead of the plastic ones normally used; the process was inspired by nature's solutions, and is able to transport the larvae of these insects from the Ladybug Factory to the vegetable gardens, facilitating the handling during the release of these insects, avoiding losses, and also protecting them.

Keywords Biomimetics · Nature-inspired · Design · Packaging · Ladybug · Cycloneda sanguinea · Pests · Environmental control

F. J. da Silva · C. F. Bremer (✉) · S. W. C. Leite · V. O. Souza
Escola de Arquitetura, Departamento de Tecnologia do Design, Universidade Federal de Minas Gerais, da Arquitetura e do Urbanismo. Rua Paraíba, 697 - Savassi, Belo Horizonte/MG. CEP 30130-141, Brazil
e-mail: cynarafiedlerbremer@ufmg.br

F. J. da Silva
e-mail: fernandojsilva@ufmg.br

F. L. Palombini and S. S. Muthu (eds.), *Bionics and Sustainable Design*, Environmental Footprints and Eco-design of Products and Processes,
https://doi.org/10.1007/978-981-19-1812-4_4

1   Introduction

For a long time now, agrotoxics, pesticides, and other agricultural chemical inputs have been used for pest control in crops and plantations all over the world, and unfortunately, recently in our country, dozens of these toxic materials are being released for use in agriculture. These inputs not only compromise the quality of the vegetables produced but also harm the health of the people who work there, the health of their families, children, the elderly, and anyone else who comes to consume the products, in addition to contaminating the groundwater of the region with the infiltration of the products by rainwater.

It is also observed that as people become aware of this problem of food with pesticides, the demand for vegetables produced in organic gardens has increased, and many of them are community gardens near the concentration of residential neighborhoods in the cities [5].

With the challenge of these community gardens to produce organic products without the use of pesticides, a way out has been the use of insect larvae that predate pests without damaging the vegetables. These are the so-called "good insects", such as the well-known Ladybugs (Cycloneda sanguinea), and the Chrysopids (Chrysopidae) [2].

According to these authors, the Institute Ladybug Factory, an agency linked to the Prefecture of the city of Belo Horizonte/MG, produces and distributes the larvae of these insects, to communities and vegetable gardens, in order to protect the production of vegetables, respecting the environment and the health of the people involved and the final consumers, since they do not use pesticides for this environmental control of pests.

However, observing the traditional prismatic packaging system and model used, part of the larvae cannot survive the transport and the way of release in the garden itself. Thus, the goal of this project is the development of a packaging, inspired by nature's solutions, able to transport the larvae of these insects from the Ladybug Factory to the vegetable garden, facilitating the handling during the release of these insects, reducing losses and protecting them, being produced with biodegradable materials, avoiding environmental problems.

2   Pesticide Use Versus the Health of Living Beings

With the increase in population all over the planet, there is the need to generate food for all, in order to ensure the continuity of life. However, this need to have food has caused producers around the world to use pesticides of all kinds to control pests and, consequently, increase production [7].

In Brazil, it has been no different: Grigori [16] shows that in the Bolsonaro government, in 2018, there was an increase in the authorization of pesticide use, an addition of 998 new items, totaling 3,064 types currently marketed in the country, which represents a growth of more than 50% compared to the previous government.

It is noteworthy here that the indiscriminate use of artificial toxins in an attempt to combat pests brings immeasurable damage to adjacent populations and consumers. The high exposure to pesticides can bring symptoms such as poisoning at various levels, depending on the proximity to the pesticide, type and amount, causing allergies; gastrointestinal, respiratory, endocrine, reproductive, and neurological disorders; neoplasms; accidental deaths; even influencing suicides. It affects mainly agricultural workers, professionals who apply the pesticides, children, women of reproductive age, pregnant, and lactating women, the elderly and individuals with biological and genetic vulnerability [9]. In addition to the factors directly impinged, consumers are also prevented from accessing with transparency the chemical quantum present in their daily food [8].

Minas Gerais ranks as the state with the highest recorded growth in organic agriculture [27] a milestone to be celebrated and expanded. In this sense, it is of extreme importance and relevance the investment in research and development of more sustainable alternatives that assist in combating the advance of such systemic pests, in an effective, cheap, and accessible way. Some considerable advances have been achieved; as pointed out by Primavesi [24] and the EMBRAPA [14] report, the biological control of pests in human environments such as the ladybug plant.

Currently, there is great governmental fomentation in the form of subsidies for the diffusion of agrotoxics in the country, an investment choice with a very high environmental and social cost. The current list of these products is immense: according to MAPA [19], 2300 pesticides are registered in Brazil, being one of the largest consumers in the world.

Thus, with the large use of pesticides in agriculture, Aires [1], warns of the problems related to damage to people's health, as well as to insects and nature itself, because "one of the most common problems is the contamination of soil, groundwater, and rivers and lakes. With the rain or the irrigation used, the chemical reaches the soil, and infiltrates, reaches the local water, intoxicating the environment and the life present there.

As for the poisoning of small animals, Sánchez-Bayo and Wyckhuys [25] warn that currently, 40% of all insect species are at risk of extinction due to pesticides—especially neonicotinoids, since they are the most widely used insecticides on the planet. And without insects, much of the pollination of food-producing plants ceases. In the case of bees, the most common effects identified are disorientation in flight, nervous and digestive system disorders, and consequent death (Fig. 1).

The huge use of pesticides by large producers ties into the approval of the Pesticide Regulatory Framework [12], which included proposals to change terminologies to soften the perception of the toxicity involved in these chemicals:

> ...the project provides, for example, the change of the name "agrotoxics" to "pesticides", which should facilitate the registration of products whose formulas, in some cases, are composed of substances considered carcinogenic by regulatory agencies. Previously, the proposal was to change the nomenclature to "phytosanitary product". [10]

Fig. 1 Dead bees. *Source* https://www.flickr.com/photos/32454731@N00/3804300714

Fig. 2 Manual and mechanized application of agrochemicals in crops. *Source* **a** https://encurtador.com.br/bpwDL; **b** https://encurtador.com.br/jstL5

Figure 2a, b show manual and mechanized application of agrochemicals on farms:

In large cultivable areas of agriculture, the mechanical application of pesticides is an activity analyzed as a major villain of pollution and poisoning in the production of soybeans, corn, rice, citrus, with about 63 million hectares (2017), and 158 million hectares of pastures [3].

Linked to this huge agribusiness is the advance of deforestation, land concentration, the depletion of soil and water resources, and the threat to traditional and native regional populations. It is also observed that high exposure to pesticides can bring symptoms such as poisoning at various levels, depending on the proximity to the pesticide, type and amount of the toxin, causing various allergic reactions; respiratory, endocrine, gastrointestinal, reproductive and neurological disorders; neoplasms, accidental deaths, even influencing suicide cases [6].

3   Organic Gardens and Natural Pest Control

As an alternative for access to good quality food, without the added costs of logistics and the dangerous and uncontrolled presence of pesticides, organic gardens have been spreading in urban regions and population proximities, presenting themselves as an option for a healthy return to the relationship between man–nature, besides promoting subsistence conditions and human dignity to the families involved, combating unemployment and social exclusion [5].

There has been an increase in demand for garden crops and in small spaces [14], and discussions regarding public health and access to food and water become frequent, especially when it comes to increasing immunity by favoring healthy habits. Thus, urban agriculture becomes ecological and sustainable, and has been gaining space and incentives by government sectors, including regulatory standards. São Paulo had its first regulation concerning urban agriculture in 2004 (Law 13.727/2004), regulated by Decree 51.801, of 09/21/2010 [23]. Other metropolises, as highlighted by Lima [18], already had this context, as is the case of Lisbon, which since 2007 has been encouraging community gardens and currently has 14 sets of urban gardens. Still according to Lima, Madrid currently has 39 of these gardens, and Barcelona 14 gardens with community empowerment that resists real estate speculation.

In Belo Horizonte, the local government [21] currently has 51 productive urban community units (Fig. 3), divided into nine regions. In addition to these units, the city also has School Agroecological Systems, involving more than 160 projects in UMEIS (Municipal Education Units), besides the private network [5].

In this sense, the Prefecture of Belo Horizonte [22] has been developing since 2017, a project called Biofábrica, to raise ladybugs (Cycloneda sanguinea) and Chrysopids (Chrysopidae), Fig. 4, aimed at donation to community gardens and small farmers. This project was inspired by a similar one in Caen, France, which started in 1981 in the city's Botanical Garden, having the first official distribution to

Fig. 3 Community gardens. *Source* https://encurtador. com.br/apE15

(a)   **(b)**

Fig. 4 Ladybugs/crisopids. *Source* **a** https://encurtador.com.br/joyMX, **b** encurtador.com.br/ghjIS

the local population in 1984 and which remains as a public policy to this day and years later served as the basis for the implementation of federal policy for total elimination of pesticide use in France [11]; in Belo Horizonte, the project also helped to control the population of the whitefly, which attacked the ficus present in the capital in the year 2019.

In Caen, until 2009, the insects were packaged for distribution in their larval form in plastic jars with lids, using wheat husk and sawdust as substrate for accommodating the insects. In the following years, the insects were distributed in tissue paper and/or cut into squares, which proved to be effective. But it still presents the issue discussed in this article of creating future problems such as inadequate disposal and the safety of the small insects during transportation.

These insects, while in the larval stage, are natural predators of aphids, enemies of vegetables, acting as natural controllers of these pests. In 2020, even with the pandemic (Covid-19), the Biofábrica's production and donation reached 40 thousand ladybugs (number initially planned only for 2021), and the tendency for demand is to increase, due to the success with the educational projects developed by them, aimed at raising awareness among the population and schoolchildren. Unfortunately, with the increase in the number of COVID cases, Biofábrica was paralyzed and had its activities restarted only in October 2021 with a forecast of expansion, having as a goal for the next year to produce about 50 thousand ladybugs and deliver more than 5 thousand kits.

This type of natural biological control of environmental pests has been known since the late nineteenth century. Amaral et al. [2] show that in 1887, ladybugs were used to control pests in citrus plantations in California (imported from Australia). And, in Brazil, there are records of natural control of whitefly and mealybugs since 1921. They also explain that the chrysopids act more efficiently still in the larval stage, when they need proteins and carbohydrates, feeding mainly on aphids, mealybugs, whiteflies, psyllids and mites. Both ladybugs and chrysopids have a life cycle with complete metamorphosis (egg, larva, pupa, and adult), and consume their prey throughout their cycle, besides having fast development, thus becoming great options for the regulation of pests.

4 Need for Appropriate Packaging for Transport and Release of Larvae

Observing the growing dissemination of sustainable and nature-protective techniques, naturally controlling pests in vegetable gardens with the use of predatory insects (as is the case of ladybugs and chrysopids), several entities have distributed packages containing larvae of these insects, in programs that have advanced for decades [2].

However, the packages used are taken from plastic objects available in the market, and that were not designed for such use, not ergonomically suitable, not appropriate for the actions of insertion of inputs, insertion of the larvae, ease of aeration to not suffocate them, and ease of handling in the release of the larvae in the appropriate places in the gardens.

Figure 5a–c show examples of the use of plastic containers, which end up polluting the garden, after their inappropriate disposal.

Figure 6a shows the activity of inserting inputs and larvae inside the packaging; and Fig. 6b shows another type of packaging affixed to plants, opened to release larvae, which can also be disposed of inappropriately near the garden, causing pollution because it is non-biodegradable plastic material.

Fig. 5 Plastic packaging. *Source* The authors

(a)   **(b)**

Fig. 6 **a** Insertion of elements into the packaging. **b** Availability of packaging next to the plant.
Source The authors

5   Principles of Biomimetics

Biomimetics is an old and recently rediscovered research area, which deals with the analysis of natural systems by applying their resolution principles to projects, be they in Design, Architecture, Engineering, or any other area of knowledge. In the late 1990s, this subject gained new life, with the publication of Janine Benyus' work in 1997, and in the last decade there have been forums, congresses, and symposia spreading with this theme. Benyus shows that adaptations make possible the creation of forms and uses of principles of functions or behaviors present in nature and that are used as analogies and patterns, be they geometric or mathematical. These principles are identified as models, as measures or as mentors, "being a new way of seeing and valuing nature (…) whose foundations lie not in what we can extract, but in what we can learn from it" ([4], p. 8).

As an elementary form in nature we can observe the figure of the circle, which can insert within it the figures of the triangular plane, square, pentagon, and hexagon. Of these natural figures, the hexagonal figure and prism are shown to have better use of space, less expenditure of material and greater volume, with golden ratios in their interior proportions ([13], p. 9). Bees build their alveoli, adopting this shape (Fig. 7), optimizing the space and having economy of wax, without loss of space.

Observing the protective aspects of seeds and fruits in nature, one can also notice the geometric configuration that, besides spatially organizing the elements present, originally saves material resources in its formation. In Fig. 8, some geometric and protective forms in Brazilian seeds can be seen.

(a)   **(b)**   **(c)**

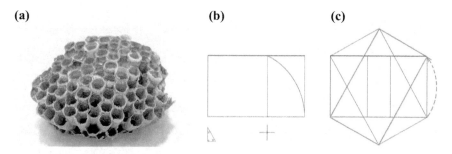

Fig. 7 a Hexagonal shape of the honeycombs; **b** and **c** Golden rectangle in the formation of the hexagon. *Source* **a** https://www.encurtador.com.br/jnuvG, **b** and **c** elaborated by the authors

(a)   **(b)**   **(c)**

Fig. 8 Geometric elements and protective forms in Brazilian seeds. *Source* **a** https://www.encurt ador.com.br/eDLM6, **b** encurtador.com.br/avGHJ, **c** encurtador.com.br/ghosF

6 Process and Packaging Design

For the new packaging proposal, with the principles and objectives already described, observing some conventional design processes, it was chosen the development of alternatives to the packaging for transportation and release of larvae based on the concepts of Munari [20]. In this methodology, starting from a problem, knowing its components, analyzing data related to it, and developed alternatives with creativity and various experiments, from drawings to three-dimensional models. Thus, after analyzing the developed generations, the solution presented in this work was reached.

Figures 9 and 10 show a series of alternatives developed in the proposal for the packaging of Ladybug larvae, or Chrysopids, to control vegetable plagues.

From the alternatives generated, three-dimensional models were built (Figs. 11 and 12) in order to better evaluate the possibilities of implementing the proposal, and then sent to Biofábrica for initial testing and prototype adjustments.

Figure 13 shows the model of the chosen proposal, in test simulation for the package opening system and release of the ladybug larvae.

Figure 14 shows details of the simulation in the final alternative model, regarding the internal movement system for the release of the ladybug larvae.

Fig. 9 Generation of alternatives. *Source* Elaborated by the authors

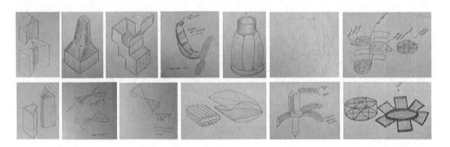

Fig. 10 Generation of alternatives and planning. *Source* Elaborated by the authors

Fig. 11 Modeling of alternatives. *Source* Elaborated by the authors

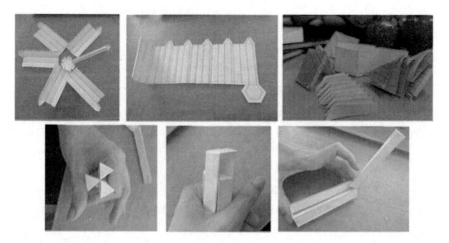

Fig. 12 Modeling of alternatives. *Source* Elaborated by the authors

Fig. 13 Final proposal model. *Source* Elaborated by the authors

In this case, with an internal element that can be made of the same material as the packaging, in biodegradable paper, and that can have a vertical movement in the module for removal and release of the larvae directly on the foliage of vegetables with aphids that will be preyed on by the ladybugs.

7   Final Considerations

In this work, the development process of packaging was characterized, a product based on the Science of Biomimetics, demonstrating to have several possibilities for favorable solutions, involving formal and functional characteristics present in nature.

As part of this process, it was observed the generation of various alternatives and then the choice of an idea being developed also with the use of three-dimensional models, exploring construction and detailing, enabling simulation of use.

Fig. 14 Final proposal model. *Source* Elaborated by the authors

It is noteworthy, however, that this proposal is currently being evaluated and adjusted according to the demands of the professionals at Biofábrica, an agency linked to the Municipality of Belo Horizonte. These adjustments are related to the size, mobile, and functional elements of the proposal, as well as the quantity and types of larvae/insects to be transported, for subsequent insertion in the production activities and donation to the vegetable producers, favoring the natural control of pests without the use of pesticides.

The partnership Design/Biomimetics also enables a range of opportunities and applications with the sectors of society, involved in the search for practical, fast, and efficient solutions to the challenges encountered, whether in the public or private sector, and that become examples to future professionals, designers, architects, engineers, agronomists, among others.

Acknowledgements The authors' team would like to thank the collaboration of Professor Dany Silvio Amaral, from the Management of Actions for Sustainability of the Environment Department of the Belo Horizonte City Hall, and especially, the professor and biologist Wagner da Costa Resende, graduated in Biological Sciences at the UFMG, administrative servant of the Belo Horizonte City Hall and one of the responsible persons for the Biofábrica project, who unfortunately died of COVID-19 in April 2021 and did not follow the final version of the package developed. Our sincere thanks and condolences to his family and friends at PBH.

References

1. Aires L (2013) Os problemas causados pelos agrotóxicos justificam seu uso? https://www.ecy cle.com.br/component/content/article/35/1441-os-problemas-causados-pelos-agrotoxicos-jus tificam-seu-uso.html Access in fev 11 2021
2. Amaral DSSL, Venzon madelaine, Barbosa E, Abreu N, Resende WC (2019) Biofábrica de insetos predadores. Belo Horizonte, Informe Agropecuário, Tecnologia para manejo sustentável de pragas e doenças. v 40, n 305
3. Altas do Agronegócio (2018) fatos e números sobre as corporações que controlam o que comemos. Maureen Santos, Verena Glass, organizadoras. Rio de Janeiro: Fundação Heinrich Böll
4. Benyus JM (2003) Biomimética: Inovação Inspirada pela Natureza. São Paulo, Cultrix, Original title: Biomimicry
5. BHVERDE (2020) Hortas comunitárias. https://bhverde.com.br/hortas-comunitarias/ Access on fev 11 2021
6. BRASIL (2018) Ministério da Saúde. Secretaria de Vigilância em Saúde. Departamento de Vigilância em Saúde Ambiental e Saúde do Trabalhador. Agrotóxicos na ótica do Sistema Único de Saúde. Brasília: Ministério da Saúde, vol 1, Tomo 2. https://bvsms.saude.gov.br/ bvs/publicacoes/relatorio_nacional_vigilancia_populacoes_expostas_agrotoxicos.pdf. Access on fev 11 2021
7. Camponogara AS (s.d.) Pesticidas, Herbicidas e Agrotóxicos no contexto da agricultura. São Paulo. https://siteantigo.portaleducacao.com.br/conteudo/artigos/biologia/pesticidas-her bicidas-e-agrotoxicos-no-contexto-da-agricultura/57630 Accessed 11 fev 2021
8. Carneiro FF, Pignati W, Rigotto RM, Augusto LGS, Rizollo A, Muller NM, Alexandre VP, Friedrich K, Mello MSC, Abrasco D (2012) Um alerta sobre os impactos dos agrotóxicos na saúde. Parte 1: Agrotóxicos, Segurança Alimentar e Saúde. ABRASCO, Rio de Janeiro, abril de. www.mprs.mp.br/media/areas/ambiente/arquivos/agrotoxicos/dossie_abrasco_agrotoxico. pdf. Access 01 Mar 2021
9. Carneiro FF, Rigotto RM, Augusto LGS, Friedrich K, Búrigo AC (2015) Dossiê ABRASCO: um alerta sobre os impactos dos agrotóxicos na saúde. Rio de Janeiro: EPSJV; São Paulo: Expressão Popular. https://www.arca.fiocruz.br/bitstream/icict/26221/2/Livro%20E PSJV%20013036.pdf Access 08 Dec 2021
10. Cristaldo H (2018) Comissão da Câmara aprova projeto que flexibiliza uso de agrotóxico. Brasília: EBC, Empresa Brasil de Comunicação. https://agenciabrasil.ebc.com.br/politica/not icia/2018-06/comissao-da-camara-aprova-projeto-que-flexibililza-uso-de-agrotoxico Access 11 fev 2021
11. DEVPB (2019) Direction des Espaces Vert, du Paysage et de la Biodiversité. La Protection Biologique a Caen. Jardin des plantes de la ville de Caen
12. DOU (Diário Oficial da União) (2019) Ministério da Saúde, Agência Nacional de Vigilância Sanitária. Critérios para avaliação e classificação toxicológica. Brasília, DOU, Edição 146, Seção 1, p. 78. https://www.in.gov.br/web/dou/-/resolucao-da-diretoria-colegiada-rdc-n-294-de-29-de-julho-de-2019-207941987 Accessed 11 fev 2021
13. Elam K (2001) Geometry of design. Princeton Architectural Press, New York
14. EMBRAPA (2020) Hortaliças em Revista. Brasília: EMBRAPA, Ano IX, nº 30. https://www. embrapa.br/documents/1355126/2250572/revista_ed30+web+links.pdf/afe0cf1b-06e8-6df3-2e87-e2ad29a65098 Accessed 11 fev 2021
15. Fontes EMG, Valadares-Inglis MC (2020) Controle Biológico de Pragas da Agricul-tura. Brasília, EMBRAPA. https://ainfo.cnptia.embrapa.br/digital/bitstream/item/212490/1/ CBdocument.pdf Accessed 11 fev 2021
16. Grigori P (2021) Bolsonaro bate o próprio recorde: 2020 é o ano com maior aprovação de agrotóxicos da história. https://reporterbrasil.org.br/2021/01/bolsonaro-bate-o-proprio-rec orde-2020-e-o-ano-com-maior-aprovacao-de-agrotoxicos-da-historia/ Accessed 02 Dec 2021
17. Hsuanna, T (2002) Sementes do Cerrado e Design Contemporâneo. Goiânia, UCG

18. Lima MT (2020) Por que agricultura na cidade? A importância da Agricultura Urbana em contexto de emergência climática e sanitária. Campinas: UNICAMP, Boletim DPCT/IG nº 20. https://www.unicamp.br/unicamp/sites/default/files/2020-08/Boletins%20DPCT%20IG%20n20.pdf Accessed 11 fev 2021
19. MAPA: Ministério da Agricultura, Pecuária e Abastecimento (2019) Anvisa vai reclassificar defensivos agrícolas que estão no mercado. Brasília. https://www.gov.br/agricultura/pt-br/assuntos/noticias/anvisa-vai-reclassificar-todos-os-agrotoxicos-que-estao-no-mercado Accessed 11 fev 2021
20. Munari B (1998) Das Coisas Nascem Coisas. São Paulo, Martins Fontes
21. PBH, Prefeitura de Belo Horizonte (2021a) Unidades Produtivas Coletivas e Comunitárias. https://prefeitura.pbh.gov.br/smasac/susan/fomento/sistemas-de-producao/coletivas-e-comunitarias Accessed 11 fev 2021
22. PBH, Prefeitura de Belo Horizonte (2021b) Biofábrica. https://prefeitura.pbh.gov.br/meio-ambiente/biofabrica Accessed 11 fev 2021
23. Prefeitura de São Paulo (2010) Programa de Agricultura Urbana e Periurbana - PROAURP - no Município de São Paulo e suas diretrizes. Decreto 51.801. http://legislacao.prefeitura.sp.gov.br/leis/decreto-51801-de-21-de-setembro-de-2010 Accessed 11 fev 2021
24. Primavesi A (1994) Manejo Ecológico de Pragas e Doenças: técnicas alternativas para a produção agropecuária e defesa do Meio Ambiente. São Paulo, Nobel
25. Sánchez-Bayo F, Wyckhuys KAG (2019) Worldwide decline of the entomofauna: a review of its drivers. Biolo Conserv 232:8–27. https://www.sciencedirect.com/science/article/pii/S0006320718313636 Accessed 11 fev 2021
26. Santos M, Glass V (2018) org. Atlas do agronegócio: fatos e números sobre as corporações que controlam o que comemos. Rio de Janeiro: Fundação Heinrich Böll. Available from: https://br.boell.org/pt-br/2018/09/04/atlas-do-agronegocio-fatos-e-numeros-sobre-corporacoes-que-controlam-o-que-comemos Access 11 fev 2021
27. Tallmann H, Zasso J (2019) Em alta, agricultura orgânica reúne todos os elementos da produção sustentável. Revista Retratos. https://agenciadenoticias.ibge.gov.br/agencia-noticias/2012-agencia-de-noticias/noticias/25126-em-alta-agricultura-organica-reune-todos-os-elementos-da-producao-sustentavel. Access 05 dec 2021

Characterization of the Gradient Cellular Structure of Bottle Gourd (*Lagenaria Siceraria*) and Implications for Bioinspired Applications

Danieli Maehler Nejeliski⬡, Lauren da Cunha Duarte⬡, Jorge Ernesto de Araujo Mariath⬡, and Felipe Luis Palombini⬡

Abstract Cellular materials can be found in numerous ways in nature, presenting interesting properties for applications in many fields, such as engineering and design. By combining a low density with a rigid structure, their macroscopic properties can essentially be derived from their microstructure, which can be classified into open- and closed-cell arrangements. Besides directly employing them in projects, synthetic polymeric alternatives have been developed, such as those bioinspired in their natural counterpart. Plant-based materials are originated from either wild or cultivated plants, which can be used as a raw material in product design after a few processes. They are considered mostly heterogeneous, biodegradable, and renewable materials. Bottle gourd (*Lagenaria siceraria*—CUCURBITACEAE) is a plant-based cellular material used empirically. After harvesting and drying, the fruit becomes hollow, with porous mesocarp and impermeable exocarp. Present on all continents, even before human presence, the fruit was one of the first domestic plants, exhibiting the characteristics of easily adapting to any climate, presenting high productivity and its annual production cycle configures it as a renewable raw material alternative.

D. M. Nejeliski (✉)
Design School, Pelotas Campus; Federal Institute Sul-Rio-Grandense of Education, Science and Technology (IFSul), Praça Vinte de Setembro, 455, Pelotas, RS, Brazil
e-mail: danielinejeliski@ifsul.edu.br

L. da Cunha Duarte
Engineering School, Materials Department, Federal University of Rio Grande do Sul (UFRGS), Av. Osvaldo Aranha, 99, Porto Alegre, RS, Brazil
e-mail: lauren.duarte@ufrgs.br

J. E. de Araujo Mariath
Laboratory of Plant Anatomy – LAVeg, Graduate Program in Botany – PPGBot; Institute of Biosciences, Department of Botany, Federal University of Rio Grande do Sul – UFRGS, Av. Bento Gonçalves, Porto Alegre, RS 9500, Brazil
e-mail: jorge.mariath@ufrgs.br

F. L. Palombini
Design and Computer Simulation Group – DSC; Laboratory of Plant Anatomy – LAVeg, Graduate Program in Botany – PPGBot, Federal University of Rio Grande do Sul – UFRGS, Av. Bento Gonçalves, Porto Alegre, RS 9500, Brazil

Among its many applications, people use bottle gourd as vases, musical instruments, buoys, and masks; in southern Brazil—along with Argentina and Uruguay—the fruit is the main raw material for the manufacture of *cuia*, a container used for *chimarrão*, a traditional tea-like beverage *(mate)*. The application of bottle gourd in the design of new products depends on the study of its structure and properties. This chapter presents the characterization of bottle gourd cellular structure regarding the material's gradient porosity—from open to closed cells—with aims at bionics, using scanning electron microscopy and transmission light microscopy. The analyses showed the exocarp as a thin layer of compact closed cells, thus being waterproof, and the mesocarp formed by parenchyma cells that progressively increase in size towards the center, characterized by large empty spaces with thickened and lignified cell walls and intercellular communication channels, making the material water permeated. Overall, the material's microstructure is presented as a functionally gradient material, leading to newer possibilities for the development of bioinspired cellular materials.

Keywords Biomimetics · Plant anatomy and morphology · Natural materials · Sustainability · Bottle gourd · *Lagenaria siceraria* · Cellular materials · Material characterization

1 Introduction

Cellular solids are an important structural classification of various types of materials that generally combine a number of interesting physical and mechanical properties with a low density. By definition, it is essentially a porous material whose microstructure is composed of an internal arrangement of open or closed cells, which only contains solid material in the edges or walls of the unitary cell, respectively [6]. Cellular materials can be divided into two main groups: honeycombs and foams. The first one is known as a two-dimensional cellular solid, having aligned prismatic cells which usually have a hexagonal shape in materials found in nature. The second group is known as three-dimensional cellular solids, consisting of polyhedral cells arranged in a more complex way. Biomimicking of cellular materials is a key area of study for the development of more lightweight, efficient, or insulated designs [7]. Since the main properties of cellular solids are derived from their microstructure—or morphology in the case of biological materials [11], comprehending their microscopic arrangement is fundamental for studying their possible bioinspired applications.

The bottle gourd, scientific name *Lagenaria siceraria*, is the fruit of a plant of the cucurbit family, which also includes pumpkin, melon, watermelon, and other widely used in food. The structures of the plants of this family are very similar to each other, with large leaves, growing near the soil. However, in relation to the fruits, it has very peculiar characteristics [10]. When the fruit ripens and is harvested, if allowed to dry in the shade for a few months, the placenta with the seeds dries and the fruit becomes hollow. The inner part of the shell acquires characteristics similar to wood and the outer skin, extremely thin and smooth, becomes impermeable and with a brown coloration, and when polished it acquires a marked brilliance. Due to the hard shell, it is known as gourd in several regions of Brazil, being the term *porongo* most used in the southern region of the country.

The cultivation of porongo is an important agricultural activity being widely used for the manufacture of containers for *chimarrão* and handicrafts [4]. The fact of being basically associated with the production of artifacts of the local culture ends up restricting cultivation in other regions of the country, as well as its use as raw material for the manufacture of other products, yet it is the main source of income for dozens of small producers in Rio Grande do Sul, Brazil [15]. Using it on new products such as containers for food and beverages, the demand for growing the fruit would increase, encouraging family farming as a whole.

2   Bottle Gourd (*Lagenaria Siceraria*)

2.1   Origins of the Oldest of Domesticated Plants

Bottle gourd belongs to the family of cucurbits (CUCURBITACEAE), one of the most important families of plants used by man, providing food and fiber. These plants are related to the origin of agriculture and human civilizations, being among the first species to be domesticated [3]. The pumpkin, the cucumber, the melon, and the watermelon, important foodstuffs, and also the bottle gourd, used for various purposes, belong to the family. Samples dating back to 11,000 BC were found in East Asia. In the American continent, they were used from at least 10,000 BC [8]. Regarding specifically South America, the records date from 6,000 to 5,000 BC, however, there is no safe argument that can resolve the issue of the unusual bi-hemispheric distribution of the fruit [3].

In the phylogenetic tree, it can be observed that a common ancestor originates two others, A and B, from which all the other Asian and African variations are generated, respectively. A wild African specimen is the origin of everything, related in the first name of the list (wild Zimbabwe), and appeared approximately 250 thousand years ago. The bottle gourds of the American continent share a common ancestor with Africans between 60,000 and 103,000 years ago, long before the first men who migrated to the Americas, which happened about 15,000 years ago. That is, the bottle gourd arrived in the American continent without the interference of humans [8].

The theory of transport by sea is one of the explanations for the spread of the species. The seeds may have been transported by sea, as computational models of sea currents indicate that it would take only a few months for the seeds of wild gourds to reach the American continent [8]. The hard, impermeable shell would retain the seeds and, once out of the ocean, they would have sprouted. While Alaska would have been an inhospitable place for bottle gourds to grow, places with a tropical climate such as Florida, Mexico, and Brazil would have been ideal [8].

Africa was probably the first place where porongo was domesticated. [8] suggest that different groups of people may have come across different wild populations of the plant and selected those with the hardest shell to use as containers. Moreover, in Asia the fruit is among the earliest domesticated plants, being brought by the first Polynesians to settle in New Zealand thousands of years ago. The bottle gourd was the first plant propagated from seeds by people in New Zealand soils [5]. The Maoris, a native people of New Zealand, use containers made from the fruit.

In New Zealand, the bottle gourd was also valued as a food plant. The young fruits resemble a zucchini and were cooked in a hangi (a traditional cooking method that uses natural vapors from the soil of volcanic terrain). However, it also provided the natives with a wide range of containers, floats, musical instruments, ornaments, and masks. Its use spread rapidly in Eastern Polynesia because their clay sources were scarce and over time the ceramic techniques were forgotten. Hence, the importance of a plant that could annually produce a series of containers of different sizes [5].

In Africa, in addition to being used in the production of food and beverage containers, its excellent acoustic properties have been exploited for thousands of years in the making of musical instruments. The fruit is used in the manufacture of a range of flutes, drums, and stringed instruments in India, where it is used as a resonance chamber for the sitar and other musical instruments [5].

With regard to the dissemination of the bottle gourd in South America, more specifically in southern Brazil, the fruit was already known to the Guarani people in the year 1580, when the Jesuits arrived in the territory now corresponding to the Brazilian state of Rio Grande do Sul [9]. The *chimarrão* itself was a habit of the natives that the Europeans assimilated. But the Indigenous people and the first immigrants of the region did not use the fruit only for the making of containers for *chimarrão*. Instead, they used it for the production of other containers, as can be seen in the permanent collection of the Parque Gaúcho museum, in Gramado/RS. The museum goes back to the origins and habits that formed the people of Rio Grande do Sul, and together with leather and ceramics, porongo was one of the main natural materials employed in the production of everyday artifacts, such as containers for food (Fig. 2a) and water (Fig. 1b).

The bottle gourd is part of the history and evolution of countless peoples and cultures. It has maintained its main aesthetic and functional characteristics intact for thousands of years and continues to serve as an interesting raw material for the production of manufactured goods today, as it did at the beginning of civilizations. Its triumph has long impressed scientists, among domesticated species, only the dog has spread more widely on the globe [8].

2.2 From Planting to Fruit

The information regarding the planting refers to the cultivation in the central region of the RS state. For the sowing of the fruit in the region seeds are selected in the crop itself, according to the size and shape of the fruit, without any attribution to the

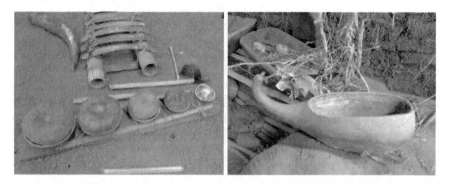

Fig. 1 Guaraní people containers made from bottle gourds: **a** Food containers; **b** water container. *Source* By the authors

10 cm

Fig. 2 Morphological variations of the bottle gourd. *Source* By the authors

plant, being a phenotypic selection of individual fruits. It is not known the origin of the populations used in the region [14]. The growth habit of the plant is climbing and remains low in the absence of a support structure, which has the advantage of greater control of the final fruit format [16].

Regarding its structures, the plant has a well-developed and branched root system, the stem is herbaceous and firm, from which originate the leaves, shoots, inflorescences, and tendrils. Regarding the leaves of the bottle gourd plant, they are usually simple, reniform, undulated, with an entire margin. It is a monoecious plant, in which the male and female gametes are separated, the flowers have a white corolla with five petals. The fruit is classified as a berry of the pepônio type, because of its hard and resistant hull, the term gourd refers to the hard shell. As for the internal morphology, the fruit presents exocarp, mesocarp, and endocarp [16]. The species

Lagenaria siceraria has a great variability of fruit shapes and sizes (Fig. 2), hull thickness, length, symmetry, width, and the number of seeds [14].

The positioning of the fruit in the soil, when small, must be vertical to position the base of the bottle gourd in the correct form. Fruits kept in the horizontal position can bend and lose quality as a *chimarrão* container [16]. The harvest is performed after the natural senescence of the plant, in which the fruits are harvested and huddled in the shade to dry slowly. The drying process in the shade may take more than 6 months to allow the ripening of the material. The green aspect of the outer epidermis will disappear and the fruit will lose 90% of its weight with the water evaporation. When the bottle gourd is completely dry, the seeds will shake inside it (Fig. 3) and the pulp will detach from the mesocarp (Fig. 3).

The importance of knowing the harvest season is directly related to the quality of the fruit, since drying in the field after this phase increases the risk of necrosis of the material caused by fungi, with external and internal browning and loss of quality. In the research conducted by [16], the end of the cycle occurred at 195 days, the author developed a technique that relates the harvest point to the color of the fruit pedicel.

According to [2], the average productivity of porongo reaches 12,000 fruits per hectare. The research carried out by [16] aimed at the management of the crop in order to increase the productivity and quality of the fruit. In the experiment performed by the author, the number of fruits per hectare varied according to the density of plants, being between 9,000 and 10,000 samples. At the end of the research, he concluded that the low yield of marketable fruits is related to the genetic variability of the species.

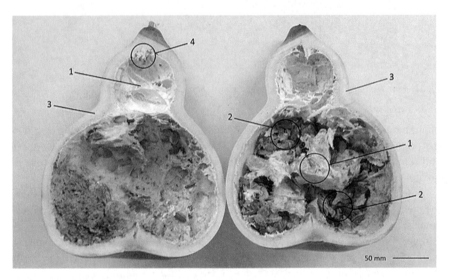

Fig. 3 Longitudinal section of the porongo after the drying process: (1) Endocarp; (2) Seeds; (3) Mesocarp; (4) Encounter of the mesocarp with the endocarp. *Source* By the authors

3 Characterization of Bottle Gourd and Implication for Bionics

3.1 Materials and Methods

The samples of the fruits used in this research were collected in the city of Santa Maria (state of Rio Grande do Sul, Brazil), in 2015. Samples were prepared out of four different fruits, based on the NBR 7190–1997 standard: Design of wooden structures [1]. The standard establishes that wooden specimens must have a rectangular cross section measuring 20 × 30 mm, and length along the fibers of 50 mm. As the bottle gourd has concave shapes and irregular skin thickness, measures were adapted to 50 mm in the transversal direction of the fruit, 30 mm in the longitudinal direction, and 5 mm in thickness. In total, 36 samples were made.

In order to analyze the microstructure of the bottle gourd, characterization was performed by scanning electron microscopy (SEM) and transmitted light microscopy. SEM is a non-destructive method that allows visualizing the surface morphology of materials from solid samples. TM3000 equipment (Hitachi® High-Technologies Corp., Tokyo, Japan) was used, located in the Design and Materials Selection Lab (LDSM) at the Federal University of Rio Grande do Sul (UFRGS). The acceleration of the electron beam used was 15 keV, and the equipment operates with image magnification of up to 30,000 times. Backscattered electron images were obtained.

To observe the morphology and histology of the bottle gourd, a botanical procedure was followed, in which semi-permanent slides were prepared for visualization under light microscopy, at the Laboratory of Plant Anatomy (LAVeg/UFRGS). To prepare the sections, porongo samples were previously immersed in 10% ethylene-diamine for 48 h. Samples taken from the solution were dried in a controlled environment at 20 °C with 50% humidity. For the procedure, they were hydrated in water for 24 h at room temperature. The hydrated samples were then sectioned with a Ranvier microtome and subsequently hand-cut with blades with a thickness between 15 and 8 μm, approximately.

Afterward, the slices underwent the staining process on the Kline plate. Toluidine Blue dye 0.05% was used to observe the presence of lignin in the sample, and Safranin, a reddish dye, to better visualize the parenchyma walls and phenolic compounds, such as lignins. The images were generated in BX41 microscope (Olympus® Co., Tokyo, Japan), using a digital camera model AxioCam ERc5s, at LAVeg/UFRGS.

3.2 Characterization of the Gradient Cellular Structure

Some characteristics of the porongo can be visualized without the aid of magnification equipment, such as the clear distinction between the bark and the inner part (Fig. 4). One of the main aspects is the coloration, which is darker on the outside than

Fig. 4 Observable differences between the exocarp and the mesocarp of the bottle gourd: **a** Exocarp (dark coloration) and mesocarp (light coloration); **b** detail of the pores after the sanding process. *Source* By the authors

that seen on the inside, which is light in color. The difference in thickness between these two zones is also observable, as the shell is constituted by a layer of a micrometric magnitude, in which its proportions can only be quantified using microscopy techniques, the inner porous layer covers the almost totality of the material, reaching several millimeters in thickness. Another striking difference is regarding textures. The skin is extremely smooth and homogeneous to the touch, in contrast to the internal part, which is irregular and has a rough surface (Fig. 4a), as it has the function of holding the placenta with the seeds when fruit.

The characterization of the bottle gourd initiated with the analysis of the microstructure with aims at identifying the differences between the structure of the shell and those observed inside. The observation of structures on a micrometric scale helps the visualization and identification of the basic units that compose the material's microstructure and thus contributes to understanding how these units provide the material its behavior and interesting properties. One of the techniques used was scanning electron microscopy, as described above.

In the fruits of angiosperms, the wall of the fertilized ovary is transformed during development into the pericarp of the mature fruit, which normally develops into three distinct layers: exo-, meso-, and endocarp [13]. The exocarp is the outermost layer, while the mesocarp develops into a fleshy layer and the endocarp has the function of protecting the seeds. In the bottle gourd, the mesocarp is the dominant porous part, and the endocarp is the innermost layer, which is located between the mesocarp and the seeds, which is usually sanded to make products.

The difference between the exocarp and the mesocarp in the bottle gourd is easily noticeable, the outer wall of the fruit is darker and more compact, while the inner layer is lighter in color and porous (Fig. 5a). These differences become more accentuated when the microstructure is observed. The exocarp is seen as a compact structure with the presence of clear spots, which are inorganic materials (Fig. 5a). The mesocarp, on the other hand, is presented as a network of pores, with variable size and shape and different elongation directions (Fig. 5b), which defines it as a less compact

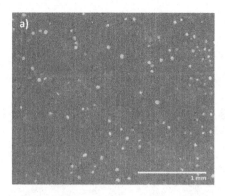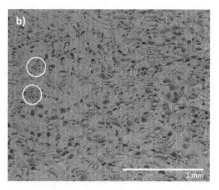

Fig. 5 SEM of the bottle gourd: **a** Exocarp, compact and homogeneous layer, with bright spots of inorganic materials; **b** mesocarp, open-cell foam-like structure formed by parenchymatic cells. Vascular bundles are highlighted in white circles. *Source* By the authors

and less dense structure than the exocarp. As a cellular solid, the mesocarp can be characterized as a three-dimensional cellular material, defined as an open-cell foam.

According to [13], the mesocarp often develops as a fleshy ground tissue, the parenchyma, being made up of large isodiametric cells, with more or less thin walls and large vacuoles. In the case of the bottle gourd, the observable porous structures in the mesocarp are the parenchyma cells of the fruit (Fig. 5b). In young fruits, the peripheral parenchyma is usually rich in chlorophyll and carries out photosynthesis. In the bottle gourd, when the fruit is collected and passes through the drying process, the parenchyma cells die and their walls are lignified while the cells internal space is emptied. The vascular bundles are scattered along the mesocarp (highlighted by white circles in Fig. 5b).

Larger magnifications (Fig. 6) of the exocarp showed that the inorganic components that constitute the spots mapped in lighter tones in the BSE-type images indicate

Fig. 6 Exocarp of the bottle gourd, where the homogeneity of the outer layer, dense and compact, is emphasized. Inorganic materials enclosed in the exocarp cavities are highlighted in white circles. *Source* By the authors

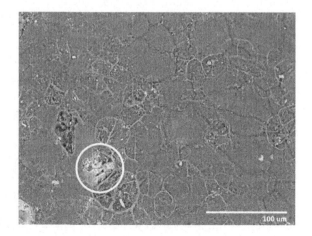

Fig. 7 Detail of the
parenchyma elements: (1)
Parenchyma cell, where the
pits are the conducting
elements; (2) Intercellular
space; (3) Orifice through
which the phloem passes; (4)
Conductor cells. *Source* By
the authors

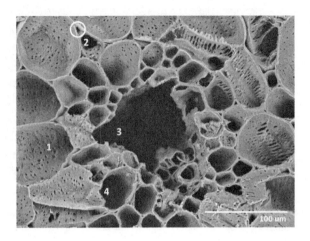

that they are constituted by chemical elements distinct from those of the matrix and
that they are enclosed inside the cells and dispersed along the entire length of the
outer wall.

Parenchyma cells are distributed around the vascular bundles (Fig. 7) and in most
of the mesocarp. It can be recognized that parenchyma cells have an isodiametric
structure while vascular bundle cells are elongated. Intercellular spaces are frequent
among these cell types [13], being of micrometric order of magnitude. Parenchyma
cells are constituted by a thickened wall and have pits through which each cell
connects with neighboring ones. This communication mechanism makes the bottle
gourd an extremely porous material, which results in the permeability of the material
to fluids.

Regarding the vascular bundles, they are composed of phloem (Fig. 7) and xylem
(Fig. 7). In the image, the phloem appears as an empty space, as it is constituted by thin
primary cell walls, which degrade when in contact with the 10% ethylenediamine
solution, used for sample preparation. The phloem is responsible for transporting
the elaborated sap. The xylem cells, which are opposed to the phloem, transport
water and mineral salts. The vascular bundles are collateral and are spread over the
mesocarp [13]. This explains the fact that, in a single cut, vascular bundles can be
sectioned in different orientations, transverse, longitudinal, or intermediate.

Parenchyma cells have a different structure from those that constitute the
conductive cells, with the parenchyma cells having pits between neighboring units
(Fig. 8), while the inner wall of the xylem elements presents cell-wall thickening,
characteristic of tracheal elements (Fig. 8).

To identify whether a pattern of cell growth and orientation can be noticed along
the pericarp of the bottle gourd, BSE-like images of a sample were grouped, covering
different portions of the fruit, from the outermost part of the exocarp to the inner
part closer to the endocarp. The selection of images includes the exocarp (Fig. 9a),
the central portion (Fig. 9b), and the innermost portion (Fig. 9c) of the mesocarp.
As the cells' location moves to the inner part, the dimensions of the parenchyma

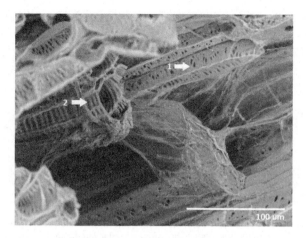

Fig. 8 Outer walls of mesocarp cells: (1) Parenchymal cell, with cavities that connect the cells; (2) outer wall of conductive cells with tracheal features. *Source* By the authors

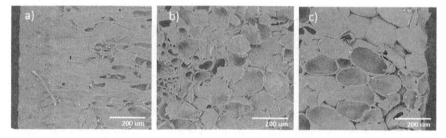

Fig. 9 Progressive increase in the dimensions of the cells: **a** Exocarp; **b** center of the mesocarp and; **c** inner part of the mesocarp. *Source* By the authors

cells and the density of void spaces increase. In the outermost portion of the exocarp, which includes the outer skin, there is a very compact layer of cells, where no intercellular space is visible (Fig. 9a). Those cells are showed very densified and packed together, which explains the impermeability of the skin. This situation is different as the innermost portions are observed, with a progressive increase in cell dimensions (Fig. 9a–c), with different orientations.

Differences in cell dimensions are easily noticed (Fig. 9), which results in different relative amounts between cellular material mass and empty volumes in each analyzed sector. This reason is related to the density of the material, which can lead to it being considered as a material with heterogeneous density. Regarding the study of cellular solids, this relationship is known as relative density, or the ratio between the density of a cellular region to that of the material of which the cell wall is made; therefore, it can also be defined as a volume ratio or the inverse of the porosity [12]. In addition, as observed in the SEM images, the closer to the exocarp, the denser the material, whereas in the inner portions the cell dimensions increase, thus reducing the density

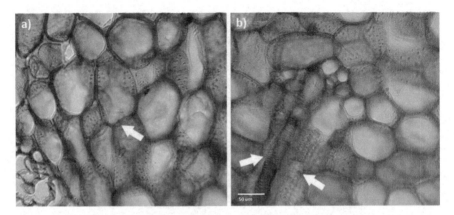

Fig. 10 Thickened walls of the parenchyma: **a** In detail, the communication channels can be seen, which are the dark horizontal lines; **b** on the left, a vascular bundle in the longitudinal direction, with its tracheal structure, and on the right, a parenchyma cell in a longitudinal section, with the communication channels. *Source* By the authors

of the material. The diameter of the cells was assessed with the BSE-type images, and in the outermost portion, closer to the exocarp, pores have values ranging between 24 and 55 μm, and the ones of the pores located in the innermost part of the mesocarp are between 67 and 93 μm.

In order to better visualize aspects that do not appear clearly in the MEV, such as cell details and lignin in the walls, transmitted light microscopy was employed. Parenchyma cells and the pits that serve as communication between neighboring cells (Fig. 10) are observed. These cells have a thickened wall, impregnated with lignin, which is responsible for the hardness and strength of the material. Characteristics of the structures observed via transmitted light microscopy are like those analyzed via SEM, however, some details such as thickened walls are well evidenced here.

The distribution and arrangement of cells from the exocarp to the mesocarp are also well evident (Fig. 11). In this situation, the outermost part, the exocarp, can be seen as a thin, impermeable micrometric layer that is difficult to visualize under a microscope magnification (Fig. 11). The first layer of the mesocarp, formed by sclereids or stone cells, which are cells with a very thicker wall, are very closely arranged so that there is no intercellular space between them (Fig. 11). In the second layer of the mesocarp, the parenchyma cells appear larger, with less thicker cell walls, and with a greater presence of vacuoles, increasing progressively as they are located inwardly to the endocarp (Fig. 11). Regarding the endocarp, it was not possible to visualize it, as it is a very thin layer, between the mesocarp and the seeds, formed by very large and dispersed cells, which ended up deteriorating in the preparation of the samples.

Lignin can be observed by adding dye (as described in the methods), which favors visualization. In the exocarp (Fig. 11), a very dark greenish color is observed, indicating the greater presence of lignin. As the mesocarp position moves inwardly, the cell walls get less thicker, and the greenish hue becomes lighter, indicating a

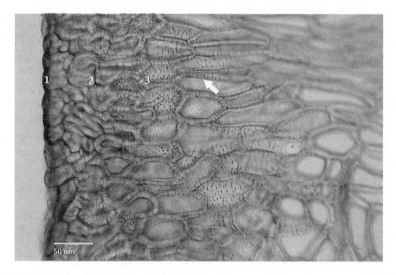

Fig. 11 Distribution and arrangement of cells from the exocarp (left) to the mesocarp (right): (1) Outer layer of exocarp, thin and compact; (2) Mesocarp sclereids; (3) Mesocarp, where a progressive increase in the dimensions of the parenchymal cells can be observed towards the inner part. The arrow indicates the intercellular communication channels, known as pits. *Source* By the authors

progressive reduction in lignin in the innermost layers. The amount of lignin in the dried fruit is relative, depending on the moment it matures.

3.3 Implications for the Study of Bionics

From the analysis of the bottle gourd microstructure, it was possible to better understand the behavior of the material in different situations. Based on the characteristics observed, one can define parameters that could contribute to the development of newer materials, according to a methodology of bionics. Two factors are fundamental for defining the main properties of the material: the difference in density between the exocarp and the mesocarp, as well as the gradient changes in the network that forms its structure. The compact cell distribution in the exocarp explains the impermeability of the skin, while the large empty cells of the parenchyma justify the high porosity of the highly hydrophilic material, as well as the reduced density of the inner part of the fruit.

According to the characteristics analyzed, the bottle gourd can be classified as a cellular material with a unique organization of its unitary components. On the outermost part, the cells are so tightly packed together that they present an elevated level of impermeability, whereas in the inner part cells are much larger and sparse, allowing the development of a complex network of intracellular communication via cell pits, leading to a permeable surface. Moreover, due to the gradient distribution of lignin in the cell walls of the material, the fruit develops a much harder and stiffer periphery, while its interior remains softer. This gradient in the structural characteristics also contributes to the maximization of the weight of the fruit, since more material—especially lignin in the cell wall—is deposited in the regions where mechanical loads are greater.

In terms of bioinspired designs, the fruit being the solution developed by nature can lead to some key features that could be explored. Foremost, not only the gradient distribution of cells can have an effect on the structural efficiency of a bioinspired material but can also propitiate distinct levels of water absorption or impermeability. This is particularly interesting for the design of panels for architecture and civil engineering, where moister is desired to be kept on one side of a surface while the other remains waterproof. Synthetic bottle stoppers as cork replacements can also be developed with this gradient distribution of water absorption and mechanical resistance capabilities. Even small kitchen accessories, like potholders, cutting boards, and oven mitts can benefit from these characteristics.

4 Conclusions

The bottle gourd can be found in different countries and continents, each place with a different vernacular name. Because the fruit originates from a plant that has spread to the continents before men, its history as a raw material for the production of artifacts goes back a long time. Due to its almost ubiquitous presence, the diversity of products made from is quite variable. In the Brazilian state of Rio Grande do Sul, Argentina, and Uruguay the dried fruit is used for the production of containers for the *chimarrão*, while in Africa, it is employed in the creation from food containers to musical instruments, to name a few examples. The diversity of surface design techniques is also diverse, as is the list of artifacts made from it.

Due to its peculiar characteristics of shape, structure, and surface, it can be considered a differentiated natural cellular material. Although it is an extremely ancient crop, from which numerous artifacts are produced, there is still little scientific research for its use as a material. The objects made from the fruit are handcrafted, based on empirical knowledge. In the places where it is cultivated and processed, it is a source of income for the farmers who cultivate the plant and also for the artisans who manufacture the products from it. With better use of the fruit in the making of other artifacts, the entire production chain increases, as well as the number of people benefited by the process.

The use of natural materials as a raw material in product design, in general, is a more complex and variable labor when compared to the use of materials developed and produced industrially. When working with natural materials, automatically a network of people is involved, starting with those responsible for planting or even extracting the material from nature, through the initial processing, the artisans who create products from the raw material, and by those who sell their products. They are also required to adapt to the natural variabilities of the material, where changes in shape, structure, colors, and textures can be expected. It is an intricate process, but inherently more sustainable, because it creates a value chain that benefits all involved, and often minimizes the production of contaminant waste. In this sense, the valorization of the characteristics of natural materials from different regions, as well as the community involved, is just another possibility of design operation.

The use of natural materials from renewable sources as well as promoting local development is an alternative to non-renewable resources that are doomed to extinction. However, the feasibility of applying biological materials in the manufacture of products like containers for food and beverages, depends on some factors, such as the identification of basic properties, so that it can be characterized as material and potentiate its use in other products; and the application of treatments that may increase durability and extend the useful life of the material by virtue of its organic, degradable nature.

Another important application of natural materials, like the bottle gourd, is by means of analyzing and exploring its characteristics as a source of inspiration. The methodology of design based on the study of bionics allows the transposition of features found in nature to the development of newer products, structures, and materials. As observed in the case of this chapter, the properties of the fruit can be explored as its natural gradient anatomy leads to regions from high stiffness and impermeability to soft and permeable surfaces.

References

1. ABNT (1997) Associação Brasileira de Normas Técnicas. NBR 7190: Projeto de estruturas de madeira. Rio de Janeiro
2. Bisognin DA, Marchesan E, Aude MI, DA S (1992) Densidade de semeadura e produtividade do porongo. Ciência Rural, Santa Maria, 22(1):15–19
3. Bisognin DA (2002) Origin and evolution of cultivated cucurbits. Revista Ciência Rural: Santa Maria 32(5):715–723
4. Bisognin DA, Silva ALL (2004) A cultura do porongo. Informe técnico, ed. Pelo Departamento de Fitotecnia do Centro de Ciências Rurais (CCR) da Universidade Federal de Santa Maria (UFSM)
5. Burtenshaw M (2003) The first horticultural plant propagated from seed in New Zealand: Lagenaria siceraria. New Zealand Garden J
6. Gibson LJ, Ashby MF (1999) Cellular solids: structure and properties, 2nd edn. Cambridge University Press, Cambridge, UK
7. Gibson LJ, Ashby MF, Harley BA (2010) Cellular materials in nature and medicine. Cambridge University Press, Cambridge, UK
8. Kistler L, et al (2014) Transoceanic drift and the domestication of African bottle gourds in the Americas. PNAS 111(8)
9. Lessa B (1986) História do chimarrão. Sulina, Porto Alegre
10. Nejeliski D, Duarte LDC (2019) Caracterização do Porongo (Lagenaria siceraria): análise termogravimétrica, determinação do teor de umidade, da densidade básica e da densidade aparente. DAT J 4(1):14–26. https://doi.org/10.29147/dat.v4i1.108
11. Palombini FL, Pestano V, Kindlein Jr., W, da Cunha Duarte L (2020a) Biônica e Seleção de Materiais Celulares para projetos de Design. Des e Tecnol 10(20):01–10. https://doi.org/10.23972/det2020iss20pp01-10

12. Palombini FL, Lautert EL, Mariath JE, de Oliveira BF (2020b) Combining numerical models and discretizing methods in the analysis of bamboo parenchyma using finite element analysis based on X-ray microtomography. Wood Sci Technol 54(1):161–186. https://doi.org/10.1007/s00226-019-01146-4
13. Roth I (1977) Fruits of angiosperms: encyclopedia of plant anatomy. Universidad Central de Venezuela, Caracas
14. Silva ALL et al (2002) Coleta e caracterização morfológica de populações de porongo – Lagenaria siceraria (Mol.) Standl. – Cucurbitaceae. Revista Ciência & Natura: Santa Maria, pp 91–100
15. Silva ALL (2005) Germinação in vitro de sementes e morfogênese do porongo (Lagenaria siceraria (Mol.) Standl.) e mogango (Curcubita pepo L.). 2005. Dissertação (Mestrado em Agronomia) – Universidade Federal de Santa Maria, Santa Maria/RS
16. Trevisol W (2013) Morfologia e fenologia do porongo: produtividade e qualidade da cuia. Tese (Doutor em Ciências). Escola Superior de Agricultura "Luiz de Queiroz", Piracicaba

Bamboo-Based Microfluidic System for Sustainable Bio-devices

Omar Ginoble Pandoli⬡, Sidnei Paciornik⬡, Mathias Strauss⬡, and Murilo Santhiago⬡

Abstract Using conventional microfabrication processes to obtain well-aligned arrays of microfluidic channels is very challenging and costly. Nature, on the other hand, is unique in creating complex hierarchical architectures. For instance, some wood-derived materials have aligned microchannels that may be explored to add new functions to these biological templates, expanding their uses toward greener electronic, biological, and energy devices. To explore novel hierarchical architectures, the 3D anisotropic structure of bamboo has been recently used as a bio-template for the fabrication of functional bio-devices, adding new functionalities to this natural material. Bamboo is a monocotyledon plant that shows high growth speed and that is widespread in tropical regions. It is considered an abundant and low-cost lignocellulosic natural resource that possesses fast microfluidic dynamics, good mechanical strength, lightweight, and high content of crystalline cellulose. Moreover, it can be pyrolyzed to become thermally and electrically conductive. From the anatomic point of view, bamboo is an anisotropic gradient functional material with an atactostele microarray channel system constituted by a complex of vascular bundles

O. Ginoble Pandoli (✉)
Department of Chemistry, Pontifícia Universidade Católica do Rio de Janeiro (PUC-RIO), Rua Marquês de São Vicente, Rio de Janeiro, Brazil
e-mail: omarpandoli@puc-rio.br; omar.ginoblepandoli@unige.it

Dipartimento di Farmacia, Università degli Studi di Genova, Viale Benedetto XV, Genova, Italy

S. Paciornik
Chemical and Materials Engineering Department, Pontifícia Universidade Católica do Rio de Janeiro (PUC-RIO), Rua Marquês de São Vicente, Rio de Janeiro, Brazil
e-mail: sidnei@puc-rio.br

M. Strauss · M. Santhiago
Brazilian Nanotechnology National Laboratory (LNNano), Brazilian Center for Research in Energy and Materials (CNPEM), Campinas, São Paulo, Brazil
e-mail: mathias.strauss@lnnano.cnpem.br

M. Santhiago
e-mail: murilo.santhiago@lnnano.cnpem.br

F. L. Palombini and S. S. Muthu (eds.), *Bionics and Sustainable Design*, Environmental Footprints and Eco-design of Products and Processes,
https://doi.org/10.1007/978-981-19-1812-4_6

141

(metaxylem, protoxylem, and phloem) protected by lignocellulosic fibers (scle-renchyma) embedded into a matrix of living cells tissue (parenchyma). The vascular vessels are radially distributed from the inner to the outer wall of the internode culm with diameters ranging from 50 to 200 μm. As the bamboo microchannel arrays allow the flow of different types of fluids, passively or actively, through capillarity, vacuum, or pumping, this opens a plethora of possibilities. The lignocellulosic walls of the microchannels and the parenchymatous living cells can be functionalized to build up novel devices for environmental, health, chemical, and energy applica-tions. Natural bamboo bio-templates decorated with plasmonic nanoparticles (Ag and Pd-NPs) have been used as a plasmonic system for solar steam generation. Conductive silver ink was used to achieve a regioselective coating of the 3D hollow channel for the prototyping of electric circuits, microfluidic heaters, and fully inte-grated micro electrochemical cells. Finally, new chemical functionalities have been added to the bamboo bio-template to obtain a chemical platform for analytical appli-cations and click chemical reactions. Bamboo carbonized by pyrolysis was used as a 3D solar vapor-generation device for water desalination and also as a monolithic air cathode for microbial fuel cell applications. Therefore, bamboo stands as a promising natural template for devices that demand and take advantage of hierarchical archi-tectures and microarray channels. It can be explored as raw or as carbonized material for scalable production of eco-friendly, sustainable, low-cost, and portable chemical, electronic, and electrochemical bio-devices. These bioinspired solutions could fulfill industrial demands for greener chemical, electronic, and energy applications.

Keywords 3D structure · Lignocellulose polymer · Vapor-generation device · Monolithic air cathode · Modeling analytical platform · Catalytic microreactor · Electrochemical cell · Microfluidic heater · Bambootronic

1 Introduction

Bamboo is considered a biological structural material that has inspired designers, engineering, and chemists to understand its structural-property relationships. The complexity of this natural material is based on 3 key factors: anatomy, chem-ical composition, and hierarchical architecture from nano to macroscale level [1]. Recently, bamboo has attracted scientific interest in the emerging fields of bioin-spired design and biomimetics. The main goal is to discover a practical biomim-icking fabrication of a biological composite with a defined 3D structure for the scalable production of engineered products. Up to now, we are far away from repro-ducing synthetically a similar man-made multifunctional "superior" material, which combines biological and/or mechanical functions. In this case, what we can do is to study from a different point of view: the single constituents at a molecular level, and the whole hierarchical architecture at a macroscopic scale. With a whole overview, we will understand the efficiency and effectiveness of bamboo 3D structure, considered a biologically evolved structural system [2].

Bamboo belongs to the grass family *Gramineae (Poaceae)*, subfamily *Bambusoideae*, tribe *Bambuseae*. *Bambusoideae* include two tribes: *Bambuseae* (wood bamboo) and *Olyreae* (herbaceous bamboo). All *Bambuseae* are native in tropical and sub-tropical countries and are classified into 80 genera and an estimated 1200 species [3]. The most important giant bamboos include *Phyllostachys pubescens* (Moso), *Guadua angustifolia* (Guadua), and *Dendrocalamus giganteus* (Dendrocalamus). Depending on the species, bamboo reaches its maximum height between 3 and 8 months and its complete maturity in 4 years. After that, it starts slow biodegradation. 36 million hectares of bamboo forests are distributed worldwide in China (24 million hectares, 65%), America (10 million hectares, 28%), and Africa (2 million hectares, 7%).[4] With a high yield per hectare, fast harvesting, without deterioration of the soil, bamboo is an exceptional candidate as abundant, sustainable, and renewable lignocellulosic polymers resources with a potential economic impact for undergoing economic developing countries in Asia, Latin America, and Africa. The traditional bamboo use in indigenous cultures has been replaced (housing and utensils) with new industrial and economic opportunities as a substitute of wood for agroforestry, textile, reinforced composite materials, charcoal, pulp, paper, laminated board, and new engineered materials.

Bamboo macrostructure is constituted by a hollow tube-like structure named culm that is divided into transversal nodes and longitudinal internodes. The diameter and the thickness of the culm decrease along the bamboo length. From the ground to the top head of the culm the thickness of the woody wall is constituted by longitudinally vascular bundles (metaxylem, protoxylem, and phloem) protected by lignocellulosic fibers (sclerenchyma) embedded into a matrix of living cells tissue (parenchyma). The morphology and the spatial distribution of vascular bundles into the longitudinal axis and radial axis are not homogenous and well ordered, indicating an atactostele distribution [5]. Indeed, the bamboo structure is more heterogeneous than wood [6]. The density of the vascular bundles with its sclerenchyma tissue is graded and the shape changes as well. The volume fraction of the vascular bundles and their solid fraction of the cellulose fiber increase radially from the inner to outer part across the transversal section of the culm. More gradually the density of the fiber increases from the bottom to the upper part of the bamboo. Contrarily, the density of parenchymatous matrix cells increases from the outer to the inner region of the bamboo culm.

Concerning stiffness and strength, the mechanical properties show significant variations depending on the measurement positions, both radially and longitudinally [7]. The exceptional anisotropic mechanical properties are determined by a more evident radial density gradient in the bamboo culm [8]. The distribution of Young's modulus and tensile strength along the transverse section of bamboo culm varies from about 5 to 25 GPa, and from about 100 to 800 MPa, respectively [9]. This asymmetric behavior similar to a cantilevered structure is useful to absorb the highest stress at the bottom, and lowest stress at the top head of the bamboo [10]. For these reasons, bamboo is well-known as an intelligent functional gradient material [9]. The gradients and the heterogeneities in all living organisms allow obtaining different high-performance biological materials. The gradients are the combinations of different factors: (i) variations of chemical composition and constituents; (ii) dimensions

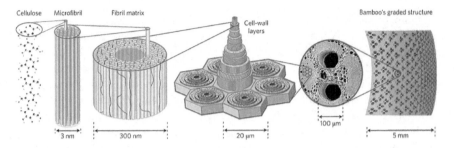

Fig. 1 The hierarchical structure of bamboo. Reprinted with permission from [1]. Copyright 2014, Nature Publishing Group

and orientations of the molecular building blocks; (iii) complex arrangements of hierarchical structures from nano to macroscale level (Fig. 1) [11].

The chemical composition varies depending on the species, growth conditions, bamboo age, and part of the culm [12]. The main constituents of bamboo are holocellulose 65–75% (cellulose and hemicellulose); lignin 20–30%; extractive 5% (resins, fatty acids, essential oil, tannins, etc.) and ashes 2% (inorganic substances such as potassium, sodium, calcium, silicon, magnesium, iron, aluminum, zinc, phosphor) [13]. The D-glucose unit ($C_6H_{10}O_5$) is a molecular building block of the cellulose polymer. The linear long-chains composed of β-1,4 glycosidic bonds are nanostructured in microfibrils alternating amorphous and crystalline regions packed inside fibril aggregates (macrofibrils). The fibrils are enveloped in a complex lignin and hemicellulose matrix, called lignin-carbohydrate complex (LCC) [14]. LCC works as a binder material inside the macrofibrils. The fibril matrix is a constituent of the cell-wall layers that are assembled in different orientations to obtain a protective sclerenchymatous tissue (fiber) organized in a honeycomb-like structure around the hollow channels. The microarray of channels is straight and parallel to the longitudinal axis of the internode. The internal diameter of the channels increases from the outer to the inner wall, from 50 to 200 μm, across the transversal section of the culm. Each vascular bundle is embedded in the parenchyma cells tissue with variable square-like shape and volume of 1000–2000 μm^3.

The hierarchy of cellulose from the molecular level to the higher structure is of paramount importance for the anisotropic mechanical properties of bamboo [15]. The Young's modulus of crystalline cellulose is estimated at 200 GPa, but experimentally the value for the microfibrils decreases to 50–70 GPa, depending on the species. The consequence of the hierarchical organizations of the fiber and parenchyma structures is the loss of stiffness with an increase in toughness. The origin of the strength and stiffness of bamboo is hidden at a molecular level and was revealed with different theoretical atomistic/molecular modeling [14, 16]. The impressive mechanical behavior, particularly its flexibility, depends on the supramolecular interactions of hydroxyl groups between stacked biopolymers. The dynamic hydrogen bonding between the cellulose polymers plays an important role to absorb energy avoiding fracture and preserve the structural integrity [17]. The integration of the supramolecular structures [18] within its hierarchical material is the key strategy of nature to balance strength, stiffness, and toughness [19, 20].

Similar to a bone, bamboo shows another interesting characteristic concerning the ability to adapt its growth to external stimuli. This adaptability is a feature of a living organism to build up its hierarchical structure depending on the loading charges [21]. The optimization of the chemical constituents for self-assembling into determined arrangements is a key factor to obtaining a determined structural design [2]. Bamboo is considered a self-optimizing graded material built by a cell-based mechanosensory system. It uses loading external stimuli to shape the volume density of the fibers, their distribution, as well as the thickness of the culm. The remarkable piezoelectric behavior of the bone to model its skeleton structure was demonstrated for bamboo as well [9].

Several 2D techniques have been used to get information about the nanoscale topography, chemical and physical properties of different bamboo species, such as peak-force quantitative nanomechanics atomic force microscopy (PF-QNM-AFM) [22], scanning thermal microscopy (SThM) [23], scanning electron microscopy (SEM) and transmission electron microscopy (TEM),[24] mapping confocal Raman spectroscopy [25] and confocal laser scanning microscopy (CLSM) [26]. With the aims to study the width and length of cellulose crystallites, the crystallinity and the orientation of the microfibril angle, X-ray diffraction (DRX), small-angle X-ray (SAXS), and wide-angle X-ray scattering (WAXS) have been used to reveal the anisotropic arrangement of crystalline cellulose fibers [10, 27, 28]. More recently, X-ray computed microtomography (μCT) has been employed as a non-invasive approach to visualize and analyze the biological structure of bamboo, such as the vascular bundle system [29], and the degree of fibril orientation [30], High-resolution X-ray μCT with nonlinear finite element analysis (FEA) was used for the reconstruction of a complex vascular system of the node [31] and to evaluate the strength of parenchyma and sclerenchyma structures [32]. μCT images have been useful to rebuild the physical 3D structure of the parenchyma by stereolithography printing, to characterize its mechanical properties and structure-property relationships [33]. μCT has also been used as a powerful tool to identify a regioselective metal coating into bamboo biological structures, such as vascular bundles[34, 35] and parenchyma cells [26, 36].

This chapter will present several 3D devices obtained from two kinds of engineered bamboo-based biomass: natural bamboo and carbonized bamboo. The functionalized natural bamboo-based is built adding new functionality through chemical and physical functionalization of the lignocellulosic internal channels. First, we report lignocellulosic-based chemical platforms for copper-catalyzed organic reactions[34] and analytical detection for safe water and urinalysis [37]. Second, we describe the possibility to graft plasmonic nanoparticles (Ag and Pd-NPs) for self-floating bamboo solar steam generation [38]. Third, with a simple and low-cost silver ink internal coating it is possible to transform an insulating natural bamboo into a high conductive biocomposite material for electric and electrochemical applications.

Bambootronics devices will be described, such as electrical 3D circuits, a microfluidic heater, and a fully integrated micro electrochemical cell [35]. To improve the electrical and photothermal properties of bamboo, pyrolytic processes are explored to transform the crystalline phase of cellulose into conductive graphitic carbon. Two devices based on carbonized bamboo will be presented: a solar vapor-generation device for water desalination [39] and a monolithic air cathode for microbial fuel cell applications [40].

2   Natural Bamboo Bio-templates for Chemical Platform, Steam Generation, and Electric and Electrochemical Devices

The world of plants is full of examples that allow us to create biomimetic materials for specific purposes and spin-off a new global market for advanced functional materials, such as superhydrophobic surfaces with lotus leaves effect or greener composite material for harvesting solar energy [41]. Plants, as a biological evolution system, use hydrophilic microenvironment channels to flow, in and out, water, nutrients, and synthetic bioproducts. Similarly, a man-made microfluidic reactor (Lab-on-Chip) allows to carry out chemical reactions that can be tuned by designing the size, shape, and polarity of the microreactor channels [42, 43]. In the last 20 years microreactor fabrication technology (MRT) was able to design and prototype several artificial microfluidic systems with several aims: increasing diffusion mixing of reagents and heat transfer; decreasing of reaction-time and waste; high-throughput chemical synthesis; increasing productivity and scalability of chemical production in flow mode [44].

The deposition of catalyst in the flow reactor design opened up an emergent research field named heterocatalysis in microfluidic reactors [45, 46]. At the same time MRT was applied in the field of sensors incorporating sensitive material for bioanalytical applications [47]. The bottleneck of MRT is the cost of fabrication technology that in several cases includes the use of clean rooms and expensive facilities. Another problem is the cost of the materials for the chip fabrication [48]. Glass, silicon, thermoplastic, elastomers, and paper have been used for different designs of microfluidic chips [44, 49]. The idea of exploring the natural 3D microarray channels of bamboo to create a bio-microfluidic system will avoid some expensive microfabrication steps and will minimize the costs of the laboratory facilities. With this aim, a proper functionalization of channels' walls of the vascular bundles should be addressed for the prototyping of two kinds of chemical platforms: (i) a lignocellulose-based microreactor for continuous flow organic reaction [34], and (ii) a lignocellulose-based analytical device for chemical detection [37].

2.1 Copper-Functionalized Lignocellulosic Microreactor (Cu-LµR)

A functionalized biological microreactor device will be a novel class of cheaper and sustainable bio-microfluidic devices compared to current artifact flow chemical systems. If bio-microfluidic is an emergent research field that applies biomaterials and biologically inspired design (biomimetics) to microfluidic devices [50], another approach to developing a microfluidic device is using the renewable and biodegradable hierarchical structural directly imprinted into a 3D natural specimen of bamboo. Without the need to fabricate artificially microsized polar channels, the natural bamboo structure with its microarray of vascular bundles can be directly prototyped as chemical flow reactors to continuously inject reagents for aqueous reactions or in other greener solvents (Fig. 2a) [45, 51].

Figure 2b depicts the fabrication sequence for lignocellulose-based microreactors (LµRs) from bamboo internodes consisting of:

(i) cutting of the internode with different types of rotary saws to obtain a cylindrical shape of 3 cm length and 5 mm diameter;
(ii) cleaning of the channels with water penetration under negative pressure and drying in an oven;
(iii) reduction of the diameter extremities; and
(iv) gluing and insertion of the two syringe needles as inlet and outlet for polyethylene tubes.

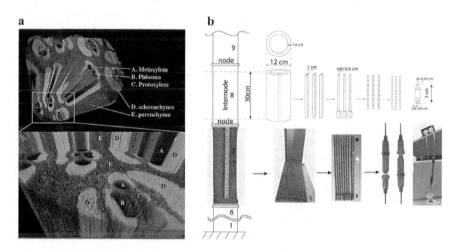

Fig. 2 **a** µCT image of the internal structure of bamboo with its vascular bundles (metaxylem, phloem, and protoxylem), sclerenchyma, and parenchyma tissue. Reprinted with permission from [36]. Copyright 2016 Royal Society Chemistry; **b** Fabrication of lignocellulose-based microreactors (LµRs) from bamboo internode 7. A tutorial video is available in the supplementary information of reference [34]. Reprinted with permission from [34] Copyright 2019 American Chemical Society

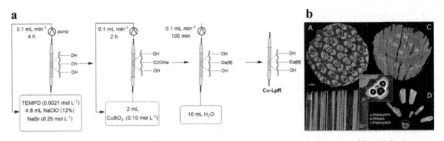

Fig. 3 a Functionalization with copper ions of LμR. **b** μCT images of Cu-LμR: transversal and longitudinal cross-sections of bamboo culm (A–B); 3D image with and without the vegetal biomass (C–D). A dotted circular yellow line with an internal diameter of 4.2 mm corresponds to the internal area of flow injection. Insight of the metal deposition onto microchannels (metaxylem, phloem, protoxylem) are highlighted with red lines. Reprinted with permission from [34]. Copyright 2019 American Chemical Society

A peristaltic pump is used to inject and recycle pure water or other solutions from a reservoir at different flow rates (from 0,1 to 2,0 mL min^{-1}). Up to 2,0 mL min^{-1} a very low back pressure (5 psi) and no leaching at the bamboo-needle connector nor along the external walls of the bamboo-based device was observed.

The chemical functionalization of the LμRs is presented in Fig. 3a. First, chemoselective oxidation of the hydroxyl carbon 6 with the formation of carboxylate functionalities is carried out recirculating an oxidative solution containing TEMPO/NaClO/NaBr for 4 h at 0,1 mL min^{-1}. Second, the cation exchange and the complexation of copper ions are obtained with the injection of CuSO$_4$ solution (2 h at 0,1 mL min^{-1}). Third, water (100 min at 0,1 mL min^{-1}) was used to wash the microchannels and be ready for the use of the Copper-functionalized lignocellulosic microreactor (Cu-LμR). The metal coating of the microsized wall was characterized by X-ray μCT. The transversal and longitudinal sections in Fig. 3b show a brighter contrast due to the metallization of the internal wall of the vascular bundles. In Fig. 3b, image processing steps allowed us to segment and observe the hollow channels with a deposited copper layer with 21 μm thickness.

As a proof-of-concept study, copper(I)-catalyzed 1,3-dipolar cycloaddition between azide and terminal azide (CuAAC) was selected to test the effectiveness of the fabricated bamboo-based microreactors. A series of 1,4-disubstituted triazole derivatives were yielded in flow regime with an aqueous-methanol solvent with good efficiency (60–96%) and minimal leaching of copper (5 ppm) (Fig. 4a). To evaluate the contribution of the leached copper to the formation of the product 3aa in homogenous catalysis instead of the heterocatalytic regime, the following experiment was considered (Fig. 4b). The reaction mixture was pumped through the Cu-μLR for 2 h with a product yield of 16%. Then the system was stopped and the reaction mixture was maintained under stirring for 6 h without contact with the Cu-μLR, leading to a small increment of the yield of ca. 2%. Finally, the reaction mixture was recirculated through the Cu-μLR for 6 h with a final product yield of 80%. This result indicated the effective contribution of the heterocatalysis in the flow regime and the minimal

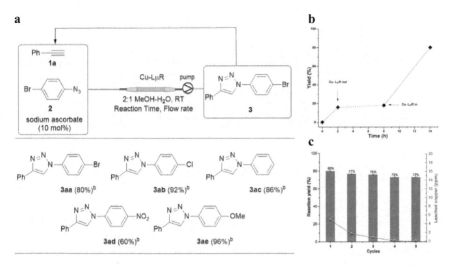

Fig. 4 **a** Scope of the CuAAC flow procedure with Cu-LμR. [a]1a (0.14 mmol), 2 (0.12 mmol), sodium ascorbate (10 mol%), 2:1 MeOH-H₂O (3 mL). [b]Isolated Yield. **b** Evaluation of the contribution of homogeneous catalysis (leached copper) to the reaction outcome of the optimized flow process with Cu-LμR. **c** Reusability of Cu-LμR and quantification of leached copper (ICP-MS analysis). Reprinted with permission from [34]. Copyright 2019 American Chemical Society

contribution of the leached copper to the catalytic process. The reusability of the Cu-LμR up to 5 cycles was tested keeping a good reaction yield (80–73%) (Fig. 4c). The maintenance of the process efficiency demonstrated the long-term operation of the novel copper-functionalized lignocellulosic microreactor. Jointly with a cheaper and sustainable scalability production of the bamboo-based bio-microfluidic devices, a new biological chemical platform was demonstrated as a potential alternative to the established market of microreactor fabrication techniques.

2.2 Lignocellulose-Based Analytical Devices (LADs)

Alternatively, to pump solution inside the vascular bundles system, it is possible to explore a passive absorption of analytes into the 3D bamboo biomass for analytical detection [37]. Lignocellulose-based analytical devices (LADs) were demonstrated for several biochemical analyses. Bamboo biomass was impregnated with colorimetric indicators to analyze food and water safety (i.e. nitrite assay and bacterial detection in water), and complete urinalysis (i.e., nitrite, urobilinogen, and pH assays in human urine). Figure 5 shows, in sequence: (I) the fabrication of the device drilling a V-shaped reaction zone; (II) the immobilization of the detection reagents into the reaction zone; and (III–IV) easy operability of the LADs for colorimetric recognition.

A pre-treatment of bamboo with deionized water at 100 °C for 5 h was necessary to remove unknown chemicals and starch granules. Then the LADs were dried at

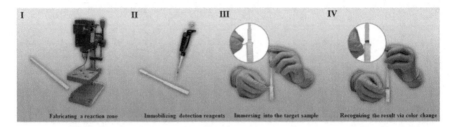

Fig. 5 Schematic fabrication process of LAD: (I) drilling a reaction zone; (II) immobilization of the colorimetric reagents; (III) immersion of the LAD into the target sample; (IV) colorimetric recognition of the results into the reaction zone. Image by [37] licensed under CC-BY

50 °C for 8 h. After the optimization of the fabrication process and the colorimetric signal output, it was chosen to build the device with a reaction zone 3 cm from the absorption end. For each kind of analysis, the reaction zone was functionalized with a specific reagent. The LADs were soaked into the specific solution target for several minutes (5–15 min) waiting for a different time for each specific reaction in the reaction zone (7–55 min). Then, the images of the reaction zone were captured with a digital camera, and the color intensity was analyzed by ImageJ software to create a calibration curve and to determine the limit of detection (LOD) for each analysis.

The nitrite analysis was based on the Griess reaction. The reagent solution (3 μL of sulfanilamide, citric acid, and N-(1-naphthyl)-ethylenediamine dihydrochloride) was deposited into the reaction zone and the LADs were dried at room temperature for 15 min. The device was immersed in the sample target for 7 min and the images were taken after a 20-min wait to complete the reaction (Fig. 6a).

The bacterial detection of *E. Coli* was based on the oxidation-reduction of the indicator (3-[4,5-dimethylthiazol-2yl]2,5-diphenyltetrasodium bromide, MTT) in the presence of the electron mediator (phenazine methosulfate, PSS). The device was dried for 2 min and then was immersed in the target sample for 5 min. Before image acquisition, the LAD was kept for 35 min until the chemical reaction was completed (Fig. 6b). The schematic representation of the PMS-MTT assay for bacterial detection in water is shown in Fig. 6c. The LOD for the nitrite assay in deionized water was 0.06 mM and compatible with the typical paper analytical device (PAD). The LOD for the bacterial detection in drinking water was in the range of 1.8×10^4–9.3×10^4 cfu/mL. In this case, the LOD for this assay is higher compared to the LOD of available point-of-care of lab-on-chip (POC-LOC).

The Urobilinogen assay was based on the Ehrlich reaction using 4-dimethylaminobenzaldehyde. The reagent solution (3μL) was deposited into the reaction zone of the LADs followed by drying at 25 °C for 5 min. For the pH assay, a reagent solution (3 μL of bromothymol blue and resazurin sodium salt) was immobilized in the reaction zone, followed by 5 min of drying at 25 °C. For both the urinalysis, urobilinogen, and pH, the LADs were immersed for 7 min, and then it was necessary to wait 20 min for the chemical reaction process to take place and observe the colorimetric results shown in Fig. 7a–d. The LODs for the urinalysis of

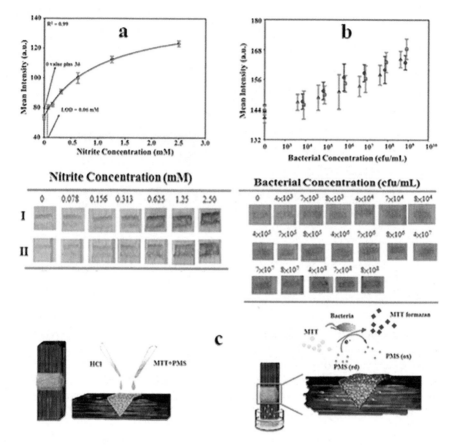

Fig. 6 Colorimetric results for **a** nitrite assay and **b** bacterial detection in drinking water (PMS-MTT assay). **c** Schematic representation of the PMS-MTT assay. Image by [37] licensed under CC-BY

nitrite and urobilinogen were respectively 0.06 mM and 0.16 mg/mL. The LADs were reliable for a urine pH range from 4.0 to 8.0 (Fig. 7c). Since there is no transversal flow between parallel vascular bundles, it was possible to include two reaction zones for multiple detections on different sides of the LADs. Figure 7e shows multiple detections of nitrite and urobilinogen from human urine. The red box indicates that the measurement was in a normal human urine sample; the blue box indicates that the measurement was in human urine spiked with 1.25 mM nitrite and 1.25 mM urobilinogen.

Recently, the anisotropic behavior of water and moisture sorption at various structural levels—cell walls and bamboo blocks—was demonstrated. The study revealed different sorption directions, longitudinal (L) and transversal (T), for a saturated salt solution and water vapor. These results will be very useful to understand the water transport in bamboo, aiming at more efficient drying and impregnation processes

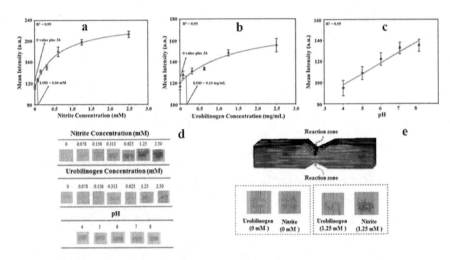

Fig. 7 LADs for urinalysis of **a** nitrite, **b** urobilinogen, **c** pH with **d** the colorimetric results of different assays. **e** Multiple detections of nitrite and urobilinogen on two sides of the LDA with the colorimetric results in red and blue boxes. Image by [37] licensed under CC-BY

for future applications [52]. Because of their abundant and independent longitudinal microfluidic channels, LADs analytical platforms show rapid passive transportation with the capillary action of different solutions in selective reaction zones. These new lignocellulosic-based analytical devices have been demonstrated as potential low-cost engineered materials to advance microfluidic development compared to PADs [53], wood-microfluidic systems [6, 54], and PDMS-based microfluidic devices [55].

2.3 Solar Steam Generation

Another application exploring water transportation through the capillary effect is the production of a plasmonic bamboo device for solar steam generation (Fig. 8) [38]. Inspired by the transpiration of the natural bamboo culm, oriented bamboo was decorated with plasmonic metallic nanoparticles (Ag and Pd-NPs) for solar energy-based seawater desalination. Light energy was converted into heat energy under local surface plasmon resonance. A nest of bamboo rings floating on the water allowed an upward flow by capillary action and the heat vaporized water to form steam.

The synthesis and the deposition of the plasmonic NPs (Pd and Ag) were executed in-situ through a chemical reduction of salt precursors, $PdCl_2$ and $AgNO_3$ respectively. The function of the NPs is to convert the sunlight into heat. The photothermal behavior showed a high light absorption efficiency with a solar energy conversion of 87% under the intensity of 10 suns. The plasmonic bamboo performance for water desalination under different seawater conditions was stable for 140 h under the intensity of 5 suns. Considering the long-term operation of the engineered bamboo

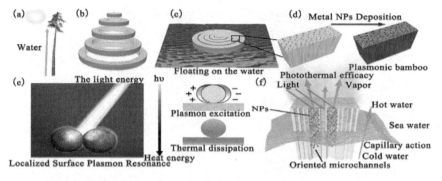

Fig. 8 **a–d** Schematic fabrication of plasmonic bamboo. **e** Light energy was converted into heat energy by local surface plasmon resonance. **f** Photothermal effect under irradiation and water transportation into the microchannel converted to vapor. Reproduced with permission [38]. Copyright 2019, Elsevier Ltd.

plasmonic device, the natural oriented hydrophilic microchannel, good water transportation, biocompatibility with the NPs and water, self-floating property, and the low cost of the support material, bamboo is a promising material for an efficient solar steam generation without any support system [38].

2.4 Electrical 3D Circuits

The fabrication of electrical circuits is a crucial aspect in the development of electrical and electrochemical devices on lignocellulosic substrates, such as bamboo [35]. Bamboo offers a unique platform to create a plethora of devices with unique properties when compared to traditional substrates, such as paper. For instance, the paper has been widely used in the fabrication of flexible electronic and electrochemical devices. However, the fabrication of three-dimensional (3D) devices is very challenging and there are only a few routes to enable this process [56, 57]. These routes are based on creating hydrophobic walls on paper followed by delivering precursors or conductive materials by using capillary flow. As can be observed in Fig. 9a, b, the width of conductive channels is in the mm range, which negatively impacts the miniaturization of devices. In comparison, bamboo-based devices offer two main advantages against classical paper substrates, (i) high lateral resolution and (ii) high density of channels per area. For instance, we can find more than 40 microchannels in a diameter of 5 mm [35]. The aligned channels can be viewed as templates for the fabrication of conductive tracks. Figure 9c shows an optical microscopy image of a transversal cut of a bamboo culm. The vascular bundles (phloem and metaxylem) are isolated from each other by crystalline cellulose fibers (sclerenchyma tissue). The brownish-yellow color represents the living cells of bamboo, namely the parenchyma tissue.

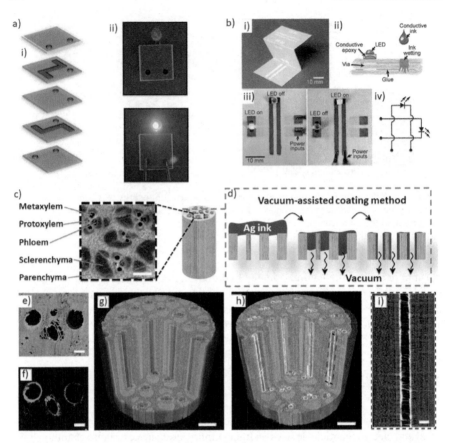

Fig. 9 **a** Three-dimensional organic conductive tracks on paper using polypyrrole. (i) schematic view of the conductive tracks (black regions), (ii) LED device being activated. Reprinted (adapted) with permission from [57]. Copyright (2016) American Chemical Society. **b** Three-dimensional conductive paper-based device. (i) Picture of the paper sheet containing hydrophilic channels. (ii) Cross-sectional view of the device and flowing direction of PEDOT:PSS. (iii) Picture of the 3D device lighting up a LED and iv) Circuit diagram. Reproduced with permission [56]. Copyright 2016, WileyVCH. **c** Optical microscopy of bamboo. The scale bar is 500 μm. **d** Schematic view of the Ag-coating method. **e–f** SEM image of the transversal cut of bamboo modified with Ag highlighting the channels and its respective Ag mapping using EDS. Scale bars are 100 μm. **g, h** X-rays microtomography (μCT) 3D images of the pristine bamboo template before and after silver coating. Scale bars are 500 μm. (i) 2D μCT detail of a single vascular channel with its internal wall metal modified (thickness 10.3 ± 2.2 μm). Scale bar is 100 μm. Reproduced with permission [35] Copyright 2020, RSC

In theory, the channels can be filled with many different conductive materials to create highly conductive tracks. For instance, low melting temperature metal alloy (Sn-Bi) can be impregnated inside the channels to obtain high conductivity (5.4 × 10^4 S m^{-1}) [58]. One of the issues of this route is that the resulting channels are clogged after impregnation, thus limiting many applications that require hollow channels. In this regard, one of the possibilities to keep the channels hollow while adding conductivity consists in the deposition of conductive materials only at the walls of the channels.

Figure 9d shows a schematic figure of a vacuum-assisted coating method that ultimately results in highly conductive walls. In brief, the method consists of flowing silver ink inside the microchannel arrays by vacuum pumping, as schematically shown. In the next step, the excess solvent is removed by flowing N_2 inside the channels. This simple yet effective method results in the formation of silver coating of the microchannels' inner walls. The presence of hollow channels after modification with silver ink can be assessed by SEM-EDS and μCT. Figure 9e, f shows the SEM image and its respective Ag mapping using EDS. As can be observed, the entrance of the channels contains silver only at the inner walls. To confirm that the entire length of the channels is hollow, μCT 3D images of the same bamboo specimen before and after silver coating were obtained, as shown in Figs. 9g–i. In addition, the white contrast observed in the images is due to the presence of silver coatings formed at the inner walls along the entire length of microchannels. The thickness of the silver coating is ~10 μm.

Since silver coatings are only formed at the inner walls of bamboo channels and these channels are aligned, one can expect the electrical conductivity to be extremely anisotropic. For example, electrical measurements showed that Ag-coated bamboo is highly conductive along the microchannel direction, as schematically shown in Fig. 10a. The conductivity of the channels was 9.3 ($\pm$4.0) $\times$ 10^5 S m^{-1}, which is the highest value reported so far for cellulose-based ordered materials. When the electrical test probes are placed at the bottom and top surfaces, the resistance achieves its minimum values. On the contrary, when the measurement is performed orthogonally to lateral surfaces the electrical resistance is ultra-high even with a short distance between probes.

A remarkable feature of these highly conductive structures is the absence of short-circuit between microchannels, making it possible to create several electrical and electrochemical devices. The high anisotropy of bamboo combined with electrically addressable microchannels enables the fabrication of many functional 3D devices. By creating contact pads on single channels or groups of channels it is possible to fabricate complex electrical circuits in bamboo.

For instance, Fig. 10b shows an electric circuit in which the current starts at the bottom face of bamboo, reaches the top surface by flowing through the microchannel, lights up the LED, and then returns to the bottom face. Even though the lateral surface of bamboo is not conductive, small cuts in the lateral surface can locally assess the conductive microchannels, as shown in Fig. 10c. Depending on how deep the cuts are, inner channels can be assessed by this simple strategy.

Another example illustrated in Fig. 10d shows even more complex 3D electrical circuits. In this case, a serpentine-like circuit was fabricated, showing the potential of connecting multiple faces, such as the bottom, top, lateral and inner regions. In addition, two pieces of bamboo can be combined to light up specific LEDs by rotating vertically aligned bamboo pieces. Figure 10e shows pictures of the devices. A triangular area was patterned with silver ink at the bottom of bamboo (i) and

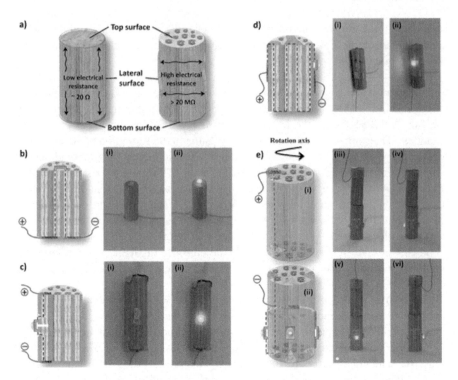

Fig. 10 **a** Schematic layout of the Ag-coated bamboo ($\emptyset = 6$ mm and $L = 20$ mm) and its anisotropic electrical properties. **b–e** Schematic figure of different circuits and photos of the devices at ON and OFF states. Reproduced with permission [35]. Copyright 2020, RSC

bamboo (ii) contains three independent circuits that can be activated when the Ag contact pads perfectly match during rotary stepwise movement. Figure 10b–e also shows the ON and OFF states.

2.5 *Microfluidic Heaters*

In the previous section, we showed the fabrication of 3D electrical circuits that have a hollow structure, which means it is possible to flow gases and liquids through the microchannels and explore the electrical circuits for new applications [35]. 3D fluidic devices can be fabricated by using other routes like soft photolithography and 3D printing, for instance. Fabrication of 3D devices on PDMS is time-consuming and the layered patterning process limits the final structure that can be obtained. The continuous advance in 3D printed routes enables the fabrication of fluidic arrays with excellent features.

On the other hand, we can mention bamboo as a uniquely sustainable and raw bio-template for microfluidic applications. For instance, most 3D printers and micro-fabrication routes have issues achieving centimeter-long highly oriented arrays of microchannels (up to 50 cm) that are naturally found in the full extent of the bamboo culm. The conductive Ag-coated bamboo can be used as a Joule heater to boost the kinetics of chemical transformations or to increase the temperature of fluids. Figure 11a shows a picture of Ag-coated bamboo where all the microchannels were short-circuited. By increasing the current flows through the microchannels it is possible to tune the bamboo temperature, as shown in the infrared thermographic images in Figs. 11b–d. The homogeneous thermal images at different currents are also a good indicator of the uniform metal coating distribution in the microchannels.

Figure 11e shows a bamboo-based microfluidic device that was used to heat water. An external thermocouple was placed at the end of the device to measure the temperature. As also shown in Fig. 11e, the water temperature becomes constant after approximately 4 min. The equilibrium temperature can also be tuned by adjusting the flow rate and current, as shown in Fig. 11f. It is possible to increase the water outlet temperature by applying higher electric currents. In addition, by increasing the flow rate the temperature measured at the end of the channels decreases. In this configuration, the temperature can be set from 25 to 55 °C, which is suitable for several organic reactions in flow mode [56].

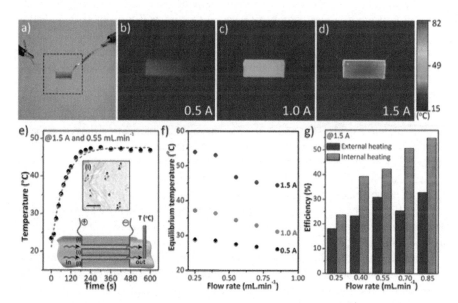

Fig. 11 a Microfluidic heater using bamboo (Ø = 6 mm and L = 20 mm). **b–d** Infrared thermographic images during application of different currents. **e** Water temperature versus time. The inset photo shows the open channels. Scale bar is 1.0 mm. **f** Water temperature at different currents and flow rates. **g** Joule heating efficiency. Reproduced with permission [35]. Copyright 2020, RSC

Figure 11g shows energy efficiencies (EE) for two different microheater setups to highlight the unique advantages of Ag-coated bamboo systems. In one example, current flows through the array of channels (internal heating) while in the other example the current flows on the outer Ag-coated surface (external heating). As can be viewed in Fig. 11g, in both cases EE increases since higher flow rates led to an increase in forced convection that ultimately results in a higher heating transfer coefficient between water and the heating source. However, by comparing both conditions, EE can be 20% higher for internal heating at higher flow rates. Such differences observed in EE can be explained by the low thermal conductivity of lignocellulosic materials. Thus, in the case where the heating step is performed at the external region, bamboo poses higher resistance to deliver energy to heat-up water inside microchannels. By using external heating, some part of the heat that should be delivered to the inner regions of bamboo is lost to the environment. This heat transfer issue can be even more problematic for thicker pieces of bamboo, thus highlighting the importance of preparing conductive coating on the walls of the microchannels.

2.6   Fully Integrated Electrochemical Cells

Another additional advantage of the hollow conductive channels of bamboo is the possibility of incorporating conductive pastes to fabricate fully integrated electro-chemical cells [35]. Figure 12a shows the step-by-step fabrication process. The first step consists of spreading an epoxy resin in the bamboo face where the electrodes will be fabricated. Before resin curing, nitrogen gas is flowed through bamboo channels as schematically shown in Fig. 12a. Most of the channels remain unclogged and can be filled with carbon black nanostructures to fabricate an array of microelectrodes. In this way, carbon pastes (carbon particles + mineral oil) are excellent alternatives to fabricate electrodes. They are low cost, have simple fabrication, have low back-ground current, and many modifiers can be added in paste formulation. Figure 12b–d show the face of bamboo filled with carbon black paste at different magnifications. Figure 12e shows the cyclic voltammogram obtained using a bamboo sample modi-fied with carbon black paste as the working electrode. In this experiment, Pt wire was also used as counter electrode and saturated calomel electrode (SCE) as reference. The carbon electrodes showed a well-defined electrochemical response when tested using established redox probes, such as ferri/ferrocyanide.

As demonstrated in the previous section, the Ag-coated microchannels are not in short-circuit and can be individually addressed. Thus, by creating electrical contact pads in groups of channels it is possible to fabricate a fully integrated electrochemical cell, as shown in Fig. 12f. The setup illustrated in this figure is only possible due to the unique architecture of bamboo. As mentioned before, carbon pastes can be easily modified with catalysts, redox mediators, and many other compounds. As a proof-of-concept, carbon paste was modified with Prussian blue, which is known to work as a redox mediator to detect hydrogen peroxide at low potentials. Figure 12g shows

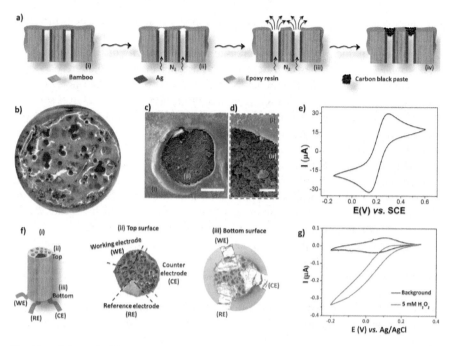

Fig. 12 **a** Schematic fabrication process. **b** Stereomicroscope image of the bamboo after modification with carbon paste. **c–d** SEM Images of one microelectrode. Scale bars in Figures c–d are 50 and 5 μm, respectively. **e** Cyclic voltammograms using modified bamboo as working electrodes. **f** Schematic layout of a bamboo-based electrochemical cell with working, counter, and reference electrodes fully integrated into one single bamboo specimen. **g** Cyclic voltammogram of the integrated electrochemical bamboo cell. The carbon paste was modified with Prussian blue (Black curve). The red curve shows the voltammogram in the presence of hydrogen peroxide (5 mM). Reproduced with permission [35]. Copyright 2020, RSC

the curves in the absence and presence of hydrogen peroxide. As can be observed, a large current is obtained at potentials around −0.15 V versus Ag/AgCl.

3 Devices Based on Carbonized Bamboo

As mentioned before, bamboo vegetable anatomy, structure, and hierarchical architecture, together with the chemical and mechanical properties given by its composition based on biopolymers and biomolecules such as cellulose, lignin, and hemicellulose are critical characteristics that make it stand as a unique bio-template for the fabrication of functional bio-devices. However, raw bamboo has very deficient electrical, thermal and solar radiation absorption properties that hinder its full exploration in devices for electro, electrochemical and photothermal applications. One way to overcome these limitations is the carbonization of bamboo, which leads to the transformation of its structural organic components into carbonaceous materials that present electrical and photothermal properties associated with the formation of conducting graphitic domains and increased light absorptivity respectively.

Carbonized bamboo has historically found uses as an energy resource [59, 60], raw material for the production of activated carbon [61, 62] and other porous materials for environmental remediation [63], gas adsorption [64, 65], catalyst supports [66, 67], and materials for batteries and supercapacitors [68–70]. However, none of the applications makes rational use and takes great advantage of the hierarchical structure and arrangement of the bamboo channels. One of the most efficient ways to obtain carbonized bamboo without the loss or drastic alterations of this functional architecture is through pyrolysis. Pyrolysis is a well-explored thermochemical route to convert several types of solid organic matter into carbonaceous materials through several simultaneous and sequential chemical reactions on which organic matter chemical bonds are broken and formed toward its final carbonized product.

The use of controlled pyrolysis parameters, such as adequate heating rates and target temperatures, applied to non-meltable organic precursors, such as cellulose and lignin from bamboo, are likely to result in carbonaceous materials which mimic the morphology and architecture of the starting organic material. Febrianto and collaborators [71] have studied the physical and anatomical characteristics of different types of bamboo carbonized at 200, 400, 600, 800, and 1000 °C. An increase in carbonization temperature led to a decrease in the volume and weight of carbonized samples. The greater change occurred between 200 and 400 °C as at this temperature range cellulose, hemicellulose, and lignin from bamboo experience the highest degradation and mass-loss rates. The density of all carbonized bamboos decreased after carbonization between these temperatures and then became almost constant for higher temperatures. Anatomical observation of carbonized bamboo showed that the vascular diameter decreased with the increase of carbonization temperature, and shrinkage in radial and tangential directions followed the same trend. The author found by statistical analysis that there was a significant correlation between physical contraction and anatomical contraction.

In a similar study involving pyrolyzed bamboo samples at temperatures between 300 and 950 °C, Krzesinska discussed in more detail the structural anisotropy of carbonized bamboo. He explored a macro scale description based on physical parameters measured along basic directions of samples and microscopic analysis of the skeleton structure [72]. The temperature of pyrolysis and the heat rate were found again as critical parameters affecting the structure of carbonized bamboo [73]. An increase in pyrolysis temperature led to increased samples' true density as observed also by Febrianto et al. As pyrolysis temperature evolved the highly anisotropic structure of raw bamboo was transformed into low-temperature carbons with a disordered nearly isotropic structure which developed into ordered turbostratic carbon clusters at temperatures above 600 °C. Krzesinska concluded that the electrical and elastic modulus anisotropy of carbonized bamboo structure is mostly due to its pore system, rather than from some microstructural arrangement of the solid carbon skeleton, which is nearly isotropic.

3.1 Solar Vapor-Generation Device for Water Desalination

As mentioned before, there is still plenty of room to be explored in applications that take advantage of the hierarchical structure and arrangement of bamboo that remains intact after carbonization by pyrolysis under controlled conditions. A very interesting example is presented by Shulin Gu and collaborators who have used carbonized bamboo to fabricate a 3D solar vapor-generation device (Fig. 13) [39]. Solar-driven vapor-generation devices have become a smart alternative and an energy-efficient way to produce clean water and tackle the global challenges of water scarcity. So far, several devices based on 2D structures have been reported for this purpose on which improvement of the vapor-generation efficiency is achieved, enhancing the

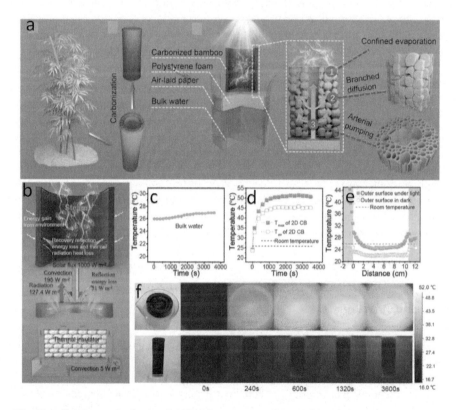

Fig. 13 **a** Bamboo carbonization for 3D solar vapor generator and water pathways at the natural structure, **b** schematic of the heat behavior in the 3D carbonized bamboo-based solar vapor generator. Temperature changes of **c** bulk water **d** and 2D carbonized bamboo over time under 1 sun, **e** temperature distribution of the outer surface of 3D carbonized bamboo in dark and after illumination under 1 sun for 3600 s. **e, f** Images of carbonized bamboos with 2D and 3D structure in a wet state and the corresponding infrared photos under 1 sun for 1 h. The photos, in order from left to right, correspond to $t = 0$, 240, 600, 1320, and 3600 s after illumination. Reproduced with permission [39]. Copyright 2018, Wiley-VCH

solar absorbance, heat confinement, continuous water replenishment, and quick vapor transportation. Moreover, other alternatives have been explored considering materials with lower cost, better mechanical stability, and larger active area, well fulfilled by 3D carbonized wood generators [74–76].

Authors have demonstrated that the 3D solar vapor-generation device based on carbonized bamboo offers an extremely high vapor-generation rate of 3.13 kg m^{-2} h^{-1} under 1 sun illumination [39]. Such a high evaporation rate is achieved due to the unique natural structure, properties, and hierarchical cellular architecture of carbonized bamboo such as.

(i)   hydrophobicity;
(ii)   numerous aligned microchannels that enable rapid water transport;
(iii)   high light absorptance which covers a broad range of the Solar spectrum;
(iv)   reduced thermal radiation heat losses;
(v)   lower average temperature than the environment;
(vi)   reduced vaporization enthalpy of water confined in the cellular structure;
(vii)   remarkable mechanical properties;
(viii)   the ability of salt self-cleaning;
(ix)   good scalability and low cost.

From a mechanistic point of view, good performance occurs because the inner wall of bamboo recovers the diffuse light energy and thermal radiation heat losses from the 3D bamboo bottom, while the outer wall captures energy from the warmer surrounding environment (Fig. 13a–c). Moreover, water confinement effects in the bamboo cellular structure mesh reduce its vaporization enthalpy increasing the evaporation rate. In this sense, carbonized bamboo presented great characteristics and favorable overall performance as a solar vapor generator standing as an attractive alternative for desalination and industrial and domestic wastewater treatment [39].

3.2   Monolithic Air Cathode for Microbial Fuel Cell Applications

As mentioned, in addition to maintaining almost unaltered morphological and anatomical characteristics, bamboo pyrolysis leads to the formation of carbonaceous materials that contain conductive graphitic domains that can be explored in electrical and electrochemical devices [77]. Such devices play a pivotal role in biomedical and environmental technologies as well as in the transition to sustainable technologies for energy generation and storage.

Microbial fuel cells are examples of bio-electrochemical devices that use the degradation of organic matter by bacteria metabolism and simultaneously generate bioelectricity. They are considered both wastewater treatment and electricity generation technologies. Microbial fuel cells based on air cathode have great potential for practical utilization due to their simple design and direct use of oxygen in the air as electron acceptors. However, commercial applications of such devices still depend on lowering fabrication costs for scaling up, high electrocatalytic activity as well as easily fabricated design and porous structure. Some of these issues were tackled by Qiang Liao and collaborators who have made use of a carbonized bamboo tube as the air cathode of a microbial fuel cell taking advantage of its anatomical and structural characteristics (Fig. 14) [40].

They used a bamboo charcoal tube obtained by carbonization at 900 °C in N_2 atmosphere followed by a heat treatment at 350 °C for 2 h under air to increase porosity. This bamboo charcoal tube was coated with PTFE and then directly used as the air cathode of a microbial fuel cell. This device has achieved a maximum power density (P_{max}) of 40.4 ± 1.5 W m^{-3} which is slightly higher than 37.7 ± 2.5 W m^{-3} found for a standard noble metal-based electrocatalyst (Pt/C), at the same operating conditions. The abundant porous structure of the carbonized bamboo led to an increase in triple-phase interfaces (TPIs) for the oxygen reduction reaction (ORR), contributing to its higher performance. Besides, the lowest oxygen mass transfer coefficient (K_{O2}) values observed for the bamboo charcoal tube compared to the Pt/C cathode provided a lower oxygen leakage from the cathode to the anode resulting in a comparable power generation with a high coulombic efficiency of 56%. Due to these results, the bamboo charcoal tube cathode stands out as a low-cost, simple fabrication, and high-performance alternative for the cathode in microbial fuel cells and other electrochemical devices [40].

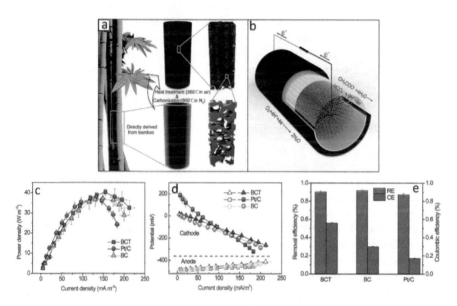

Fig. 14 a Schemes of the preparation process of the bamboo charcoal tube cathode and the microstructure formed, and **b** a microbial fuel cell built with a bamboo charcoal tube cathode. Devices performance tests: **c** Power density curves, **d** polarization curves, and **e** removal and coulombic efficiencies (at 50 Ω). Image by [40] licensed under CC-BY

4 Remarks and Potential Future Applications of Bamboo Bio-templates

Bamboo-based electrochemical devices are still at a very early stage in their development. In this chapter, we showed some of the applications that can be explored by coating the inner channels with silver ink, for instance, 3D circuits, joule heaters, and fully integrated electrochemical cells. Regarding the electrochemical cells, the interface where the redox reaction occurs is still at the external region of bamboo. There is a lot of room to perform electrochemical reactions that take place inside of the inner channels. For instance, other metal coatings (Au, Pt, Ni) and carbon could be used to create hollow channels to conduct electrochemical reactions in flow conditions. Since the channels are highly-aligned, the product of these electrochemical reactions can be directed to specific compartments. This could be interesting for application in the field of energy. For instance, water splitting will generate H_2 and O_2 at the cathode and anode, respectively, and such products could be transported by the bamboo channels.

We foresee the integration of bamboo-based electrodes with the plethora of 2D materials available. These materials are very promising for hydrogen evolution reactions and the fabrication of supercapacitors. Their integration into bamboo channels can be achieved by flowing a dispersion of these materials. In addition, by applying a potential onto the conductive channels it is possible to reduce some 2D materials, such as graphene oxide, and these materials will be deposited at the walls of the conductive channels. In general, there are many possibilities to functionalize the walls of the conductive channels by many different routes and such modifications may help to push the field for many new technologies using conductive bamboo.

5 Conclusion

Bamboo is a hierarchical, anisotropic, graded, and hygroscopic biocomposite material. Its complex 3D structure has inspired designers and engineers to reproduce a similar bioinspired structural material. Still far to mimic its biological and functional properties, many works have been presented exploring one of the most important characteristics of bamboo: the microarray of hydrophilic, parallel, and longitudinally oriented microchannels.

The bottom-up approach of nature to build hierarchical 3D structures from D-glucose molecular building blocks is very difficult to mimic. To obtain a man-made multifunctional material that encompasses extraordinary mechanical properties, thermal stability, biocompatibility, and fast fluid dynamic behavior is not trivial. One way to bypass these difficulties is to use biomass as a bio-template to build engineered bamboo-based devices for chemical, electric, and electrochemical applications.

Lignocellulosic-based chemical platforms, with metal catalysts and specific colorimetric reagents, have been presented for organic synthesis and analytic detections. A regioselective coating of the vascular bundles' system with commercial conductive silver ink allowed to convert insulating biomass into a highly anisotropic conductive composite material to prototype electric 3D circuits, microfluidic heaters, and integrated electrochemical cells.

Decorating the bamboo matrix with plasmonic metal nanoparticles or through pyrolysis of the bamboo, different designs of solar steam generation for water desalination were developed. The transformation of the lignocellulosic biomass into carbonaceous conductive material has added new electric and photothermal properties with important applications for energy transformation such as microbial fuel cells. We think that the use of bamboo, such as other wood-based materials, can lead to new applications by fine-tuning a selection of catalysts, indicators, redox mediators, metals, and conductive polymers which could be incorporated into its highly oriented 3D hollow structure.

References

1. Wegst UGK, Bai H, Saiz E, Tomsia AP, Ritchie RO (2015) Bioinspired structural materials. Nat Mater 14:23–36. https://doi.org/10.1038/nmat4089
2. Srinivasan AV, Haritos GK, Hedberg FL (1991) Biomlmetlcs: Adwancing man-made materials through guidance from nature. Appl Mech Rev 44:463–482. https://doi.org/10.1115/1.3119489
3. Judziewicz E, Clark L (2007) Classification and biogeography of new world grasses: anomochlooideae, pharoideae, ehrhartoideae, and bambusoideae. Aliso 23:303–314. https://doi.org/10.5642/aliso.20072301.25
4. INBAR, FAO (2005) World bamboo resources. A thematic study prepared in the framework of the global forest resources
5. Schweingruber FH, Börner A (2018) The plant stem. Springer International Publishing, Cham. https://doi.org/10.1007/978-3-319-73524-5
6. Jia C, Jiang F, Hu P, Kuang Y, He S, Li T, Chen C, Murphy A, Yang C, Yao Y, Dai J, Raub CB, Luo X, Hu L (2018) Anisotropic, mesoporous microfluidic frameworks with scalable, aligned cellulose nanofibers. ACS Appl Mater Interfaces 10:7362–7370. https://doi.org/10.1021/acsami.7b17764
7. Dixon PG, Gibson LJ (2014) The structure and mechanics of Moso bamboo material. J R Soc Interface 11. https://doi.org/10.1098/rsif.2014.0321.
8. Tan T, Rahbar N, Allameh SM, Kwofie S, Dissmore D, Ghavami K, Soboyejo WO (2011) Mechanical properties of functionally graded hierarchical bamboo structures. Acta Biomater 7:3796–3803. https://doi.org/10.1016/j.actbio.2011.06.008
9. Nogata F, Takahashi H (1995) Intelligent functionally graded material: bamboo. Compos Eng 5:743–751. https://doi.org/10.1016/0961-9526(95)00037-N
10. Salvati E, Brandt LR, Uzun F, Zhang H, Papadaki C, Korsunsky AM (2018) Multiscale analysis of bamboo deformation mechanisms following NaOH treatment using X-ray and correlative microscopy. Acta Biomater 72:329–341. https://doi.org/10.1016/j.actbio.2018.03.050
11. Liu Z, Meyers MA, Zhang Z, Ritchie RO (2017) Functional gradients and heterogeneities in biological materials: design principles, functions, and bioinspired applications. Prog Mater Sci 88:467–498. https://doi.org/10.1016/j.pmatsci.2017.04.013
12. Liese W, Grosser D (1971) On the anatomy of Asian bamboos, with special reference to their vascular bundles. Wood Sci Technol 5:290–312

13. Rusch F, Wastowski AD, de Lira TS, Moreira KC, de MoraesLúcio D (2021) Description of the component properties of species of bamboo: a review. Biomass Convers Biorefinery. https://doi.org/10.1007/s13399-021-01359-3

14. Youssefian S, Rahbar N (2015) Molecular origin of strength and stiffness in bamboo fibrils. Sci Rep 5:11116. https://doi.org/10.1038/srep11116

15. Silva ECN, Walters MC, Paulino GH (2006) Modeling bamboo as a functionally graded material: lessons for the analysis of affordable materials. J Mater Sci 41:6991–7004. https://doi.org/10.1007/s10853-006-0232-3

16. Hao H, Tam LH, Lu Y, Lau D (2018) An atomistic study on the mechanical behavior of bamboo cell wall constituents. Compos Part B: Eng 151:222–231. https://doi.org/10.1016/j.compositesb.2018.05.046

17. Habibi MK, Samaei AT, Gheshlaghi B, Lu J, Lu Y (2015) Asymmetric flexural behavior from bamboo's functionally graded hierarchical structure: underlying mechanisms. Acta Biomater 16:178–186. https://doi.org/10.1016/j.actbio.2015.01.038

18. Pandoli O, Spada GP (2009) The supramolecular chemistry between eastern philosophy and the complexity theory. J Incl Phenom Macrocycl Chem 65:205–219. https://doi.org/10.1007/s10847-009-9643-5

19. Sarikaya M (2002) Biomimetics: materials fabrication through biology. Proc Natl Acad Sci 96:14183–14185. https://doi.org/10.1073/pnas.96.25.14183

20. Gibson LJ (2012) The hierarchical structure and mechanics of plant materials. J R Soc Interface 9:2749–2766. https://doi.org/10.1098/rsif.2012.0341

21. Chen PY, McKittrick J, Meyers MA (2012) Biological materials: functional adaptations and bioinspired designs. Prog Mater Sci 57:1492–1704. https://doi.org/10.1016/j.pmatsci.2012.03.001

22. Ren D, Wang H, Yu Z, Wang H, Yu Y (2015) Mechanical imaging of bamboo fiber cell walls and their composites by means of peakforce quantitative nanomechanics (PQNM) technique. Holzforschung 69:975–984. https://doi.org/10.1515/hf-2014-0237

23. Shah DU, Konnerth J, Ramage MH, Gusenbauer C (2019) Mapping thermal conductivity across bamboo cell walls with scanning thermal microscopy. Sci Rep 9:1–8. https://doi.org/10.1038/s41598-019-53079-4

24. Lian C, Liu R, Zhang S, Yuan J, Luo J, Yang F, Fei B (2020) Ultrastructure of parenchyma cell wall in bamboo (Phyllostachys edulis) culms. Cellulose 27:7321–7329. https://doi.org/10.1007/s10570-020-03265-9

25. Wang X, Ren H, Zhang B, Fei B, Burgert I (2012) Cell wall structure and formation of maturing fibres of moso bamboo (Phyllostachys pubescens) increase buckling resistance. J R Soc Interface 9:988–996. https://doi.org/10.1098/rsif.2011.0462

26. GinoblePandoli O, Martins RS, De Toni KL, Paciornik S, Maurício MH, Lima R, Padilha NB, Letichevsky S, Avillez RR, Rodrigues EJ, Ghavami K (2019) A regioselective coating onto microarray channels of bamboo with chitosan-based silver nanoparticles. J Coat Technol Res 16:999–1011. https://doi.org/10.1007/s11998-018-00175-1

27. Wang Y, Leppänen K, Andersson S, Serimaa R, Ren H, Fei B (2012) Studies on the nanostructure of the cell wall of bamboo using X-ray scattering. Wood Sci Technol 46:317–332. https://doi.org/10.1007/s00226-011-0405-3

28. Ren W, Guo F, Zhu J, Cao M, Wang H, Yu Y (2021) A comparative study on the crystalline structure of cellulose isolated from bamboo fibers and parenchyma cells. Cellulose 28:5993–6005. https://doi.org/10.1007/s10570-021-03892-w

29. Peng G, Jiang Z, Liu X, Fei B, Yang S, Qin D, Ren H, Yu Y, Xie H (2014) Detection of complex vascular system in bamboo node by X-ray μCT imaging technique. Holzforschung 68:223–227. https://doi.org/10.1515/hf-2013-0080

30. Ahvenainen P, Dixon PG, Kallonen A, Suhonen H, Gibson LJ, Svedström K (2017) Spatially—Localized bench—Top X-ray scattering reveals tissue—Specific microfibril orientation in Moso bamboo. Plant Methods 1–12. https://doi.org/10.1186/s13007-016-0155-1

31. Palombini FL, Nogueira FM, Kindlein W, Paciornik S, Mariath JEDA, De Oliveira BF (2020) Biomimetic systems and design in the 3D characterization of the complex vascular system

of bamboo node based on X-ray microtomography and finite element analysis. J Mater Res 35:842–854. https://doi.org/10.1557/jmr.2019.117

32. Palombini FL, Kindlein Jr W, de Oliveira BF, de Araujo Mariath JE (2016) Bionics and design: 3D microstructural characterization and numerical analysis of bamboo based on X-ray microtomography. Mater Characteriz 120:357–368. https://doi.org/10.1016/j.matchar. 2016.09.022

33. Dixon PG, Muth JT, Xiao X, Skylar-Scott MA, Lewis JA, Gibson LJ (2018) 3D printed structures for modeling the Young's modulus of bamboo parenchyma. Acta Biomater 68:90–98. https://doi.org/10.1016/j.actbio.2017.12.036

34. de Sá DS, de Andrade Bustamante R, Rodrigues Rocha CE, da Silva VD, da Rocha Rodrigues EJ, DjenneBuarque Müller C, Ghavami K, Massi A, Ginoble Pandoli O (2019) Fabrication of lignocellulose-based microreactors: copper-functionalized bamboo for continuous-flow CuAAC click reactions. ACS Sustain Chem Eng 7:3267–3273. https://doi.org/10.1021/acssus chemeng.8b05273

35. Pandoli OG, Neto RJG, Oliveira NR, Fingolo AC, Corrêa CC, Ghavami K, Strauss M, Santhiago M (2020) Ultra-highly conductive hollow channels guided by a bamboo bio-template for electric and electrochemical devices. J Mater Chem A 8:4030–4039. https://doi.org/10.1039/C9T A13069A

36. Pandoli O, Martins RD, Romani EC, Paciornik S, Maurício MH, Alves HD, Pereira-Meirelles FV, Luz EL, Koller SM, Valiente H, Ghavami K (2016) Colloidal silver nanoparticles: an effective nano-filler material to prevent fungal proliferation in bamboo. RSC Adv 6:98325–98336. https://doi.org/10.1039/C6RA12516F

37. Kuan C-M, York RL, Cheng C-M (2016) Lignocellulose-based analytical devices: bamboo as an analytical platform for chemical detection. Sci Rep 5:18570. https://doi.org/10.1038/sre p18570

38. Sheng C, Yang N, Yan Y, Shen X, Jin C, Wang Z, Sun Q (2020) Bamboo decorated with plasmonic nanoparticles for efficient solar steam generation. Appl Thermal Eng 167:114712. https://doi.org/10.1016/j.applthermaleng.2019.114712

39. Bian Y, Du Q, Tang K, Shen Y, Hao L, Zhou D, Wang X, Xu Z, Zhang H, Zhao L, Zhu S, Ye J, Lu H, Yang Y, Zhang R, Zheng Y, Gu S (2019) Carbonized bamboos as excellent 3D solar vapor-generation devices. Adv Mater Technol 4:1800593. https://doi.org/10.1002/admt. 201800593

40. Yang W, Li J, Zhang L, Zhu X, Liao Q (2017) A monolithic air cathode derived from bamboo for microbial fuel cells. RSC Adv 7:28469–28475. https://doi.org/10.1039/C7RA04571A

41. Aldersey-Williams H (2004) Towards biomimetic architecture, Nat. Mater 3:277–279. https:// doi.org/10.1038/nmat1119

42. Lizana L, Konkoli Z, Bauer B, Jesorka A, Orwar O (2009) Controlling chemistry by geometry in nanoscale systems. Annu Rev Phys Chem 60:449–468. https://doi.org/10.1146/annurev.phy schem.040808.090255

43. Karlsson M, Davidson M, Karlsson R, Karlsson A, Bergenholtz J, Konkoli Z, Jesorka A, Lobovkina T, Hurtig J, Voinova M, Orwar O (2004) Biomimetic nanoscale reactors and networks. Annu Rev Phys Chem 55:613–649. https://doi.org/10.1146/annurev.physchem.55. 091602.094319

44. Ren K, Zhou J, Wu H (2013) Materials for microfluidic chip fabrication. Acc Chem Res 46:2396–2406. https://doi.org/10.1021/ar300314s

45. Mosadegh B, Bersano-Begey T, Park JY, Burns MA, Takayama S (2011) Next-generation integrated microfluidic circuits. Lab Chip. 11:2813. https://doi.org/10.1039/c1lc20387h

46. Elvira KS, Solvas XCI, Wootton RCR, de Mello RCR (2013) The past, present and potential for microfluidic reactor technology in chemical synthesis. Nat Chem 5:905–915. https://doi. org/10.1038/nchem.1753

47. Rivet C, Lee H, Hirsch A, Hamilton S, Lu H (2011) Microfluidics for medical diagnostics and biosensors. Chem Eng Sci 66:1490–1507. https://doi.org/10.1016/j.ces.2010.08.015

48. Walsh DI, Kong DS, Murthy SK, Carr PA (2017) Enabling microfluidics: from clean rooms to makerspaces. Trends Biotechnol 35:383–392. https://doi.org/10.1016/j.tibtech.2017.01.001

49. Tsao C-W (2016) Polymer microfluidics: simple, low-cost fabrication process bridging academic lab research to commercialized production. Micromachines 7:225–236. https://doi. org/10.3390/mi7120225

50. Domachuk P, Tsioris K, Omenetto FG, Kaplan DL (2010) Bio-microfluidics: biomaterials and biomimetic designs. Adv Mater 22:249–260. https://doi.org/10.1002/adma.200900821

51. Duncombe TA, Tentori AM, Herr AE (2015) Microfluidics: reframing biological enquiry. Nat Rev Mol Cell Biol 16:554–567. https://doi.org/10.1038/nrm4041

52. Chen Q, Fang C, Wang G, Ma X, Luo J, Chen M, Dai C, Fei B (2021) Water vapor sorption behavior of bamboo pertaining to its hierarchical structure. Sci Rep 11:1–10. https://doi.org/ 10.1038/s41598-021-92103-4

53. Cate DM, Adkins JA, Mettakoonpitak J, Henry CS (2015) Recent developments in paper-based micro fl uidic devices. Anal Chem 87:19–41. https://doi.org/10.1021/ac503968p

54. Andar A, Hasan MS, Srinivasan V, Al-Adhami M, Gutierrez E, Burgenson D, Ge X, Tolosa L, Kostov Y, Rao G (2019) Wood microfluidics. Anal Chem 91:11004–11012. https://doi.org/10. 1021/acs.analchem.9b01232

55. McDonald JC, Whitesides GM (2002) Poly(dimethylsiloxane) as a material for fabricating microfluidic devices. Acc Chem Res 35:491–499. https://doi.org/10.1021/ar010110q

56. Hamedi MM, Ainla A, Güder F, Christodouleas DC, Fernández-Abedul MT, Whitesides GM (2016) Integrating electronics and microfluidics on paper. Adv Mater 28:5054–5063. https:// doi.org/10.1002/adma.201505823

57. Santhiago M, Bettini J, Araújo SR, Bufon CCB (2016) Three-dimensional organic conductive networks embedded in paper for flexible and foldable devices. ACS Appl Mater Interfaces 8:10661–10664. https://doi.org/10.1021/acsami.6b02589

58. Wan J, Song J, Yang Z, Kirsch D, Jia C, Xu R, Dai J, Zhu M, Xu L, Chen C, Wang Y, Wang Y, Hitz E, Lacey SD, Li Y, Yang B, Hu L (2017) Highly anisotropic conductors. Adv Mater 29:1–9. https://doi.org/10.1002/adma.201703331

59. Yang W, Wang H, Zhang M, Zhu J, Zhou J, Wu S (2016) Fuel properties and combustion kinetics of hydrochar prepared by hydrothermal carbonization of bamboo. Biores Technol 205:199–204. https://doi.org/10.1016/j.biortech.2016.01.068

60. Hu W, Feng Z, Yang J, Gao Q, Ni L, Hou Y, He Y, Liu Z (2021) Combustion behaviors of molded bamboo charcoal: Influence of pyrolysis temperatures. Energy 226:120253. https:// doi.org/10.1016/j.energy.2021.120253

61. Hameed B, Din A, Ahmad A (2007) Adsorption of methylene blue onto bamboo-based activated carbon: kinetics and equilibrium studies. J Hazard Mater 141:819–825. https://doi.org/10.1016/ j.jhazmat.2006.07.049

62. Liu Q-S, Zheng T, Wang P, Guo L (2010) Preparation and characterization of activated carbon from bamboo by microwave-induced phosphoric acid activation. Ind Crops Prod 31:233–238. https://doi.org/10.1016/j.indcrop.2009.10.011

63. Kearns JP, Wellborn LS, Summers RS, Knappe DRU (2014) 2,4-D adsorption to biochars: effect of preparation conditions on equilibrium adsorption capacity and comparison with commercial activated carbon literature data. Water Res 62:20–28. https://doi.org/10.1016/j.watres.2014. 05.023

64. Shang G, Shen G, Liu L, Chen Q, Xu Z (2013) Kinetics and mechanisms of hydrogen sulfide adsorption by biochars. Biores Technol 133:495–499. https://doi.org/10.1016/j.biortech.2013. 01.114

65. Kumar A, Singh E, Khapre A, Bordoloi N, Kumar S (2020) Sorption of volatile organic compounds on non-activated biochar. BioresourTechnol 297:122469. https://doi.org/10.1016/ j.biortech.2019.122469

66. Chen W, Fang Y, Li K, Chen Z, Xia M, Gong M, Chen Y, Yang H, Tu X, Chen H (2020) Bamboo wastes catalytic pyrolysis with N-doped biochar catalyst for phenols products. Appl Energy 260:114242. https://doi.org/10.1016/j.apenergy.2019.114242

67. Zhang C, Fu Z, Liu YC, Dai B, Zou Y, Gong X, Wang Y, Deng X, Wu H, Xu Q, Steven KR, Yin D (2012) Ionic liquid-functionalized biochar sulfonic acid as a biomimetic catalyst for hydrolysis of cellulose and bamboo under microwave irradiation. Green Chem 14:1928. https://doi.org/10.1039/c2gc35071h

68. Gong Y, Li D, Luo C, Fu Q, Pan C (2017) Highly porous graphitic biomass carbon as advanced electrode materials for supercapacitors. Green Chem 19:4132–4140. https://doi.org/10.1039/C7GC01681F
69. Jin C, Sheng O, Luo J, Yuan H, Fang C, Zhang W, Huang H, Gan Y, Xia Y, Liang C, Zhang J, Tao X (2017) 3D lithium metal embedded within lithiophilic porous matrix for stable lithium metal batteries. Nano Energy 37:177–186. https://doi.org/10.1016/j.nanoen.2017.05.015
70. Gu X, Wang Y, Lai C, Qiu J, Li S, Hou Y, Martens W, Mahmood N, Zhang S (2015) Microporous bamboo biochar for lithium-sulfur batteries. Nano Res 8:129–139. https://doi.org/10.1007/s12274-014-0601-1
71. Park SH, Jang JH, Wistara NJ, Hidayat W, Lee M, Febrianto F (2018) Anatomical and physical properties of Indonesian bamboos carbonized at different temperatures. J Korean Wood Sci Technol 46:656–669. https://doi.org/10.5658/WOOD.2018.46.6.656
72. Krzesińska M (2017) Anisotropy of skeleton structure of highly porous carbonized bamboo and yucca related to the pyrolysis temperature of the precursors. J Anal Appl Pyrol 123:73–82. https://doi.org/10.1016/j.jaap.2016.12.024
73. Oyedun AO, Gebreegziabher T, Hui CW (2013) Mechanism and modelling of bamboo pyrolysis. Fuel Process Technol 106:595–604. https://doi.org/10.1016/j.fuproc.2012.09.031
74. Xue G, Liu K, Chen Q, Yang P, Li J, Ding T, Duan J, Qi B, Zhou J (2017) Robust and low-cost flame-treated wood for high-performance solar steam generation. ACS Appl Mater Interfaces 9:15052–15057. https://doi.org/10.1021/acsami.7b01992
75. Liu H, Chen C, Chen G, Kuang Y, Zhao X, Song J, Jia C, Xu X, Hitz E, Xie H, Wang S, Jiang F, Li T, Li Y, Gong A, Yang R, Das S, Hu L (2018) High-performance solar steam device with layered channels: artificial tree with a reversed design. Adv Energy Mater 8:1–8. https://doi.org/10.1002/aenm.201701616
76. Zhu M, Li Y, Chen G, Jiang F, Yang Z, Luo X, Wang Y, Lacey SD, Dai J, Wang C, Jia C, Wan J, Yao Y, Gong A, Yang B, Yu Z, Das S, Hu L (2017) Tree-inspired design for high-efficiency water extraction. Adv Mater 29:1–9. https://doi.org/10.1002/adma.201704107
77. Noman M, Sanginario A, Jagdale P, Castellino M, Demarchi D, Tagliaferro A (2014) Pyrolyzed bamboo electrode for electrogenerated chemiluminescence of Ru(bpy)32+. Electrochim Acta 133:169–173. https://doi.org/10.1016/j.electacta.2014.03.100

Sustainable Biomimetics: A Discussion on Differences in Scale, Complexity, and Organization Between the Natural and Artificial World

Valentina Perricone, Carla Langella, and Carlo Santulli

Abstract Biomimetics emerges as an effective approach to identify functional bio-inspired solutions for the development of original design applications. This approach does not necessarily result in sustainable products and processes, which are frequently made of petroleum-based materials fabricated with non-renewable and high-energy consuming technologies. Nevertheless, the inspiration from nature has a great potential in terms of sustainable innovation, taking into consideration not only analogies but also the differences between the natural and artificial world. In this regard, the present contribution aimed to highlight the differences between biological and human industrial systems in scale, complexity, and organization, encouraging new sustainable biologically inspired designs increasingly close to the construction law of organisms. The result of this comparison emphasized nature's intelligence concerning balanced source consumption and regeneration of ecosystems as well as the effective adaptation of organisms to natural cycles in time and space. A biomimetic approach that combines the use of bio-based materials with a coherent use of bioinspiration is here identified as a future sustainable and effective strategy to design a new human world, which does not impose on nature but is inspired and integrated with it.

Keywords Bioinspiration · Bio-based · Life cycle · Circularity · Heterogeneity · Organismal design · Waste valorization · Sustainable design

V. Perricone (✉)
Department of Engineering, University of Campania Luigi Vanvitelli, Via Roma 29, 81031 Aversa, Italy
e-mail: valentina.perricone@unicampania.it

C. Langella
Department of Architecture and Industrial Design, University of Campania Luigi Vanvitelli, Via San Lorenzo, 81031 Aversa, Italy
e-mail: carla.langella@unicampania.it

C. Santulli
School of Science and Technology, Università di Camerino, via Gentile III da Varano 7, 62032 Camerino, Italy
e-mail: carlo.santulli@unicam.it

1 Introduction

Nature is the best source of inspiration for designing environmentally sustainable artifacts that are compatible with the complexity of the present world dynamics [1]. Design projects in collaboration with biology can offer a valuable contribution to the evolution of sustainable design culture, eco-oriented marketing strategies, and environmental awareness with novel conceptual tools inspired by nature and its resilience [2].

Design for sustainability seeks solutions to resolve problems with minimal environmental impacts. In this regard, the transfer of logics found in nature to solve similar problems could be very useful. Artifacts designed using biological structures, materials, and working principles as models are certainly more performing and respectful of earth resources and its limits since they refer to strategies selected by nature and validated by millions of years of evolution.

Humans have always been inspired by nature to design artifacts that satisfy their needs improving their life. Many of the important achievements in technology, design, and art have been generated by imitating biological models. However, only current conditions, in terms of knowledge and tools, allow the creation of products and artifacts that could conceptually and concretely reproduce some of the most complex biological qualities hidden in the natural world. Indeed, the intersection between the progress of contemporary biological knowledge, together with new production technologies, proposes innovative and unique perspectives on the relationship between design and biology.

The interdisciplinary approach that combines the understanding of the natural world with its abstraction and translation into technological applications is known as "Biomimetics" [3, 4]. The biomimetic approach is increasingly spreading within the design culture at different dimensional scales from the development of nanotechnologies to systemic urban design. The acquisition of the biomimetic paradigm offers a valuable opportunity to draw new principles, strategies, and logics to make more environmentally sustainable products. However, it is not certain that the inspiration or imitation of nature always results in an improvement of the environmental/eco-friendly performance.

Biomimetics is not always a synonym of sustainability. Frequently, it is exclusively used as a method to increase the functional efficiency of man-made products, relying on Darwinian principles of progressive organismal functional adaptation in response to external conditions and stress. Sometimes the application of biomimetic models even implies an increase in environmental or economic costs. In well-known biomimetic products, such as *Fastskin* fabric or *Velcro*, the functional efficiency does not coincide with the increase of environmental safety using eco-friendly materials and production processes. Polymer-based self-cleaning coatings based on the lotus effect can even affect the environmental impact at the last lifecycle stage when there is no compatibility in terms of recycling.

Nonetheless, the inspiration from nature has great potential in terms of sustainable innovation. It can lead to the generation of products in which the increased performance, material innovation, productive technologies, and reduction of lifecycle environmental impact converge in a synergic manner.

The logics observed in biology, such as time cyclicity, energy efficiency, recovery, and regeneration of waste, and its difference with the artificial ones can guide design projects toward new sustainability scenarios. The concept of degrowth is replaced by a principle of sustainable evolution in which people can rediscover more natural behavioral and consumption models in harmony with the environment. Therefore, biomimetics can also lead the industrial product dimension closer to the more natural needs of people and the environment with an outlook on a human-centered ecological transition.

The possibility of transferring from biology principles and logics to reduce the environmental impacts of artifacts is strongly linked to the relationship between the size scale of the inspiring biological system and the artificial one; particularly when physical effects are linked to size. For example, the inspiration of principles based on optical or hydraulic phenomena observed in different organisms oriented toward the use of climatic resources such as solar radiation or rainwater, e.g., light transmission enhancement of window-leaved translucent crystals [5] or water distribution ability of the thorny devil desert lizard through its skin interscalar spaces [6]. This transfer could be very complicated and sometimes result in a reduction of effectiveness. Hence, the evaluation of the effects and changes in scaling biological working principles to the final project dimension is crucial for any biomimetic transfer process success [7]. Nonetheless, the main differences in scale, complexity, and organization between organismal design and artifacts can also provide new design perspectives in the sustainability of man-made products and productive processes.

In this context, the present contribution aims to provide a critical discussion on a series of key concepts regarding lifecycle and characteristics in which the biological world differs from the artificial one, encouraging a new cutting-edge and sustainable biologically inspired design increasingly close to the construction law of organisms. Additionally, a series of experimental case studies is provided, in which design projects are oriented toward effective integration of natural concepts, materiality, and processes.

2 Life Cycles

In nature, each organism goes through a specific life cycle, i.e., a continuous sequence of changes during its life from a primary form (gamete) to the reproduction of the same primary form. These formal transitions may involve growth, asexual or sexual reproduction. In these cycles, all "waste" turns into nutrition for other cycles creating a complex interconnection. In the food web, organisms are connected by trophic linkages and levels (autotrophs and heterotrophs): there are hierarchical organizations and conceptual scales. The first level is composed of basal species, such as algae, plants, and other vegetables, which do not feed on any other living creature on the web. Basal species can be autotrophs or detritivores. Apex predators constitute the top level and are not eaten directly by any other organisms. The intermediate levels are composed of omnivores that feed on one or more trophic levels and are themself eaten, causing a trophic energy flow. In trophic dynamics, energy transfer from one level to another is a unidirectional and noncyclic pathway with a loss of energy from the base to the top. Each organism is also characterized by a unidirectional energy flow, which typically includes ingestion, assimilation, non-assimilation losses (excrements), respiration, production (biomass), and mortality [8]. Nonetheless, the energy loss is always balanced in time by trophic relationships and organismal life cycles.

Conversely, the flow of mineral nutrients is cyclic and represents the recycling system of nature. Mineral cycles include for example the carbon, sulfur, nitrogen, and phosphorus, which are continually recycled into productive ecological nutrition. This recycling is mainly regulated by decomposition processes and relies on the biodiversity of the food web.

In human industrial systems, recycling differs from the natural one in scale, complexity, and organization. The industrial recycling systems seem to work independently from the food web without considering the waste restitution to different trophic sectors as well as source and energy regeneration time. This together with the increasing greenhouse gases concentrations owing to human combustion of fossil fuels and ecosystem degradation lead the industrial and in general the human world to be based on competitive and parasitic processes toward natural ecosystems [9, 10]. Major lifestyle and conceptual productive systems changes are needed and inevitable.

In this context, the theme of environmental sustainability applied to the design culture raises important issues centered on the difficult relationship between human activities and nature's delicate balance. Biological systems survive because of their life adaption and evolutionary processes becoming an integrating part of their environment. Organisms use local resources to build themselves (e.g., skeleton, shells, etc.) and their constructions (e.g., nests, traps, etc.), all of which are capable of complete recycling with continuous reuse and regeneration of their waste materials. They can conduct dynamic and adaptive management of both material resources and quantities of energy used for vital functions. Consumption and regeneration are always in balance, waste disposal is not necessary because everything is re-used and reintegrated into natural cycles in time and space.

2.1 Use of Resources

Organisms adapt their design and functioning to local resources and environmental biotic and abiotic characteristics, creating cascades of nutrients at the end of their lifecycle. Conversely, biomimetic products and materials such as synthetic spider silk-like materials, mechanically and optically adaptive materials, self-healing elastomers and hydrogels, and antimicrobial polymers have often been made using petrochemical origin materials, which have devastating effects on terrestrial and ocean life other and furthermore an inherently toxic life cycle from production to the final disposal. Society increasingly pushes toward ecological transition resulting in a closer look at the development of sustainable polymers from renewable natural products or biomass. Diverse bio-based and biohybrid materials are rising as greener alternatives to their petroleum-based counterparts. In particular, bio-based materials consist of substances naturally or synthetically derived from living matters [11], whereas biohybrid or living building materials are based on microorganisms and used in construction and industrial design exhibiting biological functional properties [12]. Bio-based and biohybrid materials are therefore based on inert or active natural components that produce little or no waste using small amounts of energy and producing multifunctional and adaptable systems.

In the biomimetic field, the use of these materials is however limited due to their complexity. Particularly, the non-homogeneity leads to difficulties in experimental, computational, theoretical calculation, and predictability response of these materials. Moreover, they are difficult to manage and design at a molecular level. One of the most effective biomimetic material research projects refers to the optimization of crosslinking/networking processes, dynamic interactions, and self-assembly (or phase separation) of synthetic polymers [13].

Additionally, Ganewatta et al. [13] pointed out that natural polymers or bio-based compounds do not inevitably result in materials necessarily more sustainable than those based entirely on synthetic polymers. The overall sustainability of a material can only be assessed through a life cycle analysis that considers each stage's impact, such as pre-production, production, distribution, use, disposal, and end of life. A material that has a sustainable start in life, because based on highly renewable and easily accessible raw materials or not requiring energy-intensive processing and environmental emissions, may not be durable, well-performing, or need treatments that compromise recycling.

2.2 Time and Scale

Time is a crucial factor in nature. All organismal components and constructions in the natural world are at the right time biodegradable, becoming a source and food for other ecological chains. The degradability is a function of their time and utility. For example, the difference between a paper wasp and honeybee in constructions and material choices are related to the time of their social persistence. Social paper wasps live in brief annual communities and construct their nests using wood fibers (dead wood and plant stems) mixed with saliva resulting in a paper-like material, whereas honeybees build pluriannual colonies and use beeswax to construct high-resistant and durable nests. Materials and structural configurations differ in organismal design with different life perspective duration, besides protection needs. For example, bivalves that can have a relatively long survival time protect themselves with shell composed high-structured hierarchical ceramics [14]. The shell is realized by the continuous addition of materials necessitating constant strength throughout organismal life: any breaking or cracking will always constitute a point of weakness. Conversely, lightweight, and rapidly biodegradable polysaccharide-based materials are employed in seasonal cycles. For example, deciduous tree leaves quickly disintegrate after falling ending their function.

In this regard, one of the main human technology errors lies in the unbalanced connection between material choice and time of use in products and processes. Small scale disposable products, e.g., plastic bottles and flatware, have been made of materials that require 100–1000 years to degrade. This is an example that results in the need to use biodegradable materials with short disassembling and degradation time. On the contrary, fast biodegradability emerges as a paradox in the scale of architectural design, in which case the disposal time should be extended as much as possible since the structures are intended to last.

2.3   Use of Waste

Nature is based on completely zero-waste systems: the waste of one system becomes food for another. This smart cyclicity of nature is one of the most important logics to be transferred to the biomimetic design of artifacts. It induces the recovery and regeneration of material waste after production or consumption through reuse, recycle, or upcycling strategies. From this point of view, designers could be involved in the identification of waste types most suitable to be ennobled through bioinspiration. Biomimetic design can raise the final aesthetic, economic, ethical, and environmental value of wastes conceiving attractive and desirable products such as jewelry, furniture, and fashion accessories making upcycling processes convenient and profitable [15]. In this sort of project, designers are asked to analyze production processes, with particular attention to local activities, and to interpret waste transformed into resources in terms of technical characteristics, perceptive qualities, and processability. Thus, it is possible to identify new applications that maximize their potential by reducing their limitation impacts, transforming them into factors of specificity and originality [16]. In upcycling, biomimicry is an important added value in terms of marketing because it produces attractive and desirable products for the market and, therefore, economically viable. The increasing awareness of climate and environmental issues together with the impact of lifestyle on health and well-being leads people to choose what they perceive to be most natural and akin to their biological roots, preferring products that implement a biological factor in terms of raw materials or design inspiration. In a market that is progressively inclined to choose low-impact products, bioinspiration is proposed as an effective strategic vehicle to characterize, identify, and promote eco-sustainable products and eco-oriented innovation actions [17]. For these reasons, companies and commercial organizations are now aware of the great competitive potential of bioinspiration in terms of attractiveness, perceived value, and marketing, underpinning studies on bio-oriented entrepreneurship, referred to as biopreneuring [18], resulting in a biomimicry and upcycling synergy.

2.4 Production

Nature has been criticized for not producing enough and too slowly for the industrial productive standards, which conversely require efficient, rapid, precise, calculable processes and results. The productive scale and time of human technology seem not to be comparable with natural ones. This assumption seems to be true when a ceramic object produced at 1000 °C is compared with the productive time of a bivalve shell. However, the advantages in terms of time and efficiency are taken less for granted considering that a ceramic industry requires materials extracted from worldwide caves, which need to be imported and processed at high temperatures, pressures, and energies and furthermore must be transported and delivered to clients [19]. In nature, local extraction is part of the productive process and energetic costs are notably reduced.

Organisms produce biomaterials at local pressure and temperature conditions using locally available raw materials; in industrial production, artificial temperature and pressure conditions are often obtained by using great amounts of energy as well as raw materials generally transported from remote locations. Hence, the comparison between the natural and industrial production scale processes leads to another quantitative aspect. The number of natural creations depends on physical forces respecting the environmental carrying capacity in a potentially infinite cycle, while industrial processes are based on high-energy loss and resource depletion.

3 Biological Versus Artificial

During millions of years, organisms evolved complex shapes, structures, and processes generally tending to optimize the cost-benefit ratio and minimize energy and materials to be used for their construction, development, and maintenance. In a constructional perspective, they respond to principles of lightness, resilience, flexibility, resistance, and efficient logics oriented to ensure the high performance of organisms in their environment. Organisms are in this respect of particular interest for design, architects, civil engineers, and many other technical disciplines since they provide new technical and sustainable methodological strategies that can be transferred based on analogies as well as differences between the natural and technical world.

In particular, numerous differences emerge in the comparison between organismal design and human artifacts and technologies, which can lead to a change in perspective and to a design possibility and sustainability expansion.

A primary concept of sustainability emerges in terms of energy and productive processes: humans consume a vast amount of energy (60% of the time) to develop numerous diverse materials with novel properties; whereas organisms invest minimum energy (5% of the time), using few materials and synthetic processes, wherein energy contribution is high, and utilizing more structural organization (e.g., hierarchy, strategical porosity, textures), wherein energy is negligible [20]. In this regard, Vincent et al. [20] stated that "instead of developing new materials each time we want new functionality, we should be adapting the materials we already have". Analogies and differences in scaling between organisms and artifacts can encourage cutting-edge, sustainable, and biologically inspired designs increasingly close to the construction law of organisms. The solution lies in the identification of functional strategies and properties that can add mechanical resistance and multi-functionalities to materials that have been already developed.

Numerous are the interesting differences between natural and artificial materials that can lead to other interesting insights regarding sustainability. Firstly, the genesis of biological structures is not based on a mere assembly of parts, rather it consists of a continuous growth process, i.e., a self-assemble automatism that generates structures with full functionality and integrity at all different stages of life. Secondly, organisms use basic, autochthonous, and sustainable materials that require neither excessive energy-consuming methods for their realization (working at environmental temperature and pressures) nor long-distance transportation. Lastly, biological structures are perfectly integrated into their environment and continuously interact and react to its biotic and abiotic components. In addition, many other functional features characterize biological structures, such as heterogeneity, anisotropy, hierarchy, modularity, adaptability, self-healing, and multifunctionality, which are still needed to be explored in-depth and employed as technical solutions.

3.1 Heterogeneity

Heterogeneity is a widespread characteristic in natural materials (e.g., soils, geological formations, biological tissues) occurring at different scales: from molecular to macroscopic. Organisms are characterized by a remarkable material and geometrical differentiation of their structural components as well as local adaptations of their physical and chemical properties.

Being based on biological matter, bio-based materials are generally characterized by a high heterogeneity (Fig. 1), which limits their application, particularly on large scales (e.g., building construction). Indeed, this characteristic determines notable complexity and limits in experimental, computational, theoretical calculation, and predictability response of these materials. Nonetheless, the non-homogeneity of these materials can be valorized and enhanced in some design projects, generating multiple unique features, such as light effects and transparencies, multisensorial connotation, colorful effects, thickness differentiation, and singular textures (see sector 5). On the other hand, geometrical heterogeneity combined in an organized structure (achieved for examples with controlled porosity configuration) might also result in some benefits, such as crack propagation blunting and energy dissipation in bone ceramics [21].

Fig. 1 Bio-based composite
material heterogeneity.
Bioplastic matrix based on
starches and waste liquid
from buffalo mozzarella
production and hemp fibers.
Retrieved from: https://www.
hybriddesignlab.org

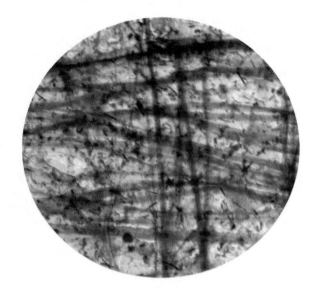

3.2 Anisotropy

Anisotropy derives from the uneven distribution and organization of materials; thus, each direction in the material has different properties and behavior. In nature, the anisotropic feature is generally exploited by combining different structural organizations that can result in emerging properties, e.g., movements, and lead to lightweight and efficient structures: e.g., the cellulose fiber orientations determine the shape and kinematics of plant cells and tissues as well as the anisotropic trabecular architecture resisting to predicted directional stresses [22]. Indeed, organismal design is adapted to forces that very seldom have the same intensity in all directions; therefore, it generally requires an adapted anisotropy. This stands for diverse natural structures: two examples are body tissues, where anisotropy is required for repair purposes [23], and rice leaves, where directional forces are related to water surface tension [24].

3.3 Hierarchy

Organisms are characterized by a multilevel organization from nano-to macroscale (Fig. 2). Increasingly sophisticated intelligence emerges from hierarchies, leading to different emerging functional properties. Emergent properties of a system arise from the interactions of its interrelated elements and cannot be reduced to or derived from the sum of the single element properties. Emergence is related to hierarchical organizations and occurs at all scales in nature, mainly: atoms, molecules, cells, tissue, organs, systems, organisms, populations, communities, ecosystems [25]. All these elements together determine unique properties and behaviors non-deductible

Fig. 2 Hierarchical porosity
in diatom valves. SEM
Micrograph made by
Valentina Perricone at
Stazione Zoologica Anton
Dohrn, Naples, Italy

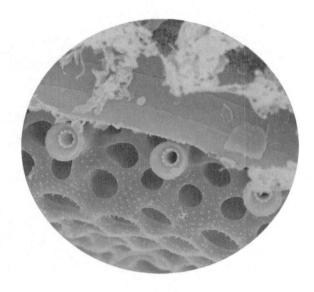

by a single component. For example, different organs constitute an organism that interacts with the environment trying to preserve itself and reproduce; while different organisms constitute a community that interacts and creates a stable environment for its members. The properties emerging from the organism and community are not deducible from their single constituents. Material science is currently working with hierarchical controlled gradients and configurations at a different scale to create high-performance materials with unique emerging properties (e.g., micro- and nano-structured materials to create structural colors). Numerous studies have been carried out at nano- and microscale, however, the introduction of these structural materials on large scales, such as building construction, is still very challenging [26].

3.4 Modularity

A common strategy of organismal design is modularity at different scales. Modules are functional units that generally implement and satisfy local needs with some degree of self-maintaining and self-controlled properties. For example, the liver system can operate controlling nutrients in the blood in relative independence from the central nervous system [27]. Moreover, the subdivision of the body in a series of segments is a common phenomenon in nature, known as metamerisms. An outstanding example is the subdivision of the worm body in multiple meters with a repetition of organs and muscles. Modules or segments have a proper problem-solving intelligence and are independent of external changing conditions. Modularity is also used in artifacts and their fabrication processes based on the assembling of units that provides differentiation of materials, reproducibility adaptability, and easy substitution. This is however based on inert modules that create a final configuration by a mere assembly

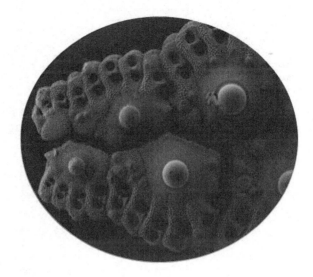

Fig. 3 Echinoid test tessellation. SEM Micrograph of *Paracentrotus lividus* ambulacral plates made by Valentina Perricone at Stazione Zoologica Anton Dohrn, Naples, Italy

of parts. There is often no intelligence, self-maintaining or self-controlled properties. Nonetheless, natural modularity can lead to functional strategies that can inspire effective biomimetic configurations. For example, diverse studies have been carried out on the modular structures that combine hard and soft materials (tessellations) that characterize numerous invertebrate and vertebrate biomaterials and structures (Fig. 3). The studies reported how tessellation can optimize mechanical configuration, e.g., maximizing mechanical material toughness with minimum expenditure of stiffness or strength [28].

3.5 Adaptability

Organisms and systems in nature vary their properties according to predictable external constraints to which they are subjected during growth and/or throughout their life cycle. For example, vertebrate bones are made of composite material (mainly hard hydroxyapatite and elastic collagen) creating a complex internal trabecular system that varies in porosity and orientation according to the main stress trajectories; this allows to create of an adaptive, lightweight, and resistant skeleton able to withstand both tensile and compressive internal and external forces.

Conversely, industrial products generally consist of repeated parts with identical properties that are not able to adapt, rapidly becoming waste. As stated by Oxman [29], design should create new concepts of formation, in which products adapt and perform, i.e., behave, rather than form and absolve a unique function.

Fig. 4 Echinoid spines.
SEM Micrograph of *Arbacia lixula* spines made by
Valentina Perricone at
Stazione Zoologica Anton
Dohrn, Naples, Italy

3.6 Self-healing

In case of fractures, amputation, or damages, organisms adopt specific self-repairing, self-healing, and regeneration mechanisms. For example, rapid self-sealing and self-healing prevent plants from desiccation and infection. The repairing properties are frequently related to the damage extension as well as organismal component and complexity: sea urchins can repair their test (endoskeleton) only if the damage is circumscribed and can entirely regenerate their spines (Fig. 4), whereas sponges can completely regenerate from fragments or even single cells.

In material science, synthetically created materials have been developed with the ability to automatically repair damages. These materials are usually polymers, metals, ceramics, or cementitious materials. A self-healing concrete has been also realized using bacteria that are able to precipitate calcium carbonate in concrete sealing micro-cracks [30]. This innovative concept has been successfully applied at small or lab-scale tests. However, some limitations have been identified in large and real-scale applications [31].

3.7 Multifunctionality

Natural materials and structures are multifunctional, i.e., they do not absolve a single role but generally provide diverse important properties that are useful to enhance organismal survival and reproduction. For example, sharkskin with its texturized denticles is able to provide, e.g., fluid drag reduction, anti-fouling, and antimicrobial functionalities. Compared to biological ones, artificial materials appear to be less effective and wasteful [32]. These materials are discrete solutions generated to absolve one or a few rigid and distinct functions. Their diversity is achieved by sizing rather than by substance variation, and is typically mass-produced and not customized [32]. Presently, material scientists are however designing and fabricating multifunctional composites for various applications taking inspiration from hierarchical micro/nanostructures and biological functions (see [33] for a review).

4 Hybrid Design Lab: Experimental Designs Closer to Nature

The Hybrid Design Lab (HDL) is a laboratory of the Department of Architecture and Industrial Design of the University of Campania "Luigi Vanvitelli", founded in 2006 and dedicated to bio-inspired design and the relationship between design and science. The HDL interdisciplinary team aims to transfer theoretical and experimental research, achieved in biosciences, new materials, and technologies, to the design innovative and sustainable products and services. The following examples show how different functional biological features can be applied, often complementing each other, in sustainable bio-inspired designs.

4.1 Designing Bio-based Products on Life Cycle Disposal Time

HDL carried out different projects aimed to develop products with natural materials coherent with their time of use. Orthopedic supports are examples of reduced life cycle products used in a limited therapy time; nevertheless, they are usually produced using conventional polymeric materials with high-temporal disposal processes. *Thumbio* emerged as an example of a promising bio-based orthopedic brace for hand and wrist immobilization in case of inflammatory, degenerative diseases, and small fractures (Fig. 5). This brace was produced using a biodegradable composite made of a bioplastic matrix based on starches and waste liquid from the buffalo mozzarella production and hemp fibers (*heterogeneity*) to modulate stiffness and elasticity according to the degree of immobilization indicated by the orthopaedist. The arrangement of the fibers in the bioplastic depends on the location of the type of lesion or inflammation and, therefore, on the movements that must be prevented and the micro-movements that can be allowed (*anisotropy, adaptability*). The bio-composite is also functionalized with natural anti-oedematous and anti-inflammatory herbal ingredients which slowly release phytotherapeutic principles during the healing process, avoiding the use of creams (*multifunctionality*). At the end of its short life, the product can be composted, releasing no harmful substances for the environment due to its fertilizing properties [34] (Fig. 6).

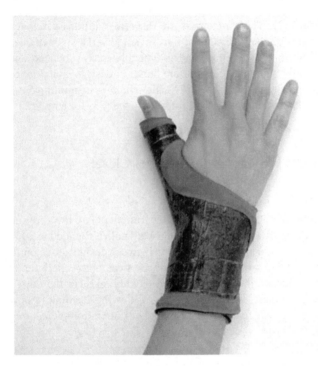

Fig. 5 Thumbio. Bio-based orthopedic brace for hand and wrist immobilization. Credits: Clarita Caliendo (Design); Carla Langella (Scientific coordination); Carlo Santulli (Material engineering); Antonio Bove (Orthopedics)

4.2 Designing and Valorizing Waste

Other than being reduced, waste can be also valorized by transforming it into a new resource enhancing its unique characteristics in an expressive way. Based on a learning from nature approach, inspired by the ability of natural systems to reuse and regenerate materials and energy, the project "+Design − Waste" carried out by HDL, was aimed to design products developed by reinterpreting different types of waste. Through a multidisciplinary approach, which involves design, material science, and biology, waste was nobilitated raising the final economic value through the project of products such as jewelry, furniture, fashion, and accessories. In the *Diaglass* project, glass waste obtained from broken building glass was upcycled and enriched with gold flakes (*heterogeneity*) through a specific heating process giving life to precious jewels inspired by diatom material and forms (Fig. 7).

By imitating nature, upcycling products should remain in the production environment in which they originated to minimize transport environmental costs. In the Flora project, waste from floriculture production was used to make biodegradable pots sold in the nurseries to contain plants (Fig. 8). The presence of coarse fragments

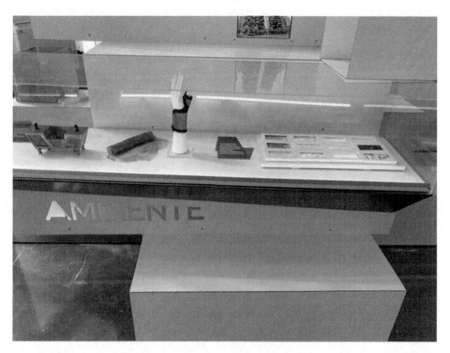

Fig. 6 Thumbio and different material solutions showed in the international itinerant exhibition "Italy: The Beauty of Knowledge", Farnesina, Rome 2018

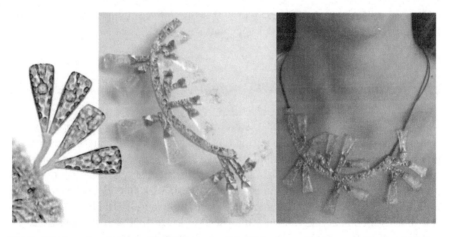

Fig. 7 Diaglass. Jewels inspired by diatom material and forms made of upcycled broken building glass enriched with gold flakes. Credits: Serena Miranda (Design); Carla Langella (Scientific coordination)

Fig. 8 Flora. Biodegradable
pots made of bioplastics with
fragments of floral petals and
leaves. Credits: Maria
Petrillo, Lorenzo Villani
(Design); Carla Langella
(Scientific coordination)

of petals and leaves in the bioplastic provides strength, color heterogeneity, clearly
communicating the ethical value of upcycling.

4.3 Designing Bio-inspired Variability

Non-homogeneity, cyclicity, and hierarchization of biological structures can be
emulated in sustainable design by using bio-based materials and alternative production processes to generate multiple unusual features, such as discontinuous light
effects and transparencies, multisensorial intensity, color variegation, thickness
differentiation, and singular textures. In this regard, HDL in collaboration with the
CNR- Institute of Polymers, Composites, and Biomaterials (IPCB) developed 60
new bio-inspired material samples with a design-driven approach, starting from raw
materials of marine origin including algae, mussel shells, and shrimp, incorporated
in biodegradable polymer matrixes (Figs. 9 and 10). The samples were developed
within the European project PIER framework led by Città della Scienza. One of its
main specificities was that new materials were created by designers and personally
produced in a chemical lab with the supervision of chemists. In this experience, the

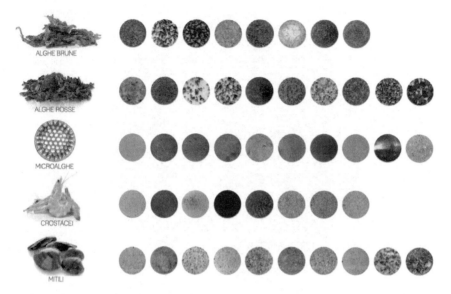

Fig. 9 Materials from the sea. Bioinspired material samples realized with raw materials of marine origin including algae, mussel shells, and shrimp, incorporated in biodegradable polymer matrixes. Credits: Francesco Amato, Clarita Caliendo (Design); Carla Langella (Scientific Design coordination); Mario Malinconico (Science material coordination)

material design, conducted from the designers' point of view, chose to favor perceptive, experimental, and functional qualities required by the application field (furniture, accessories, packaging), rather than the homogeneity and isotropy that chemists and material engineers generally give priority to. The material design was inspired by biological structures and their properties, favoring discontinuity over continuity, dishomogeneity over homogeneity, color shades and opacity gradients over chromatic and optical uniformity, and the modulation of mechanical performance in relation to expected stress. The samples were conceived by giving particular attention to the aspects of environmental sustainability, the enhancement of natural materials, and the interpretation of biological materials from a design point of view. The relationship of these projects with nature is therefore bivalent since the new materials developed contained raw materials of natural origin as well as were inspired by principles and logics studied in biology. The samples were exhibited in Città della Scienza museum in 2014, and in the itinerant exhibitions "Italy: The Beauty of Science" from 2018 to 2020 and "Italy: The Art of science" in 2021 (Fig. 9).

Heterogeneity and structural hierarchy biological features were also applied in an auxetic 3D printed collar aimed to safeguard the well-being of the neuromuscular system of the cervical spine [35]. The collar had a preventive purpose because it dissuaded the user from keeping his head tilted forward induced by the use of portable devices. It also had a therapeutic function for cervical pathologies with no serious

PCL   HIJIKI ALGAE   HIJIKI ALGAE

Fig. 10 Material sample composed of PCL and algae. Design: Francesco Amato e Clarita Caliendo. Design coordination: Carla Langella. Science material coordination: Mario Malinconico

alterations as the chin was slightly supported, partially unloading neck muscles from the mechanical stresses due to head support.

Auxetics are meta-materials observed in nature in the skins of some reptiles such as the salamander, but also in the stems of various plant species. The auxetic structure provides these tissues with greater extensibility and mechanical strength, preventing them from tearing, even when subjected to sudden and intense stress. The auxetic behavior derives from the morphological structure and not from the chemical characteristics of the material. Specifically, a meta-material is defined as auxetic when it has a negative Poisson's modulus. The Poisson's modulus is defined as the ratio of the transverse and parallel deformations with respect to a load applied to the section.

The term auxetic derives from Auxesis, a Greek word meaning to grow, which refers to the increase in cell size when structures are subject to tensile stress.

Generally, when we solicit a material with a positive Poisson's modulus to uniaxial tensile stress, it expands in the stretching direction and thins in the cross-section. Similarly, a material subjected to compression contracts in the direction of force and expands laterally.

A negative Poisson's modulus, on the other hand, means that materials also expand in the orthogonal direction when subjected to a tensile force and contract on all sides when subjected to compression.

The use of the auxetic structure in the collar, compared to conventional materials, results in more resistance, flexibility, breathability, and adaptability to the anatomy of the neck in different postures, like a second skin. The auxetic structure developed by the designer in the final project is a hybridization of two types of auxetic geometries observed in nature: indented cells and rotating cells. In the collar structure, different cell shapes were organized in a strategic position array to differentiate stiffness according to the orthopedical therapeutic indications. This structure resulted in a more effective and sustainable collar compared to the traditional one, allowing the use of less material that can be recycled to produce new 3D printing filaments at the end of its life cycle (Fig. 11).

5 Conclusions

In this complex framework, the need to design biomimetic materials and constructions based on intelligent and coherent use of resources, scale, and function emerges as a priority, including product duration, type and intensity of use, application context, and disposal choice. These parameters strongly influence the characterization of the life cycle and artifact performances such as the renewability of raw materials or biodegradability at the end of life.

Bio-based materials are encouraging for a sustainable future and their limits should be overcome by enhancing their unique properties creating new ones. New resistant and lightweight configurations, multifunctionality, regeneration properties, circularity, and sustainability can be applied to bio-based materials taking inspiration from organismal designs and working principles. Indeed, organisms also use natural materials often fragile (e.g., biogenic high-magnesium calcite of echinoid skeleton or silica of diatom valves); however, they optimize and adapt their materiality to scale, functionalities, and environmental context using more structural organization (e.g., hierarchy, strategical porosity, textures). Hence, a biomimetic approach that combines the use of bio-based materials with a coherent use of bioinspiration can be configured as a future sustainable and effective line of human design able to integrate and imitate nature through multiple dimensions.

Recent technological advances seem to have opened a new biomimetic era. Technologies such as computational design and fabrication allow the design of complex structures that can perfectly reproduce biological-like functions, whereas material

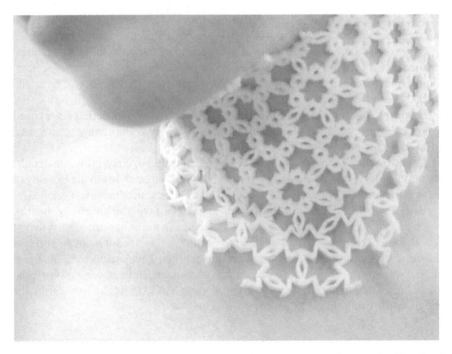

Fig. 11 Auxetic neckbrace, a detail of the heterogeneous structure. An auxetic 3D printed collar aimed to safeguard the well-being of the neuromuscular system of the cervical spine. Credits: Martina Panico (Design); Carla Langella (Scientific coordination); Carlo Santulli (Material engineering)

scientists lead to the design of new bio-based materials by integrating sustainability and biological materiality into design products. Therefore, the differences between natural and technical entities can now encourage new cutting-edge and sustainable biologically inspired designs increasingly closer to the construction law of organisms.

From a methodological point of view, sustainable biomimicry can be established on the mutual collaboration between disciplines such as biology, material science, engineering, and design [36]. The synergy between disciplines can lead to the awareness of real possibilities of transferring constructional and adaptive characteristics of organisms, scaling them up with respect to the context and dimensions of the final application. Without this synergy, conceptual and sustainable limits can emerge in the design of artifacts, processes, and services since the biological logics and the physical principles are deviated or not fully comprehended. At the same time, biologists and engineers cannot foresee with sufficient reliability the way in which products will be used, maintained, and finally discarded, because these factors are closely linked to the knowledge of market, lifestyles, user attitudes, and, finally, design strategy. An effective collaboration must be mutual [37] and synergic, and not linear and progressive, but interactive, with continuous back-and-forth, trial-and-error paths, as in nature.

This mutuality can be necessarily achieved by facing and overcoming disciplinary specificity, such as differences in objectives, tools, timeframes, and languages [38].

The possibility of developing products and materials inspired by nature and integrated with the environment also emerges from the application of biological complex logics such as cyclicity, adaptability, self-organization, redundancy, hierarchy, and non-homogeneity. These features applied to bio-based materials and artifacts can result in more resistance, durability, originality, and attractiveness, thus suitable for responding to the complex needs of contemporary living. Logics that until a few years ago were impossible to replicate due to their complexity, today can be applied to artifacts increasingly closer to biological ones. Nature can be part of the product and a source of inspiration as well as an active agent, as in the case of materials made by bacterial fermentation or cell culture [39]. In the next sustainable future, biomimetic artifacts will be shaped on the use of renewable and controlled biodegradable materials or waste-based ones. Additionally, they will probably be dis-homogeneous, anisotropic, multi-colored, and biologically produced [40].

In the history of mankind, nature has always been a source of inspiration. The future perspective foresees a new fundamental step forward, leading to the human world in overcoming the negative dichotomy between nature and artifice. In this regard, biomimetics can assume an important role in bringing new knowledge and awareness of the environment, human health, and social equity into the lives of people, offering new and concrete prospects for integrating sustainable strategies, increasing awareness, and improving life quality. Biomimetics must encourage the creation of a human world increasingly closer to nature construction laws that do not impose themselves on nature but are inspired and integrated with it.

References

1. Fisch M (2017) The nature of biomimicry: toward a novel technological culture. Sci Technol Hum Values 42(5):795–821
2. Olaizola E, Morales-Sánchez R, Eguiguren Huerta M (2020) Biomimetic organisations: a management model that learns from nature. Sustainability 12(6):2329
3. Fayemi PE, Wanieck K, Zollfrank C, Maranzana N, Aoussat A (2017) Biomimetics: process, tools and practice. Bioinspiration Biomimetics 12(1):011002
4. Vincent JF, Bogatyreva OA, Bogatyrev NR, Bowyer A, Pahl AK (2006) Biomimetics: its practice and theory. J R Soc Interface 3(9):471–482
5. Krulik GA (1980) Light transmission in window-leaved plants. Can J Botany 58(14):1591–1600
6. Sherbrooke WC, Scardino AJ, de Nys R, Schwarzkopf L (2007) Functional morphology of scale hinges used to transport water: convergent drinking adaptations in desert lizards (Moloch horridus and Phrynosoma cornutum). Zoomorphology 126(2):89–102
7. Perricone V, Santulli C, Rendina F, Langella C (2021) Organismal design and biomimetics: a problem of scale. Biomimetics 6(4):56
8. Benke AC (2010) Secondary production. Nat Educ Knowl 1(8):5
9. Huesemann MH (2003) The limits of technological solutions to sustainable development. Clean Techn Environ Policy 5:21–34
10. Rees WE (2009) The ecological crisis and self-delusion: implications for the building sector. Build Res Inf 37(3):300–311

11. Curran MA (2000) Biobased materials. In: Kirk-Othmer encyclopedia of chemical technology, pp 1–19
12. Qiu J, Artier J, Cook S, Srubar III, WV, Cameron JC, Hubler MH (2021) Engineering living building materials for enhanced bacterial viability and mechanical properties. IScience 24(2):102083
13. Ganewatta MS, Wang Z, Tang C (2021) Chemical syntheses of bioinspired and biomimetic polymers toward biobased materials. Nat Rev Chem 1–20
14. Croll RP (2009) Developing nervous systems in molluscs: navigating the twists and turns of a complex life cycle. Brain Behav Evolut 74(3):164–176
15. Santulli C, Langella C (2013) '+design–waste': a project for upcycling refuse using design tools. Int J Sustain Des 2(2):105–127
16. Binotto C, Payne A (2017) The poetics of waste: contemporary fashion practice in the context of wastefulness. Fash Pract 9(1):5–29
17. Hart SL, Milstein MB (2003) Creating sustainable value. Acad Manag Perspect 17(2):56–67
18. Ulhøi JP (2015) Framing biomimetics in a strategic orientation perspective (biopreneuring). Technol Anal Strateg Manag 27(3):300–313
19. Pauli G (2015) The blue economy version 2.0: 200 projects implemented, US$ 4 billion invested, 3 million jobs created. Academic Foundation
20. Vincent J, Bogatyreva O, Bogatyrev N, Pahl AK, Bowyer A (2005) A theoretical basis for biomimetics. In: MRS online proceedings library (OPL), p 898
21. Tai K, Dao M, Suresh S, Palazoglu A, Ortiz C (2007) Nanoscale heterogeneity promotes energy dissipation in bone. Nat Mater 6(6):454–462
22. Charpentier V, Hannequart P, Adriaenssens S, Baverel O, Viglino E, Eisenman S (2017) Kinematic amplification strategies in plants and engineering. Smart Mater Struct 26(6):063002
23. Datta P, Vyas V, Dhara S, Chowdhury AR, Barui A (2019) Anisotropy properties of tissues: a basis for fabrication of biomimetic anisotropic scaffolds for tissue engineering. J Bionic Eng 16(5):842–868
24. Wu D, Wang JN, Wu SZ, Chen QD, Zhao S, Zhang H et al (2011) Three-level biomimetic rice-leaf surfaces with controllable anisotropic sliding. Adv Funct Mater 21(15):2927–2932
25. Odum EP, Barrett GW (1971) Fundamentals of ecology, vol 3. Philadelphia, Saunders, p 5
26. Knippers J, Speck T (2012) Design and construction principles in nature and architecture. Bioinspiration Biomimetics 7(1):015002
27. Yuste R, Levin M (2021) New clues about the origins of biological intelligence. https://www.scientificamerican.com/article/new-clues-about-the-origins-of-biological-intelligence/?amp;text=New
28. Fratzl P, Kolednik O, Fischer FD, Dean MN (2016) The mechanics of tessellations-bioinspired strategies for fracture resistance. Chem Soc Rev 45(2):252–267
29. Oxman N (2020) Neri Oxman: material ecology. Museum of Modern Art, New York, USA
30. Vijay K, Murmu M, Deo SV (2017) Bacteria based self-healing concrete—A review. Construct Build Mater 152:1008–1014
31. De Belie N, Wang J, Bundur ZB, Paine K (2018) Bacteria-based concrete. In Eco-efficient repair and rehabilitation of concrete infrastructures. Woodhead Publishing, pp 531–567
32. Oxman N (2010) Structuring materiality: design fabrication of heterogeneous materials. Archit Des 80(4):78–85
33. Vijayan PP, Puglia D (2019) Biomimetic multifunctional materials: A review. Emerg Mater 2(4):391–415
34. Caliendo C, Langella C, Santulli C, Bove A (2018) Hand orthosis designed and produced in DIY biocomposites from agrowaste. Des Health 2(2):211–235
35. Panico M, Langella C, Santulli C (2017) Development of a biomedical neckbrace through tailored auxetic shapes. Emerg Sci J 1(3):105–117
36. Langella C, Perricone V (2019) Hybrid biomimetic design for sustainable development through multiple perspectives. GRID-Archit Plan Des J 2(2):44–76
37. Antonelli P (2008) Design and the elastic mind. The Museum of Modern Art, New York

38. Langella C (2021) Design and science: a pathway for material design. In: Materials experience, vol 2. Butterworth-Heinemann, pp 259–277
39. Myers W (2012) Bio design. Museum of Modern Art
40. Ferrara M, Langella C, Lucibello S (2019) Bio-smart materials for product design innovation: going through qualities and applications. In: International conference on intelligent human systems integration. Springer, Cham, pp 634–640

Bionics for Inspiration: A New Look at Brazilian Natural Materials for Application in Sustainable Jewelry

Mariana Kuhl Cidade, Janaíne Taiane Perini, and Felipe Luis Palombini

Abstract Natural material is the classification given to those retrieved from nature, whether of plant, animal, or mineral origin. Essentially, after extraction, they are characterized by requiring little or no additional processing prior to their application in a project. Since the beginning of humanity, nature-sourced elements have been used as tools and weapons, and manufacturing techniques have been perfected by civilizations, due to humans' wonder and curiosity about them. In addition to their mechanical properties of interest, natural materials are still valorized by their uniqueness and aesthetics—with their perfection through imperfection—giving individuality and character to each piece produced. Bionics is defined by the use of features from natural elements, such as shape, structure, organization, and aesthetics, in many fields such as design, engineering, and architecture, among others. This application can be realized directly or via a source of inspiration, through observations, adaptations, and parameterization. Brazil has one of the richest biodiversities in the world, with a great variety of both fauna and flora species, as well as rocks and minerals. In the country, there are many examples of different representative natural materials, such as golden grass (*S. nitens*—ERIOCAULACEAE), sisal (*A. sisalana*—AGAVACEAE), bottle gourds (*L. siceraria*—CUCURBITACEAE), mane and tail hair of horses, sheep wool, agate, opal, amethyst, Paraiba tourmaline, among others; many of which come from renewable sources, and others that are underutilized,

M. K. Cidade (✉)
Department of Industrial Design, Federal University of Santa Maria — UFSM, Av. Roraima, n° 1000, Prédio 40, Sala 1136, Santa Maria, RS 97105-900, Brazil
e-mail: mariana.k.cidade@gmail.com

J. T. Perini
Industrial Design Undergraduate Course, Federal University of Santa Maria — UFSM, Av. Roraima, n° 1000, Prédio 40, Santa Maria, RS 97105-900, Brazil

F. L. Palombini
School of Engineering, Federal University of Rio Grande do Sul — UFRGS, Av. Osvaldo Aranha 99/408, Porto Alegre, RS 90035-190, Brazil

Laboratory of Plant Anatomy – LAVeg, Institute of Biosciences, Department of Botany, Federal University of Rio Grande do Sul – UFRGS, Av. Bento Gonçalves, Porto Alegre, RS 9500, Brazil

© The Author(s), under exclusive license to Springer Nature Singapore Pte Ltd. 2022
F. L. Palombini and S. S. Muthu (eds.), *Bionics and Sustainable Design*, Environmental Footprints and Eco-design of Products and Processes,
https://doi.org/10.1007/978-981-19-1812-4_8

leading to a large amount of wasted residues. Furthermore, little has been studied and explored regarding the aesthetic value of those materials. For instance, contemporary jewelry is defined by the application of unusual materials, techniques, and creative processes, redefining concepts and exploring new ways for a sustainable luxury. This chapter proposes a new look at Brazilian natural materials from renewable sources and waste, focusing on their use in two ways, (i) as raw material and (ii) as a source of inspiration. Brazilian natural materials from animal, plant, and mineral sources were collected and examined via light microscopy for the search of morphological characteristics and colors to be employed in a creative process. Collected samples were also employed as highlighted raw materials for the manufacturing of a jewelry collection, aiming to valorize and emphasize the usage of those materials that already comes from an inspirational origin.

Keywords Renewable materials · Waste · Biomimetics · Design · Microscopy · Natural materials · Natural fibers · Contemporary jewelry

1 Introduction—Natural Materials

Natural materials play a key role in the development of all civilizations and humankind as a whole. Among the first objects ever used by humans, we can point out woods, rocks (particularly flint), skins (furs and leathers), bones (including horns, tooth, and tusks), shells, and several types of natural fibers [12] that have been used for the manufacturing of tools, weapons, ornaments, clothing, buildings, and many other types of products [8, 55]. In terms of relative importance, they had endured for millennia as the main source of raw materials [7, 12], prior to the age of metals—with the exception of gold, which due to its chemical stability it could be found in nature as metal form, instead of ore minerals like aluminum, copper, iron, etc. [8]. Only much later did humans begin to explore chemical and physical experiments for the development of synthetic materials, like metals and alloys, ceramics and glasses, polymers, and, in the last century, composites. Despite industrialization and modern technologies which lead to the development of advanced materials and manufacturing processes, natural materials have kept significant importance in many industries throughout the years. Even though the existence of synthetic alternatives in the present day, there are still examples of those materials that are largely used and commonly found in many applications, such as latex, bovine leather, and cork, in addition to wood and bamboo as building materials in houses, scaffoldings, besides furniture and a variety of utensils. However, despite the fact that they are almost omnipresent at every home or workplace, the very definition of the term "natural material" can sometimes be confusing and misleading.

Classic material selection in engineering and design literature and textbooks, like the pioneer works of William John [75] to Michael Farries [5], divide materials into four main classes—Metals, Ceramics, Polymers, and Composites—as a reason of the "common underlying structural features (the long-chain molecules in polymers, the intrinsic brittleness of ceramics, or the mixed materials of composites) which, ultimately, determine the strengths and weaknesses (the "design-limiting" properties) of each in the engineering context" [9]. Later on, those classes were expanded into general families with the inclusion of separate divisions for Elastomers, Glasses, and Hybrids. This time, instead of simply assorting materials because of their structural characteristics, the families were chosen due to the similarity of properties, processing routes, and, often, applications [6]. Because of the complexity and the fine line between one class of materials to another, it is becoming more common to separate them according to their origin. Chris [56], for instance, utilizes the macro-grouping of "grown materials", "oil-based materials", and "mined materials". The author utilized these sections with aims at sustainably in a way to highlight and better control the resources for each type of material, once their availability is likely to change drastically in the upcoming decades, either from the scarcity of oil to the development of new materials from renewable sources, like fungi [43], algae [49], and plants [65].

1.1 Classification of Natural Materials

Regarding natural materials, they can be roughly classified according to their origin, whether they are mineral or biological. The first group, also known as geological materials, can be divided into rocks and gemstones, in addition to meteorites and metal alloys [27]. While rocks and gemstones can be better interpreted as natural materials, due to their straightforwardness usage, others would still require some level of processing prior to being ready for application. A traditional classification in gemology separates minerals according to their chemical composition: silicates (the largest group), sulfides, carbonates, oxides, halides, sulfates, phosphates, and native elements [62]. Among naturally occurring metals, known as native metals, gold is a noticeable exception due to its chemical stability in its pure form; therefore, despite also being found in nature alloyed with silver, this precious metal does not require the same ore processing as others that are much more reactive in nature. Gem is a classification intended for highly prized minerals, due to their beauty, durability, and rarity [62]. Usually, they are classified as gems after being cut and polished, but in archeology, gems may refer to engraved stones (cameo, intaglio, seals, etc.) [27]. A more commercial classification divides gemstones into groups that share similar characteristics [83]: "best known gemstones", traditionally traded or generally known gemstones (like diamonds, rubies, sapphires, emeralds, tour-malines, amethysts, agates, jades, and many others); "lesser known gemstones", formerly appreciated mostly by collectors, they are today made into jewelry (like fluorite, serpentine, apatite, tiger's eye matrix, etc.); "gemstones for collectors", which although having no practical use in jewelry (because they are too soft, too brittle, or too rare), collectors gather them as rarities or as small *objets d'art* (like tantalite, rutile, and others); and "rocks as gemstones", in fringe zone of gems, earlier used for decorative purposes and ornamental objects, they are becoming more important for personal jewelry (like marble, limestone, moldavite, obsidian, and alabaster). However, the boundaries between mineral and biological natural materials become blurred when we considerer the so-called organic gemstones. This group includes materials that have preserved or acquired mineral characteristics, despite their organic origin, like coral, ivory, amber, mother-of-pearl, and fossilized materials [27, 62, 83].

Concerning those of biological origins, natural materials can be classified according to their properties/characteristics as well as their composition. Apart from having notably the absence of metals in their composition—which would usually require mechanisms that are unavailable in nature, i.e., high-temperature or high-electric current processing—mechanically, they present a generally low density, rarely exceeding $3g/cm^3$ compared to most synthetic structural materials (which are often in the $4 - 10g/cm^3$ scale); however, they can vary enormously regarding their Elastic moduli (from 0.001 to 100GPa) and strength (from 0.1 to 1000MPa) [59]. Wegst and Ashby [95] present a classification of biological materials into four groups, similar to those used on synthetic materials for engineering: natural ceramics (including ceramic composites), natural polymers (and polymer composites), natural elastomers, and natural cellular materials (or foams). Natural ceramics comprehend bone, antler, enamel, dentine, shell, and coral, and are usually made up of particles, like hydroxyapatite, calcite, or aragonite, in a matrix of collagen. Natural polymers include hooves of mammals, ligaments, and tendons, and arthropod exoskeletons, including cellulose and chitin (polysaccharides), as well as collagen, silk, and keratin (proteins). Natural elastomers encompass skin, muscle, blood vessels, and most soft tissues, made from proteins like elastin, resilin, and abductin. And natural cellular materials include wood, cork, palm, bamboo, and cancellous bone. In a recent review by [29], the authors highlight the overall resemblance of biological materials to composites, and state that they are built with a limited number of building blocks, dividing them into minerals, polysaccharides, and proteins. Mineral biological materials are mostly composed of calcium phosphate (e.g. tooth), silica glass (e.g. skeleton of glass sponge), and calcium carbonate (like the shells from snails and mussels). Polysaccharide materials correspond to cellulose (from trees to cotton), and chitin (the carapace of beetles). And protein biological materials are primarily divided into keratin (found in beaks, feathers, wools, and fingernails), silk (in spider web), and collagen (in artery walls and tendons). In addition, some natural materials are also comprised of more than one type of base substance, such as lobster shells (from chitin and calcium carbonate), and bones (from calcium phosphate and collagen).

1.2   What is a Natural Material

Affirming that a given material is "natural" may imply a broad significance, from the environmental point of view to its origin and applicability. For instance, some "natural material" may falsely indicate that it is free of hazardous substances or causes no harm to humans, which is not necessarily true [53]. A large number of toxic or dangerous substances is found in nature and have been used in many applications, from consumer products to construction materials. As examples, we may cite lead, mercury, and asbestos, which were increasingly employed after World War II, and today they are forbidden in many countries for they being the direct cause of severe lung diseases and cancer [93]. Another frequent problem of classifying natural materials lies in following the literal and absolute meaning that it represents a substance that was retrieved from nature. Oversimplifying, this would correspond to almost every substance known to men. In this sense, not just the simplest material, but even the most high-end synthetic one was once part of the natural world, and it could only be generated after some base material is extracted from nature. For instance, from the plant-based commodity polylactic acid (PLA)—used in home 3D printing systems filaments—to the oil-based engineering-level polyetheretherketone (PEEK)—which can be filled with carbon nanotubes and used in the spatial and medical industry—they were all first obtained by the removal of a basic natural substance. Instead, what can actually distinguish a natural material from others is not necessarily its origin, but *how* and *how much* of it was processed before it is applied into a product as its final form. In this way [17], defines it as:

> A product that is made from materials and ingredients found in nature, with little or no human intervention. Natural materials include stone, glass, lime or mud plasters, adobe or rammed earth, bricks, tiles, untreated wood, cork, paper, reeds, bamboo, canes and grasses as well as all natural fibers.

In its native state, almost all natural materials must be prepared before application. Even on the case of a simple table, for instance, after the tree is cut down, the wood needs to be debarked, cut, dried, and planed into lumbers that could then be employed in a piece of furniture. Not mentioning additional steps like dying, branding, steaming, or heat-based molding, and coating or treating its surface with protecting agents to prevent deterioration and wearing, either by weather or by chemical or biological means. However, in a general way, wood can be considered a natural material, particularly when comparing it to other alternatives like polymers or metals, that essentially require even more complex manufacturing processes and usually involve some level of phase change. Given the circumstances, despite being cut, molded, or contracted, wood applied into a product remains the same continuum solid as it was as secondary xylem in the plant. Wood-derived materials, like MDF (medium-density fiberboard), MDP (medium-density particleboard), OSB (oriented strand board), or even plywood involve the breakage of wood into smaller particles or plies, and then binding them together to form a new solid; therefore, they should be considered a secondary material, despite not having necessarily passed through a phase change or a full recycling procedure.

In addition to the fact that most of its bulk volume is kept in one piece—from the extraction to application—another important characteristic of natural material is that destructive or invasive treatments (chemical or physical) should not be mandatory for its use. As [17] commented, natural materials essentially require "little to no human intervention", i.e., after removing them from nature, one should be able to apply them in a fairly raw state, predominantly. Still, this should be interpreted literally. For instance, hides require a number of time-consuming processes in order to become a usable sheet of leather. From the liming of fibers to the tanning, several types of equipment and chemical products are required—either by following mineral tanning (with chrome-based products) or with more traditional, vegetal-based processes. On the other hand, the steps required for the manufacturing of most polymer-based leather alternatives—like polyurethane (PU) leather—are much more numerous and complex, considering that most of their main raw substances (polyols and isocyanates) are derived from oil; and they are even farther from genuine leather when regarding the extraction-to-application transformations, as well as changes on its original bulk form or volume.

1.3   Selection of Natural Materials

The selection of natural materials for a particular project or design can be due to many reasons. Figure 1 illustrates some of the most important ones, classified according to a holistic approach of materials selection [78]. Such classification can be initially divided into "quantitative" and "qualitative" parameters, like the inner circle of Fig. 1. The first one is based on countable factors, which can be easily measured and calculated with quite an accurate level of appraisal. As for the qualitative parameters, despite also being able to be assessed by means of specialized methodologies, it is considered more subjective and therefore can be subjected to different interpretations, depending on human-related factors like culture, age, gender, social class, etc. In the middle circle of Fig. 1 there is a separation between "economic" and "technical" approaches, in the quantitative classification, and between "sensorial" and "intangible" approaches, in the qualitative one. Whereas "economic" is obviously related to the monetary reasons some natural material is selected, "technical" regards the classical material properties, commonly used in engineering. "Sensorial" is related to parameters that can be evaluated regarding the interaction between the product and the user, and "intangible" is related to product effects measured between users.

In the case of natural materials, following the outermost circle of Fig. 1, first of all, there is the key factor of performance. Natural materials are known for their outstanding behavior along with multiple and simultaneous properties, even for the standards of modern synthetic materials [35]. For instance [59], present a list of features that are unique to natural materials, including self-assembly (in which they do not need for external means to be built up), self-healing capability (by reversing damages or failures), hydration (the properties are strongly dependent on and influenced by the moister content), mild synthesis conditions (biological materials are produced in atmospheric or underwater conditions of temperature and pressure), functionality (having more than one purpose), and hierarchy (different and organized scale levels that give unique and adaptable properties). Furthermore, all those properties are derived from just a handful of elements—carbon (C), hydrogen (H),

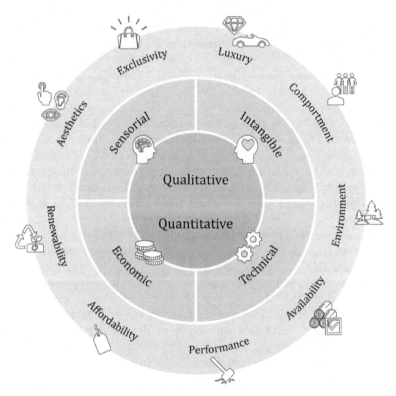

Fig. 1 Selection of natural materials: a holistic approach for some of the reasons they may be preferred as a base material in manufacturing applications

oxygen (O), nitrogen (N), calcium (Ca), and phosphorous (P)—which are organized and combined into efficient morphologies, despite environmental constraints. Another reason is aesthetics. Natural materials remained well appreciated for the feelings they invoke, not only via their look but also with their touch, sound, and scent. For instance, people seek after these materials in a way to reconnect themselves with memories (the indistinguishable scent of leather), feeling warm and cozy (with sheepskin rugs or throws), feeling comfortable or embraced (the soft touch of silk or cashmere), or experiencing deep and clear musical notes (with solid wood speakers). In some sensorial attributes, natural materials are just unrivaled [48].

Third, we can point out the factor of affordability. Despite the existence of technological manufacturing plants, capable of producing thousands of plastic consumer goods every hour, sometimes the selection of a natural material counterpart can be more reasonably priced, particularly when a lower volume manufacturing is sought, or with a craftmanship-like production. For example, the practicality of building a simple wooden stool or a bamboo scaffold is enormous [57]. There is no need for machining expensive aluminum injection molds for polymers or designing complex joints for tubular steel; the natural material is virtually ready to be employed—from harvest to use. The next decisive factor is availability. For thousands of years, local natural materials were the only existing options for fabricating almost all types of artifacts. Today with globalization any company can have access to synthetic raw materials from the most distant regions, however, in many cases selecting a local natural alternative is not only more accessible physically but also more affordable. For instance, many local and natural fibers are by-products of other applications and are still explored in the manufacture of items from reusable shopping bags to reinforcements in composites [55]. The fifth reason is one of the main benefits of biological materials: their renewability. With the clear exception of minerals, natural materials can be regenerated in a fairly brief period, compatible with human needs. Even cork used in the manufacturing of bottle stoppers, for instance, which requires around 9 years between harvests cycles, can still be managed to allow an economically sustainable supply flow [76]. And if appropriate care is taken, unlike synthetic polymer alternatives, this plant tissue is not running out, like fossil fuels eventually will. Similar to renewability, another key factor is the environment. In many life-cycle assessments (LCA), the advantages of choosing a natural material lie upon the energy costs, particularly during the extraction and manufacturing phases. Wooden products, for instance, exhibit a number of environmental benefits when compared to competing materials, such as fossil fuel consumption, contributions to the greenhouse effect, and solid waste generation tends to be much smaller [96]. Bamboo-based products, being this grass one of the most rapid growth materials, have been selected due to the reduction of carbon footprint, in addition to their carbon sequestration capabilities during cultivation [86].

The seventh factor is comportment or social behavior. It is noticeable the recent demand for products that are made using socially fair and environmentally friendly materials and processes. Contemporary trends, lifestyles, and life's philosophies, like naturalism and veganism, have been pushing companies to deliver natural materials in many markets, from phone cases to construction materials to automobile industries [51]. The eighth main factor is exclusivity. The fundamental characteristic that makes biological or mineral materials so looked-for is their individuality. Not even the most seamless and technological manufacturing process can smooth out their natural variability, which is advantageous. Every mineral, piece of leather, or wood finish is distinctive, and by that, it delivers character to a product. Including when the product is worn out, some leather goods even became more valuable with the development of a natural aging patina on the surface. Associate with this variability propriety, luxury is another reason natural materials are so often selected. If on one hand, natural materials are unique in terms of size, pattern, shine, or shape, on another they can be considered quite exclusive, which means no other consumer will have a similar product, particularly in the case of gemstones for the jewelry market [62]. In addition, many luxury consumer products, like watches and cars, are designed with natural finishes and surfaces as a way to demonstrate the brand's mindfulness and caution during the manufacturing of an artifact that is going to be exhibited as a symbol of status and wealth.

1.4 Bionics and Aesthetics of Natural Materials: From Nature to Jewelry

The fundamental characteristic of natural materials, which influences all their features and properties described before, is their variability. Principally, in biological materials, this is a consequence of both intrinsic and external factors. The first one is due to species-related differences in chemistry, polymer chain length and packing, and water content [35], and the second one is driven by environmental constraints, such as temperature, humidity, insolation, natural selection, as well as biotic effects, having an impact on their growth and development [31, 59]. However, if on one hand this variance can be seen as a disadvantage regarding strict manufacturing standards—leading to the obligation of defining a much broader range of material properties [35]—on the other, in terms of surface finish, this can be quite favorable:

> Perfection and Imperfection. Surface perfection is violated by the slightest defect – it has no hope of aging gracefully. Better, to make visual imperfection a part of the personality of the product – something that gives it individuality. It is this, in part, that makes natural materials – wood, leather, and stone – attractive [7].

As seen before, a combination of exclusivity and luxury—both dependent on the variability of natural substances—are desirable features that contribute to the perennial success of natural materials in product design. In a survey with 30 natural materials, for example, participants mentioned the reasons they express naturalness, and two important points worth highlighting is (i) their imperfect surface qualities, which were uneven or ununiform, and (ii) their "long-term use" aspect due to a worn-out appearance [47]. If for synthetic engineering materials this aging effect might be considered a sign of degradation in their physical or chemical properties, due to irreversible changes in their structure or composition, for natural materials they are perceived as a sign of maturity—modifications in scents, colors, and textures that accentuate their preciousness [79]. Despite sometimes not being valorized in the west [79], the appreciation for aging aesthetics in materials and products is better established in east cultures; for instance *Kintsugi*, the Japanese art technique of repairing broken pottery with lacquer mixed with powdered precious metals, like gold or silver. This sense of "age" and "experience" in natural materials can be reflected even when the products they are utilized on are brand new: since they were extracted from nature, their "history" is already ancient well before applied into a product [92].

When bound to artistry or craftsmanship, natural materials often are associated with imprecise aesthetics, either in shape, patterns, or colors, but in a way that harmonizes with their origin as raw materials. No two pieces are completely equal. However, even if overall measurements—like lengths and angles—are approximated and determined by the eye, this "lack of refinement not only minimizes the intervention, it also retains the qualities and attributes of the natural material" [94]. And when this inexactitude is embraced, it is transformed into an attribute of design, while reconnecting it with the variances of the natural world. In many products, this can be seen as a looked-for attribute [4], for instance, describes the development of wedding rings where one pair had the inclusion of a single piece of wood, symbolizing the "couple's unity", and involved by a rough-finished bronze, which was supposed to be naturally polished during to the long-term utilization of the jewelry, thus representing the "beauty of a growing relationship". As each individual uses and wears out a ring differently, the pieces remain distinct. More than connecting symbolic representations with valuable objects, this utilization of the variability of natural and uncommon materials represents a newer trend in contemporary jewelry [25]. Despite symbolisms in shape, colors, and textures having long been used in jewelry, nowadays special attention has been given to the application of once considered unusual—and even unvalued—materials and techniques [10, 22, 24]. From plastic waste to natural materials, contemporary jewelry follows a path towards comprehensive sustainable practices, either by the inclusion of materials that may be problematic in environmental concern, as well as social fairness, regarding the origin of each material [18]. The called "new luxury" is reached by a combination of natural materials, sustainable production, and craftsmanship, aiming at helping us to find purpose and meaning [66]. In contemporary jewelry, the liberty to choose from a wider range of possibilities allows the designer to focus more than ever on significance and meaning, rather than just the pure monetary value of the resources employed. "The applied material, regardless of whether it is gold, platinum, diamonds, emeralds, wood, polymers, rocks, or even residues from waste, is a design choice, as for the techniques, technologies, finishes, and concepts employed in jewelry" [71].

Another process for the creation of sustainable and innovative pieces that contemporary jewelry is being based on is via newer creative processes and methodologies. Among them, bionics is being explored due to its flexibility in transferring an array of features from nature to projects. More traditionally, bionics is used as a methodology for converting technical attributes—like geometries, organizational and hierarchy levels—into engineering designs ranging from simple products to large constructions, and mainly with regards to mechanical, physical, and chemical properties [52, 73, 74]. However, not only technical properties can be explored in the design of bioinspired projects. When utilized along with jewelry, for instance, aesthetical features extracted from nature became equally as important for the creative development of new patterns, geometries, and tridimensional configurations, that add value and sensorial experiences to users [23, 72]. Therefore, the combination of traditional and artisanal jewelry procedures with innovative materials, techniques, and creative methods is a way to create and develop new expressive and sustainable pieces, which not only emphasize the new meaning of contemporary luxury but also the appraisal for the aesthetics of the imperfection, valorizing the uniqueness of natural materials.

2 Brazilian Natural Materials as a Source of Inspiration

Among different sectors, natural materials exert a large influence on the amount of environmental wealth one region has at its disposal. Natural resources are among one of the main benchmarks for measuring sustainable development. In recent years, a new type of metric for assessing the gross product of countries and regions has been proposed, which takes into consideration the value of the ecosystem assets such as forests, fertile soils, and biodiversity. Ouyang et al. [70] suggested the concept of Gross Ecosystem Product (GEP), as opposed to the more traditional Gross Domestic Product (GDP), which accounts for the sums of every good and service bought and sold in a country or region, in a particular period. The authors defined GEP as the "total values of ecosystem products and services for human welfare and sustainable development" which includes "ecosystem provision value, ecological regulation services value, and ecological culture services value" [70]. Similar to the complexity of monetary economy for the definition of GDP, GEP also utilizes a number of parameters to be determined. Essentially, market prices are used, when available, and combined with estimated surrogate prices for ecosystem services, where market prices do not exist [69]. However, such ecosystem values can only be considered sustainable if the rate at which they are being extracted is lower than their level of renewability. And promoting the valorization of natural resources, especially natural materials, is a way to ensure they are going to endure for the next generations, as well as to provide means at which the population can economically benefit from their minded and prudent consumption.

Brazil is one of the world's richest countries in terms of natural resources and is known for its great diversity both in terms of mineral and biological (biodiversity) materials. However, despite the large reserves of natural resources, the country does not obtain monetary returns as relevant to its capacity. For instance, by evaluating the GEP, Brazil is the richest country in the world, estimated at 14.4 trillion USD [42], compared to a 1.4 trillion USD, at the end of 2020 [88]. In terms of mineral resources, Brazil has one of the world's largest reserves [64] and the world's sixth largest mining industry, with the production and exportation of about 80 mineral commodities [54], with emphasis on the production of niobium (87.8% of world production), iron ore (19.2%), and bauxite (7.8%) in 2019 [67]. On the other hand, the country is considered one of the main mineral exporters in the world, indicating that most of its mineral commodities does not stay in the country to be processed and developed into applied products [67]. In relative terms, despite the diversification of the mining industry in Brazil, it is estimated to be at only 1.1% of GDP [54]. Regarding gemstones, the country accounts for a variety of commercially intended gems, being the most important tourmalines, topaz, opals, varieties of quartz (agate, amethyst, and citrine) and emeralds, as well as one of the only global producers of imperial topaz and Paraiba tourmaline [11]. Regarding biodiversity, Brazil has even more resources, being grouped among the richest countries in the world, and is considered one of the megadiverse nations [90]. The country has more than 116,000 and 46,000 known animal and plant species, respectively, spread across six terrestrial biomes and three major marine ecosystems, according to the Brazilian Ministry of Environment [63]. Brazil's biodiversity comprises 70% of the world's cataloged animal and plant species, and it is estimated that it holds between 15 and 20% of the world's biological diversity, with a constantly expanding rate of 700 new animal species discovered each year, on average [19]. Nevertheless, similar to mineral resources, Brazilian biodiversity is still poorly explored in terms of products that are translated to sustainable development, covering social quality of life, environmental protection, and monetary growth. Valli et al. [91] highlights that Brazil could became a world leader in bioeconomy due to the sustainable potential of natural resources in the country. Despite the wealthiness of Brazilian natural resources, it is clear that a great part of its materials and substances is applied in low complexity products, with little added value, which consequently results in a small return for its population. In this regard, in the light of the richness of Brazilian mineral and biological diversities, this chapter intends to present a new look into some of its natural materials, in a way to incorporate them into the design of high-value products, with application in contemporary jewelry. Natural materials from a mineral, animal, and plant source were selected and used both as a base material and as inspiration for the aesthetics of a jewelry collection.

2.1 Mineral Material: Agate

As seen before, Brazil is known worldwide for the diversity and occurrence of minerals and gemstones in its soil. It is one of the main gemological provinces in the world, standing out for volume and quality of production and export of gemstones, being the second largest producer of emerald and the only producer of imperial topaz and Paraíba tourmaline [41, 83]. It produces, on a large scale, agate, amethyst, opal, aquamarine, topaz, and quartz, according to the Brazilian Institute of Gemstones and Precious Metals [41]. In the state of Rio Grande do Sul, in southern Brazil, important geode deposits filled with agate, amethyst, and opal are concentrated. Agate is a variety of chalcedony formed by successive bands, occurring in a compact form and by filling cavities, such as geodes and fractures, among other forms of incidence [20–22, 26, 36, 38, 44, 60, 61]. Geodes are cavities fully or partially allocated in volcanic rocks with rounded to ovoid shapes, having their dimensions between 20 and 60 cm in diameter, although the occurrence of larger sizes is not uncommon [20–22, 28, 46, 61, 87]. Banding can be comprised of successive layers of chalcedony or sometimes be interspersed with opal. The chalcedony bands are composed of microcrystalline and fibrous quartz, oriented perpendicularly to the surface of the individual band layers [20, 34, 83]. In terms of the chemical composition of chalcedony, 90–99% by weight is silica (SiO_2), with up to 2% of water and impurities [22, 33, 34, 37].

In addition to Brazil, agate is currently found in several places around the world, such as Botswana, South Africa, Egypt, Mexico, China, and Scotland; and fire agate only in northern Mexico and the southwestern United States [62]. Until the beginning of the nineteenth century, the most important agate deposits were located in Idar-Oberstein, Germany [62, 83], where they are currently depleted. For several years, this country was considered an excellent source of extraction of this material, divided into three major regions: Idar-Oberstein, Baumholder, and Oberkirchen [97]. The varieties of materials in these regions were highly regarded, with geodes of assorted colors and dimensions [20, 62]. The state of Rio Grande do Sul (RS) alone exported around 82 million tons of gemological materials between the years of

2017 and 2018 [41], mostly silica-rich minerals, such as agate. Both the quantity and quality of the material produced make the state not only one of the largest miners but also one of the larger suppliers to international markets, such as Asia and the United States [45, 61]. According to IBGM [41], the sum of gemological materials extracted and processed in the Rio Grande do Sul state was about US$ 200 million, in the period 2013–2016, being exported to more than 30 countries around the world. The agate extraction activity in the Salto do Jacuí Mining District took place around 1827, when German immigrants, who mastered the techniques of mining and processing, such as the glyptic art or hardstone carving, explored the area in the search for this gemological material [50, 61, 83]. Due to the fact that the mines are located on the banks of rivers, for a long time, the agate extraction activity was carried out irregularly, without environmental regulations and laws formalized by state agencies [61]. This practice brought many problems to the mining district, ranging from environmental—with inadequate exploitation—to social, including public health and human resources [61]. However, it was only around 2004 that the State Foundation for Environmental Protection (FEPAM/RS, Portuguese acronym), connected to the Secretary of Health and Environment of the Rio Grande do Sul state, implemented a resolution establishing the procedures and criteria for the extraction of mineral goods [32]. Currently, with this resolution and others that came later, the regularization of mining activity provided the existence of about 16 active mines [61].

Figure 2 illustrates the main steps of the extraction and processing of agate. The Salto do Jacuí Mining District (SJMD) is located in the central region of Rio Grande do Sul and is one of the world's largest producers of agate, covering an area of 250 km², with extraction conducted in open-pit mines. In this municipality, on the banks of the Jacuí and Ivaí Rivers, the main agate extraction is the variety known as "Umbú", which has a greyish-blue color and poorly developed banding, thus being widely used for dyeing [22]. The extraction of agate is conducted in open-pit mines, with the use of excavators to dismantle the altered rock and the manual gathering of geodes. After their extraction, agate geodes are processed, traded, and exported, either raw or in the form of simple artifacts. The main center for processing, trading, and exporting agate artifacts is located in the municipality of Soledade/RS, approximately 110 km from the SJMD. In the municipality, the processing encompasses the storage of agate geodes to their cutting into plates and many other processes with the material. The processing given to this gemological material consists of the selection and washing of geodes, cutting, and washing of plates, dyeing with the use of inorganic and synthetic organic dyes, sanding, and polishing [3, 16, 20–22, 89]. The gemological material can be found in the market in the form of plates or the cabochon cut, with or without dyeing. And the dyeing process usually involves the application of red, black, blue, pink, purple, and green colors. However, during the processing steps of plate cutting and cabochon gemstone polishing, many parts of the gemological material are rendered useless due to roughing, cutting, and shaping into the required shapes, leading to the generation of defected or broken pieces. Coming from the processing steps, these parts are classified as wasted residues and often end up having no market value. Brazil's National Policy on Solid Waste (PNRS, Portuguese acronym), defined by Federal Law n° 12.305/10 [15] states that "solid waste that, after having exhausted

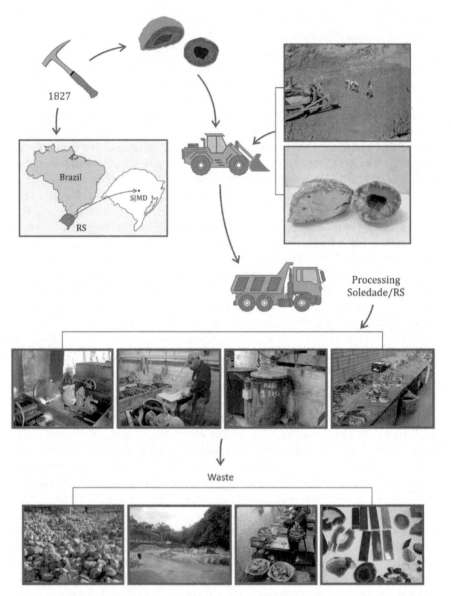

Fig. 2 Agate (Calcedony) in the Salto do Jacuí Mining District (SJMD) in the state of Rio Grande do Sul (RS/Brazil). After extraction, agate is transferred to the city of Soledade (RS) for processing, however, many types of residues are generated and wasted

all the possibilities for treatment and recovery for available and economically viable technological processes, show no other possibility than the final disposal." In recent years, with the pursuance of extraction and processing, the gemstones sector has been generating significant amounts of solid waste, mainly in the form of broken, semi-finished pieces, or with the presence of defects. Therefore, in the production process of all agate-derived products, the stored leftover materials are often discarded due to the lack of commercial interest, and thus present a potential base material to be created innovative designs upon them.

The vast majority of agate-derived products processed and sold in the city of Soledade are ornamental objects or decorative artifacts, such as plate mobiles, bookends, ashtrays, clocks, pyramids, spheres, obelisks, cutlery handles, among others. Despite these products having a simple manufacturing process, they do not present a greater level of complexity. Instead of opting for innovative projects, which would require original and new creative processes, many different industries end up always producing the same types of objects, leading to a lack of competition which, consequently, keeps the final prices equal and their profits low. Therefore, there is a noticeable demand for products with an innovative approach, to improve competitiveness. For years in Brazil, this natural material has been processed always in the same way, as half-cut geodes or in the form of plates, both dyed still in the same color shades, without highlighting the beauty of the gemstone with its bandings and appeals. Agate, regardless of the region where it is extracted, is a material with remarkable visual aspects, which its natural characteristics should be explored in projects for unconventional products, including by employing *in natura* sheets, i.e., dyeing-free. In addition, the adoption of different gemstone cuts to highlight colors, bands, shapes, and fire (play of light)—also known as luster, could be encouraged. Moreover, specifically in the jewelry sector, this material has always been linked to pieces with little intrinsic value, by which non-noble metals, such as Brass (an alloy of copper and zinc) and Zamac (the family of alloys with a base metal of zinc and alloying elements of aluminum, magnesium, and copper), with silver- or gold-platting finishing, labeling agate as a material of little value.

2.2   Plant Material: Golden Grass

Due to the great biodiversity and cultural diversity, some Brazilian regions have requested, deposited, and granted certified products as Geographical Indication. For natural materials, this process can lead to local development and significant improvements for local economies and the national agribusiness as a whole [84]. In the central region of Brazil, the Brazilian tropical savanna biome known as *Cerrado* lies an important plant which is classified as a non-timber forest product, *Syngonanthus nitens* (Bong.) Ruhland. Despite commonly being called "golden grass" due to the high shining of its flower stem, or scape, which resembles gold, the plant actually belongs to the ERIOCAULACEAE family. *S. nitens* is an everlasting plant with occurrences in all regions of Brazil, being internationally recognized as a source of handcraft products. To maintain its renewability, there are a number of restrictions for the harvesting periods of the plant, which contributes to sustainable management [68].

The golden glow of *S. nitens* is the main attribute of the plant, which was subject to some investigations to comprehend the reasons it has such a distinct visual aspect. Siqueira et al. [85] extracted cellulose microfibrils from *S. nitens* fibers, which showed birefringence due to the structural anisotropy of cellulose and a flow anisotropy, resulting from the alignment of nanocrystals under flow. Berlim et al. [14] investigated specifically the origins of the golden aspect of *S. nitens*. The authors found that the color is due to the interaction of incident light with several flavonoids present at the smooth surface of the stem. Considering that part of the light is absorbed rather than reflected, the peak spectrum of the solar light is shifted toward longer wavelengths. Besides, the flavonoids also presented fluorescence in the longer range of wavelengths, thus also contributing to the red shift, resulting in the characteristic golden brightness. In a more recent update [13], also verified that fluorophores confined into small compartments of the dry stems of *S. nitens* also lead to enhancement of the observed fluorescence, which influences the shining of the stem's surface.

Figure 3 shows the manufacturing steps of artisanal products made from golden grass. Flowering occurs between July and August, and during August and September, the production of seeds takes place. Seeds are naturally dispersed by the wind in October or manually dispersed throughout the manual harvesting period. Therefore, in the following months, seeds can germinate and grow in the form of rosettes, thus maintaining a sustainable harvesting cycle [80]. The stalks are harvested before the natural dispersion of the seeds due to the rainy season. If they remained in the fields under these humid conditions, their brightness would be reduced and, consequently, the quality of the handicraft would also be decreased [81]. When maturation occurs and the plant is dry, the stem starts to detach from the rosette. When the stalk is not yet ready for harvesting, the rosette is pulled out along with the stalk, which leads to the plant's death. Thus, if early harvesting is carried out—due to lack of knowledge about biology the rules of harvesting or due to the exploitation of raw materials—there is a great environmental impact since, during one hour of collection, more than 100 rosettes can be uprooted per a single artisan [82].

After harvesting, plant stalks are storage in bunches in a dry place out of the sun. For the craftsmanship done with the dried golden grass, a fiber extracted from the *buriti* plant (*Mauritia vinifera* Martius) is used to sew the stems [30]. By sewing the stems of *S. nitens* with fibers from *M. vinifera*, artisans weave them by twisting them into strands that are combined into a thread. Next, the tread is conformed into multiple shapes, depending on the intended product to be created. Several handicraft products are manufactured with the stem of *S. nitens*, including baskets, hats, bracelets, earrings, purses, belts, handbags, and many more items. Despite the existing creations regarding the use of golden grass in jewelry, innovative approaches can be followed in the creative process to obtain distinctive designs.

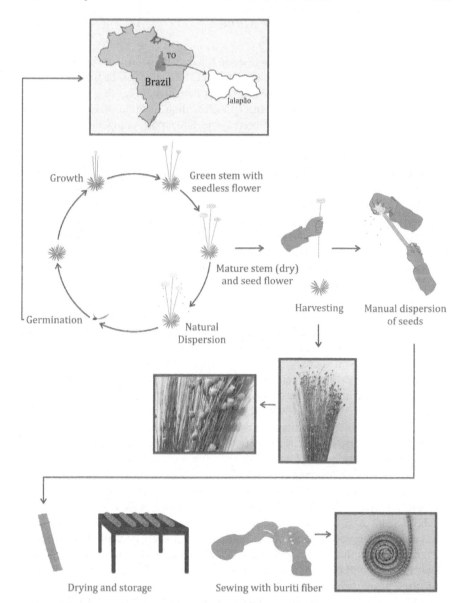

Fig. 3 Production of golden grass (*Syngonanthus nitens*) derived products, from cultivation and harvesting, to drying and sewing

2.3 Animal Material: Horse Mane

As in many countries, horses have a historic relationship with Brazil, remaining an important role for the life, culture, and work in the country [2]. In Brazil, there are approximately 5.9 million horses (*Equus ferus caballus*), according to the Municipal Livestock Production Survey, from the Brazilian Institute of Geography and Statistics [40]. Animals are mainly used for leisure, sport, competition, breeding, and the development of livestock and agriculture [58]. One of the most important breeds in southern Brazil is the Crioulo horse (or Criollo, in Spanish). The breed has its origins in the Spanish Andaluz and Jaca equines, brought from the Iberian Peninsula in the sixteenth century by the colonizers, according to the Brazilian Association of Crioulo Horse Breeders [1]. Crioulo horses are considered a traditional and super representative identity for many countries in South America, being symbols of pride and joy for millions of people in multiple countries [2]. Established mainly in Argentina, Chile, Uruguay, Paraguay, and southern Brazil, in the Pampas region, with a major representation between different breeds, many of these animals went on to live free, forming wild herds that, for about four centuries, faced extreme temperatures and adverse feeding conditions. Such adversities helped establish some of their most striking characteristics on these animals: rusticity and resistance. In the mid-nineteenth century, farmers in the south of the continent began to become aware of the importance and quality of this breed. Being this breed well defined and with its own characteristics, it started to be preserved, coming to gain worldwide notoriety from the twentieth century, when the technical selection exalted its value and proved its virtues. In the Brazilian state of Rio Grande do Sul, for an estimated 520 thousand horses [40], about 386 thousand are registered and identified as Crioulo [1], which represents about 85% of the amount of this breed in the country [40].

The Crioulo horse is used in a variety of functions, from agricultural to competitions, both related to sports of aesthetics [77]. It is the variety of coats, shades of fur and mane, and muscular strength to assist in jobs related to livestock, agriculture, and transport, which make the Crioulo horse a unique and special animal for centuries. Its morphology is characterized by a robust musculature, with a straight and broad forehead head, extensive chest; fore and hind legs rounded with strong tendons. The coats attributed to this breed are divided into approximately 24 variations, including designated names like Colorado, Picasso, Zaino, Tostado, Gateado, Tobiano, among others [1]. In the horse's mane and tail, a vast quantity of hair is found, with long structures, usually greater than 300 mm. Tail hair has a fiber diameter ranging from 75 to 280 μm, whereas mane hair is finer, varying from 50 to 150 μm [39]. Regarding these factors, the Crioulo horse differs from other breeds, which its regional identity is increased and much appreciated in terms of morphology, function, and genetics, underscoring it not only in the national but also in the international market.

Depending on the intention or job function the horse will follow, whether for competitions or to help in livestock, its mane, both on the head and tail, needs aesthetic care. Generally, in morphological competitions and in-field activities, for example, the animals are shaved monthly for reasons of hygiene and due to the heat, leaving their manes short. The hairs on both the mane and tail are trimmed by fine dimensions, sometimes even defined by millimeters, as a way to keep the cut always precise, or to specific lengths. In the morphological competitions of the Crioulo breed, the only exception for not leaving the hair with a short pattern is in the rein competitions, where the manes are required to be of large dimensions and volume. Culturally, for example, when a horse is destined to spend a specific period unfastened in the field, whether for recovery reasons from a medical issue, pregnancy, and/or for the growing phase until the exact training period, the manes tend to grow in large dimensions. In the *Cabanhas*—southern Brazil nomenclature for small farms specialized in horse breeding and maintenance—generally, two and a half years old animals are destined for training, and so the mane is trimmed with a predetermined pattern for the well-being of the animal. Another period that the mane tends to be trimmed is during the prepartum stage, which may happen once a year, most of the time, where the tail part must be prepared for the birth process. During these periods, the Crioulo breed produces a large volume of mane hair, which are discarded or destined—although not often—as fillings for the development of stuffed products. Therefore, developing new ways to incorporate such material can be a way to not only valorize it as a waste residue but also to create an aesthetic object, with a strong cultural and social meaning.

3　Experimental Procedure

Based on the evaluated natural materials from mineral, plant, and animal sources, samples were collected and employed in an experimental procedure for the development of a contemporary jewelry collection. In this experiment, the natural materials were projected to be employed in two ways in the design of the pieces. The first one is by utilizing them as raw materials in the collection. By conferring them a highlight in the pieces, they will be explored as the main featured object, as the focal point in the jewelry. The second way is to also employ them as a source of inspiration for the conceptual design. Following the principles of bionics and biomimetics, the natural materials were explored according to their potential source of information for the design of innovative pieces [23, 52, 72]. In this case, each type of material was studied regarding its aesthetic properties, and how this feature can be transported into the design of the new jewelry pieces.

For the mineral material, an agate plate was collected in the municipality of Soledade (in the Rio Grande do Sul state, Brazil). The plate consisted of a rejected piece, which was broken during the polishing process and discarded as waste due to the lack of commercial value. The plant material consisted of a mature sample of *S. nitens* collected in the municipality of Mateiros (Tocantins state, Brazil), nearby the Jalapão region. As for the animal material, a horse mane sample was collected in the Cabanha Marca do Freio, in the municipality of Tapes (in the Rio Grande do Sul state, Brazil). For the project followed in this research, each sample was observed via light microscopy. Instead of using higher resolution techniques, such as scanning electron microscopy—which would have allowed even finer topological details to be observed—we opt for light microscopy due to the capability of registering sufficient morphological details, as well as color shades and gradients. Image system equipment used were BX41 microscope (Olympus® Co., Tokyo, Japan) and M165 FC stereoscopic (Leica® Microsystems, Wetzlar, Germany), available at the Laboratory of Plant Anatomy (LAVeg) from the Federal University of Rio Grande do Sul (UFRGS). Figure 4 shows examples of light micrographs from the investigated specimens.

The agate micrographs showed the distinct aspect of bands geometries, which is highlighted due to the difference in the translucency levels of each region. Despite the color of the original variety of agate from the Salto do Jacuí Mining District being a blueish-grey, the observed sample consisted of a broken plate that was dyed

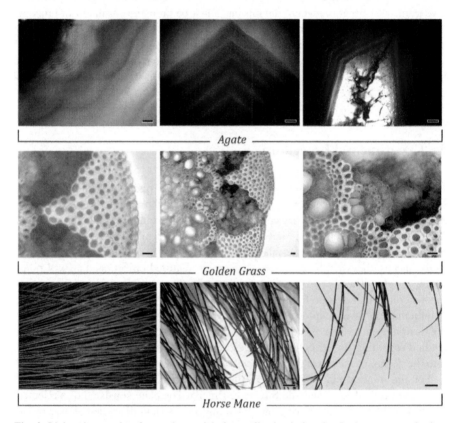

Fig. 4 Light micrographs of natural materials for application in jewelry. In the top row, a broken plate of agate shows different geometries of bands; the scale bar is 200 μm. In the middle row, unstained cross sections of the stem of golden grass (*S. nitens*) highlighting the golden aspect of the epidermis, collenchyma, sclerenchyma, alternating with the chlorenchyma, and the collateral vascular bundles surrounded by lignified cells; the scale bar is 20 μm. In the bottom row, the horse mane hair shafts are shown, with details of the length and tip portions; the scale bar is 2 mm

in purple after the cutting and before the polishing processes. Unstained *S. nitens* cross-sectional micrographs showed a matured sample with lignified cell walls and the traditional golden aspect of the tissues (epidermis, collenchyma, sclerenchyma, and chlorenchyma). As for the horse mane samples, the micrographs showed very rough and course fibrous hairs with a uniform thickness.

The creative process followed a methodology of contemporary jewelry design based on the study of bionics [21, 72], which combines traditional jewelry techniques with newer materials and processes. The creation was also based on the study of bionics, by using the micrographs of the collected natural materials as a source of inspiration for the aesthetics of the pieces. The collection consisted of three silver pendants with variable geometries based on the shapes observed in the micrographs. For the contemporary aspect, each pendant includes the addition of the original raw natural material, incorporated as an element of the design. Figure 5 shows the 3D renderings of the developed jewelry collection of pendants based on the visualized natural materials.

For the agate jewelry, the pendant included the same broken plate that was observed in the microscope, and the gemstone setting was designed based on the bands of the material observed in the micrographs. Despite its main purpose is to attach the gemstone firmly and securely into the chain, instead, the external shape was modeled, and its surface was engraved with emphasis on the geometric pattern of the bands, which influenced it aesthetically. The golden grass pendant was based on the cross-sectional views of the collateral vascular bundles that are surrounded by golden, lignified cells. Each cell was cut out from a silver plate, resulting in the cellular aspect seen on the cross sections. The region corresponding to the vascular

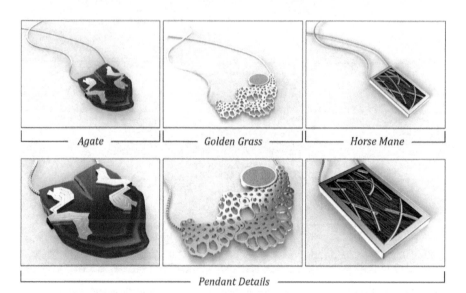

Fig. 5 Developed contemporary jewelry collection of sterling silver pendants based on mineral (agate), plant (golden grass, *S. nitens*), and animal (horse mane) materials, and details of the pendants

bundles in the micrographs was replaced by an oval setting in the jewelry with the inclusion of aligned segments of the plant's stem. The main purpose was to take advantage of the strong brightness of the color of the golden grass and to include it in the pendant as a replacement for a traditional gemstone set in the piece. The horse mane pendant was developed aiming at the valorization of the hair tips, due to their scrambled orientation in the micrograph. Mane hair samples were placed in the back of the pendant, in a silver box, but still visualized between the details of the frontal part of the piece.

In order to illustrate the production of a physical model from the developed collection, one of the pendants was chosen (Fig. 6). Among the observed samples, the micrographs of the golden grass (*S. nitens*) were the ones with the most complex geometries and details that could be explored in many ways. In addition, from the production point of view, this pendant design was the most finely detailed, thus requiring more manufacturing techniques. The process followed the traditional jewelry manufacturing techniques, and the selected material was sterling silver, an alloy with 95% Ag and 5% Cu, much utilized in contemporary jewelry due to great characteristics of mechanical strength, brightness, and general aspect. The employed silver was recycled and purified from scraps and recovered pieces. The alloy copper material was extracted from electronic waste. After the formation of the alloy, by melting the proportional Ag and Cu quantities with an oxyacetylene torch, the fabrication started the generation of a sheet with an electric rolling mill. After the desired thickness of around 2 mm, the designed pattern was cut with a jeweler's saw. Finally,

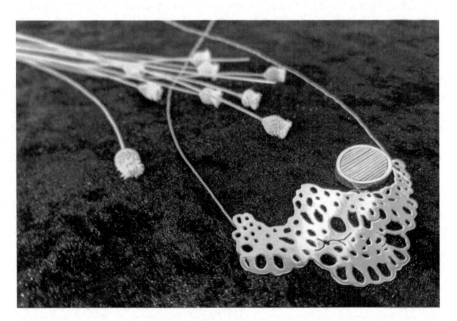

Fig. 6 Produced sterling silver pendant inspired on the micrographs of the golden grass (*S. nitens*) and utilizing the plant's stem as a replacement for a gemstone

another silver sheet was used and shaped in an oval geometry as a setting, to which the gemstone replacement would be attached. In this project, instead of a traditional gemstone, stems of *S. nitens* were cut into segments and attached parallelly inside the oval setting, which was then coupled with the pendant. Lastly, the pendant was smoothed using flies with different coarseness, sanded with abrasive paper, and polished with natural bristle and muslin buffing wheels.

4 Conclusions

Natural materials are tightly bound to humankind's history and development. From the creation of adornments to weapons and tools, their ready availability and renewability made them suitable for applications that contributed directly to our evolution towards modern civilizations. The appearance of the first known pieces of jewelry has the presence of materials retrieved from nature, being collected and applied intentionally, either for their beauty or for the pure curiosity of those who saw them. Over time, more materials were introduced into the early manufactured objects, such as rocks, woods, bones, tusks, and seeds, as well as manufacturing processes ranging from lashing to forging, among others. Analyzing this usability, we can consider that the materials themselves ended up being reused for other functions, such as the skin and bones of animals, which were the resulting waste of hunting, were employed as pendants and fibers for the attachment of ornaments. With this characteristic of reusing materials and waste, sustainability in natural materials has been traced back to the beginning of civilization, even if unintentionally, as a natural cycle of reuse and ornamentation. Over time, other materials—now considered more traditional—such as noble metals and gemstones started to become the preferred option for application in the jewelry industry.

As seen before, the boundaries between natural and synthetic materials are blurred. Differentiating them lies in the non-definitive estimation of how much they were processed prior to being able to be applied as a raw material in the materialization of a project. Nonetheless, despite their ancient history and—somewhat—primitive usage, they are still considered valuable and appreciated for several reasons. From their renewability, and relatively lower environmental footprint, to the costs of extraction, feasibility, and convenience of having a bulk, ready-to-use material, natural materials are the preferred choice for a number of applications up to this day. Moreover, their natural variability, rareness, and uniqueness also make them desirable and often contribute to associating them with wealth, status, or even social acceptance. Accompanying these characteristics is a key and inherent part of natural materials: their flaw and imperfection-related properties. Diamonds and emeralds are interesting examples. Even though synthetic versions of these gemstones can be flawlessly produced in laboratories, the very fact that they do not present defects (like their natural counterparts) makes them easily identified as "not originals", ending up acquiring a much lower market value. While defects are associated with problems in industrialized materials and processes, in natural materials they are precisely what makes them singular and precious.

Traditionally, materials used in jewelry are usually selected, manipulated, and processed in the same way for quite some time, i.e., certain gemstones are only processed with one type of cutting, and others, considered of lesser value, are destined for a relatively simpler process. Likewise, metals such as silver alloys are only intended to be used along with certain gems or pearls, and diamonds, for example, should not be incorporated in these designs. These determinations and usability requirements are being broken in contemporaneity, where new looks are being projected on in order to break these traditional standards. Take the example of unusual materials. Why couldn't they be introduced in jewelry along with gold? Agate, which is traditionally always cut in a cabochon way, why couldn't it be cut in a faceted shape, highlighting its natural bands? Why does it have to be dyed into artificial colors to be appreciated in some regions? And even the residue originated from this material, why can't they be better utilized as well as be combined along with noble materials with an innovative design with a high-quality finishing? Likewise, the application of non-conventional, renewable natural materials, such as horsehair and golden grass. Couldn't they also be applied together with more noble materials, to leave the artisanship sphere towards the high-end jewelry market? Such questions are exactly what contemporary jewelry intends to raise, debate, and incite. In addition to a greater intrinsic appreciation of these natural materials, socially due to their production process—still, we may add, quite dependent on craft practices or with direct monitoring of people—more people could benefit and profit from new applications of greater value.

Bionics, in its broad meaning, intends to present innovative ways of developing projects that take nature as a source of inspiration. As mentioned in the methodology followed, this chapter aimed at representing a new look into the development of bioinspired, sustainable contemporary jewelry collection, where unusual images—such as microphotographs of natural materials—are employed as a source of inspiration for new patterns. Moreover, by also including the same material in the design process, we proposed a way to increase the value of Brazilian traditional, biodiversity-derived natural materials, either those recovered from waste or those whose value could be more valorized. Finally, bionics should then be regarded as a powerful tool not only for the development of more efficient technical solutions, but as a means to promote sustainability, creativity, and value, even in more traditional fields, like jewelry design.

Acknowledgements The authors thank the "National Council for Scientific and Technological Development—CNPq" for supporting this study through the project "Chamada Universal MCTIC/CNPq 2018". This study was financed in part by the Coordenação de Aperfeiçoamento de Pessoal de Nível Superior—Brasil (CAPES)—Finance Code 001. The authors also thank Prof. Jorge Ernesto de Araujo Mariath and the Plant Anatomy Laboratory (LAVeg) from the Federal University of Rio Grande do Sul (UFRGS) for the availability and assistance with the light microscopy procedure; Prof. Ana Lúcia Oderich from the Federal University of Santa Maria (UFSM) for the donation of the samples of golden grass (*S. nitens*); The *Comercial De Pedras Palludo* from the city of Soledade (RS—Brazil) for the donation of the agate waste; and the *Cabanha Marca do Freio* from the city of Tapes (RS—Brazil) for the donation of the horse mane.

References

1. ABCCC—Associação Brasileira de Criadores de Cavalos Crioulos (2021) O Cavalo Crioulo. https://www.cavalocrioulo.org.br/studbook/cavalo_crioulo. Accessed 11 Nov 2021
2. Adelman M, Camphora AL (2019) Crioulos e crioulistas. In: Horse breeds and human society. Routledge, pp 104–120
3. Agostini IM, Fiorentini JA, Brum TMM, Juchem PL (1998) Ágata do Rio Grande do Sul. Ministério de Minas e Energia. DNPM, Departamento de Produção Mineral. Brasília
4. Allen J (2014) The immaterial of materials. In: Karana E, Pedgley O, Rognoli V (eds) Materials experience: fundamentals of materials and design. Elsevier, Oxford, pp 63–72
5. Ashby MF (1992) Materials selection in mechanical design. Pergamon Press, Oxford
6. Ashby MF (2011) Materials selection in mechanical design, 4th edn. Butterworth-Heinemann, Burlington
7. Ashby MF, Johnson K (2013) Materials and design : the art and science of material selection in product design. Elsevier Science & Technology
8. Ashby MF, Jones DRH (2013) Engineering materials 2 : an introduction to microstructures and processing. Elsevier
9. Ashby MF, Jones DRH (1980) Engineering materials 1: an introduction to their properties and application, 1st edn. Pergamon, New York
10. Ba'ai NM, Hashim HZ (2015) Waste to wealth: the innovation of areca catechu as a biomaterial in esthetics seed-based jewelry. In: Proceedings of the international symposium on research of arts, design and humanities (ISRADH 2014). Springer, pp 373–381
11. Barreto S de B, Bittar SMB (2010) The gemstone deposits of Brazil: occurrences, production and economic impact. Boletín la Soc Geológica Mex 62(1):123–140
12. Baysal EL (2019) Personal ornaments in prehistory: an exploration of body augmentation from the palaeolithic to the early bronze age. Oxbow Books, Oxford
13. Berlim LS, Bezerra AG, Pazin WM, Ramin TS, Schreiner WH, Ito AS (2018) Photophysical properties of flavonoids extracted from syngonanthus nitens, the golden grass. J Lumin 194:394–400. https://doi.org/10.1016/j.jlumin.2017.10.040
14. Berlim LS, Gonçalves HA, de Oliveira VS, Mattoso N, Prudente AS, Bezerra AG, Schreiner WH (2014) Syngonanthus nitens: why it looks like spun gold. Ind Crops Prod 52:597–602. https://doi.org/10.1016/J.INDCROP.2013.11.030
15. Brasil (2010) Lei nº 12.305 de 2 de agosto de 2010. Institui a Política Nacional de Resíduos Sólidos; altera a Lei no 9.605, de 12 de fevereiro de 1998; e dá outras providências. [Law Nº. 12.305, of August 2, 2010, establishing the National Policy on Solid Waste, amending Law Nº. 9,605, of February 12, 1998 and other measures]. Brasília, DF
16. Brum IAS, Silva RA (2010) Sistemas de tingimento em gemas. In: Tecnologias para o setor de gemas, joias e mineração. IGEO/UFRGS, Porto Alegre, pp 205–217
17. Burden E (2012) Illustrated dictionary of architecture, 3rd edn. McGraw-Hill, New York
18. Cappellieri A, Tenuta L, Testa S (2020) Jewellery between product and experience: luxury in the twenty-first century. In: Gardetti MÁ, Coste-Manière I (eds) Sustainable luxury and craftsmanship. Springer, Singapore, pp 1–23
19. CDB—Convention on Biological Diversity (2021) Brazil—Main details. https://www.cbd.int/countries/profile/?country=br. Accessed 4 Nov 2021
20. Cidade MK (2012) Caracterização e padronização do processo de gravação a laser em ágata aplicado ao design de joias, p 172

21. Cidade MK (2017) Design e tecnologia para a joalheria: microtomografia da gravação a laser CO_2 em ágata e implicações para projetos com desenhos vetoriais, p 106
22. Cidade MK, Palombini FL, Duarte L da C, Paciornik S (2018) Investigation of the thermal microstructural effects of CO_2 laser engraving on agate via X-ray microtomography. Opt Laser Technol 104:56–64. https://doi.org/10.1016/j.optlastec.2018.02.002
23. Cidade MK, Palombini FL, Kindlein Júnior W (2015) Biônica como processo criativo: microestrutura do bambu como metáfora gráfica no design de joias contemporâneas. Rev Educ Gráfica 19(1):91–103
24. Cidade MK, Palombini FL, Lima NFF, Duarte L da C (2016) Método para determinação de parâmetros de gravação e corte a laser CO_2 com aplicação na joalheria contemporânea. Des e Tecnol 12:54–64. https://doi.org/10.23972/det2016iss12pp54-64
25. Cidade MK, Palombini FL, Palhano AP, Melchiors A (2021) Experimental study for the valorization of polymeric coffee capsules waste by mechanical recycling and application on contemporary jewelry design. In: Muthu SS (ed) Sustainable packaging. Springer-Nature, Singapore, pp 85–110
26. Deer WA, Howie RA, Zussman J (1981) Minerais constituintes de rochas : uma introdução. Fundação Calouste Gulbenkian, Lisboa
27. DeMouthe J (2006) Natural materials: sources. Architectural Press, Oxfo, Properties and Uses
28. Duarte L da C, Kindlein Júnior W, Tessmann CS, Santos PG dos (2009) Potencialidades do design aplicado a utilização de novos materiais gemológicos no Rio Grande do Sul. In: I Seminário sobre Design e Gemologia de Pedras, Gemas e Jóias do Rio Grande do Sul. UPF, Soledade (RS)
29. Eder M, Amini S, Fratzl P (2018) Biological composites—Complex structures for functional diversity. Science 362(6414):543–547. https://doi.org/10.1126/science.aat8297
30. Eichemberg MT, Scatena VL (2011) Handicrafts from Jalapão (TO), Brazil, and their relationship to plant anatomy. J Torrey Bot Soc 138(1):34–40. https://doi.org/10.3159/TORREY-D-10-00005.1
31. Evert RF, Eichhorn SE (2013) Raven biology of plants, 8th edn. W. H. Freeman, New York
32. FEPAM Fundação Estadual de Proteção Ambiental (2004) Resolução CONSEMA n° 085/2004, de 17 de dezembro de 2004. Porto Alegre
33. Flörke OW, Köhler-Herbertz B, Langer K, Tönges I (1982) Water in microcrystalline quartz of volcanic origin: agates. Contrib to Mineral Petrol 80(4):324–333. https://doi.org/10.1007/BF00378005
34. Frondel C (1962) The system of mineralogy of James Dwight Dana and Edward Salisbury Dana, Silica materials, 7th edn. Wiley, New York and London
35. Gibson LJ, Ashby MF, Harley BA (2010) Cellular materials in nature and medicine. Cambridge University Press, Cambridge, UK
36. Götze J, Schrön W, Möckel R, Heide K (2012) The role of fluids in the formation of agates. Chem Erde 72(3):283–286. https://doi.org/10.1016/j.chemer.2012.07.002
37. Graetsch H, Flörke OW, Miehe G (1985) The nature of water in chalcedony and opal-C from brazilian agate geodes. Phys Chem Miner 12(5):300–306. https://doi.org/10.1007/BF00310343
38. Hartmann LA, da Cunha Duarte L, Massonne H-J, Michelin C, Rosenstengel LM, Bergmann M, Theye T, Pertille J, Arena KR, Duarte SK, Pinto VM, Barboza EG, Rosa MLCC, Wildner W (2012) Sequential opening and filling of cavities forming vesicles, amygdales and giant amethyst geodes in lavas from the southern Paraná volcanic province, Brazil and Uruguay. Int Geol Rev 54(1):1–14. https://doi.org/10.1080/00206814.2010.496253
39. Hunter L (2020) Mohair, cashmere and other animal hair fibres. In: Kozłowski RM, Mackiewicz-Talarczyk M (eds) Handbook of natural fibres. Elsevier, pp 279–383
40. IBGE—Instituto Brasileiro de Geografia e Estatística (2020) PPM—Municipal livestock production. https://www.ibge.gov.br/en/statistics/economic/agriculture-forestry-and-fishing/17353-municipal-livestock-production.html. Accessed 11 Nov 2021
41. IBGM—Instituto Brasileiro de Gemas e Metais Preciosos (2019) O Setor em Grandes Números 2018 [The Industry in Large Numbers 2018]. São Paulo

42. Jiang H, Wu W, Wang J, Yang W, Gao Y, Duan Y, Ma G, Wu C, Shao J (2021) Mapping global value of terrestrial ecosystem services by countries. Ecosyst Serv 52:101361. https://doi.org/10.1016/j.ecoser.2021.101361

43. Jones M, Huynh T, Dekiwadia C, Daver F, John S (2017) Mycelium composites: a review of engineering characteristics and growth kinetics. J Bionanosci 11(4):241–257. https://doi.org/10.1166/jbns.2017.1440

44. Juchem PL (1999) Mineralogia, geologia e gênese dos depósitos de ametista da região do Alto Uruguai, Rio Grande do Sul, p 225

45. Juchem PL, Brum TMM, Ripoll VM (2010) O laboratório de gemologia da Universidade Federal do Rio Grande do Sul. In: Tecnologias para o setor de gemas, joias e mineração. IGEO/UFRGS, Porto Alegre, pp 133–147

46. Juchem PL, Strieder AJ, Hartmann LA, Brum TMM de, Pulz GM, Duarte L da C (2007) Geologia e mineralogia das gemas do Rio Grande do Sul. In: 50 Anos de Geologia. Comunicação e Identidade, Porto Alegre, pp 177–197

47. Karana E (2012) Characterization of "natural" and "high-quality" materials to improve perception of bio-plastics. J Clean Prod 37:316–325. https://doi.org/10.1016/j.jclepro.2012.07.034

48. Karana E, Pedgley O, Rognoli V (eds) (2014) Materials experience: fundamentals of materials and design. Butterworth-Heinemann, Oxford

49. Kartik A, Akhil D, Lakshmi D, Panchamoorthy Gopinath K, Arun J, Sivaramakrishnan R, Pugazhendhi A (2021) A critical review on production of biopolymers from algae biomass and their applications. Bioresour Technol 329:124868. https://doi.org/10.1016/j.biortech.2021.124868

50. Kellerman CF (1990) Ágata em Salto do Jacuí: ocorrência e extração - aspectos ambientais e sócio-econômicos. Sindipedras, São Paulo

51. Kerwan K, Coles S (2018) Natural materials in automotive design. In: Designing with natural materials. CRC Press, pp 165–179

52. Kindlein Júnior W, Guanabara AS (2005) Methodology for product design based on the study of bionics. Mater Des 26(2):149–155. https://doi.org/10.1016/j.matdes.2004.05.009

53. Klaschka U (2015) Naturally toxic: natural substances used in personal care products. Environ Sci Eur 27(1):1. https://doi.org/10.1186/s12302-014-0033-2

54. Korinek J, Ramdoo I (2017) Local content policies. In: Minerals-exporting countries: the case of Brazil. OECD Trade Policy Pap. https://doi.org/10.1787/4b9b2617-en

55. Kozłowski RM, Mackiewicz-Talarczyk M (eds) (2020) Handbook of natural fibres: types, properties and factors affecting breeding and cultivation, vol 1. Elsevier

56. Lefteri C (2014) Materials for DESIGN. Laurence King Publishing, London

57. Liese W, Köhl M (eds) (2015) Bamboo: the plant and its uses. Springer International Publishing, Cham, SZ

58. MAPA—Ministério da Agricultura Pecuária e Abastecimento (2016) Revisão do Estudo do Complexo do Agronegócio do Cavalo. Brasilia

59. Meyers MA, Chen P-Y (2014) Biological materials science: biological materials, bioinspired materials, and biomaterials. Cambridge University Press, Cambridge, UK

60. Michelin CRL (2014) Ágata do distrito mineiro de Salto do Jacuí (Rio Grande do Sul, Brasil) : uma caracterização com base em técnicas estratigráficas, petrográficas, geoquímicas e isotópicas, p 167

61. Michelin CRL, Duarte L da C, Juchem PL, Brum TMM de, Mizusaki AMP (2021) Depósitos de ágata e de opala no estado do Rio Grande do Sul. In: Jelinek AR, Sommer CA (eds) Contribuições à Geologia do Rio Grande do Sul e de Santa Catarina. Compasso Lugar-Cultura, Porto Alegre, pp 355–370

62. Miller J (2016) Jewel: a celebration of earth's treasures. Dorling Kindersley Ltd., London

63. MMA—Ministry of the Environment (2021) Biodiversidade [Biodiversity]. https://www.gov.br/mma/pt-br/assuntos/biodiversidade. Accessed 4 Nov 2021

64. MME—Ministry of Mines and Energy (2020) Mineral sector bulletin 2020. Brasilia

65. Mooney BP (2009) The second green revolution? Production of plant-based biodegradable plastics. Biochem J 418(2):219–232. https://doi.org/10.1042/BJ20081769
66. Moreno Biec CL (2020) Unwritten: the implicit luxury. In: Gardetti MÁ, Coste-Manière I (eds) Sustainable luxury and craftsmanship. Springer, Singapore, pp 45–59
67. OECD (2021) Trade in raw materials. https://www.oecd.org/trade/topics/trade-in-raw-materials/. Accessed 4 Nov 2021
68. Oliveira MNS de, Cruz SM, Sousa AM de, Moreira F da C, Tanaka MK (2014) Implications of the harvest time on Syngonanthus nitens (Bong.) Ruhland (Eriocaulaceae) management in the state of Minas Gerais. Rev Bras Bot 37(2):95–103. https://doi.org/10.1007/S40415-014-0049-2/FIGURES/4
69. Ouyang Z, Song C, Zheng H, Polasky S, Xiao Y, Bateman IJ, Liu J, Ruckelshaus M, Shi F, Xiao Y, Xu W, Zou Z, Daily GC (2020) Using gross ecosystem product (GEP) to value nature in decision making. Proc Natl Acad Sci 117(25):14593–14601. https://doi.org/10.1073/pnas.1911439117
70. Ouyang Z, Zhu C, Yang G, Xu W, Zheng H, Zhang Y, Xiao Y (2013) Gross ecosystem product: concept, accounting framework and case study. Acta Ecol Sin 33(21):6747–6761. https://doi.org/10.5846/stxb201310092428
71. Palombini FL, Cidade MK (2021) Possibilities for the recovery and valorization of single-use EPS packaging waste following its increasing generation during the COVID-19 pandemic: a case study in Brazil. In: Muthu SS (ed) Sustainable packaging. Springer-Nature, Singapore, pp 265–288
72. Palombini FL, Cidade MK, Magris DA, Ghedini JVS (2021) Práticas projetuais transdisciplinares entre design e biologia: metodologia prática para o ensino de biônica. Rev Educ Gráfica 25(2):245–257
73. Palombini FL, Cidade MK, Oliveira BF de, Mariath JE de A (2021b) From light microscopy to X-ray microtomography: observation technologies in transdisciplinary approaches for bionic design and botany. Cuad del Cent Estud en Diseño y Comun 149:61–74
74. Palombini FL, Lautert EL, Mariath JE de A, de Oliveira BF (2020) Combining numerical models and discretizing methods in the analysis of bamboo parenchyma using finite element analysis based on X-ray microtomography. Wood Sci Technol 54(1):161–186. https://doi.org/10.1007/s00226-019-01146-4
75. Patton WJ (1968) Materials in industry, 1st edn. Prentice-Hall, New Jersey
76. Pereira H (2007) Cork: biology, production and uses. Elsevier, Amsterdan
77. Pimentel AMH, de Souza JRM, Boligon AA, Moreira HLM, Rechsteiner SM da EF, Pimentel CA, Martins CF (2018) Association of morphometric measurements with morphologic scores of Criollo horses at Freio de Ouro: a path analysis. Rev Bras Zootec 47:20180013. https://doi.org/10.1590/RBZ4720180013
78. Piselli A, Simonato M, Del Curto B (2016) Holistic approach to materials selection in professional appliances industry. In: Proceedings of international design conference, design, pp 865–874
79. Rognoli V, Karana E (2014) Toward a new materials aesthetic based on imperfection and graceful aging. In: Karana E, Pedgley O, Rognoli V (eds) Materials experience: fundamentals of materials and design. Elsevier, pp 145–154
80. Schmidt IB, Figueiredo IB, Scariot A (2007) Ethnobotany and effects of harvesting on the population ecology of syngonanthus nitens (Bong.) Ruhland (Eriocaulaceae), a NTFP from Jalapão region, central Brazil. Econ Bot 61:73
81. Schmidt IB, Figueiredo IB, Ticktin T (2015) Sustainability of golden grass flower stalk harvesting in the Brazilian savanna. In: Shackleton CM, Pandey AK, Ticktin T (eds) Ecological sustainability for non-timber forest products: dynamics and case studies of harvesting. Routledge, London, pp 199–214
82. Schmidt IB, Sampaio MB, Figueiredo IB, Ticktin T (2011) Fogo e artesanato de capim-dourado no Jalapão—usos tradicionais e consequências ecológicas. Biodiversidade Bras - BioBrasil (2):67–85. https://doi.org/10.37002/BIOBRASIL.V
83. Schumann W (2009) Gemstones of the world. Sterling Publishing Company Inc., New York

84. Silva LCS, Kovaleski JL, Gaia S, Back L, Piekarski CM, de Francisco AC (2013) Geographical indications contributions for Brazilian agribusiness development. African J Agric Res 8(18):2080–2085. https://doi.org/10.5897/AJAR12.2188

85. Siqueira G, Abdillahi H, Bras J, Dufresne A (2010) High reinforcing capability cellulose nanocrystals extracted from syngonanthus nitens (Capim Dourado). Cellulose 17(2):289–298. https://doi.org/10.1007/s10570-009-9384-z

86. Song X, Zhou G, Jiang H, Yu S, Fu J, Li W, Wang W, Ma Z, Peng C (2011) Carbon sequestration by Chinese bamboo forests and their ecological benefits: assessment of potential, problems, and future challenges. Environ Rev 19(NA):418–428. https://doi.org/10.1139/a11-015

87. Strieder AJ, Heemann R (2006) Structural Constraints on Paraná Basalt Volcanism and their Implications on Agate Geode Mineralization (Salto do Jacuí, RS, Brazil). Pesqui em Geociências 33(1):37–50

88. The World Bank (2021) GDP (current US$)—Brazil. https://data.worldbank.org/indicator/NY.GDP.MKTP.CD?locations=BR. Accessed 4 Nov 2021

89. Tubino LCB (1998) Tratamento Industrial da Ágata em Bruto no Estado do Rio Grande do Sul, p 177

90. UNESCO (2021) Biodiversity in Brazil. https://en.unesco.org/fieldoffice/brasilia/expertise/biodiversity-brazil. Accessed 4 Nov 2021

91. Valli M, Russo HM, Bolzani V da S (2018) The potential contribution of the natural products from Brazilian biodiversity to bioeconomy. Acad Bras Cienc 90(1):763–778. https://doi.org/10.1590/0001-3765201820170653

92. van Hinte E 1997) Eternally yours : visions on product endurance. 010 Publishers, Rotterdam

93. Vezzoli CA, Manzini E (2008) Design for environmental sustainability. Springer, London

94. Walker S (2014) Designing sustainability: making radical changes in a material world, 1st edn. Routledge, London

95. Wegst UGK, Ashby MF (2004) The mechanical efficiency of natural materials. Philos Mag 84(21):2167–2186. https://doi.org/10.1080/14786430410001680935

96. Werner F, Richter K (2007) Wooden building products in comparative LCA. Int J Life Cycle Assess 12(7):470–479. https://doi.org/10.1065/lca2007.04.317

97. Zenz J (2005) Agates. Haltern, Bode Verlang, Germany

Pherodrone1.0: An Innovative Inflatable UAV's Concept, Inspired by *Zanonia Macrocarpa's Samara* Flying-Wing and to Insect's *Sensillae,* Designed for the Biological Control of Harmful Insects in PA (Precision Agriculture)

Massimo Lumini

Abstract Pheromones are chemicals used by living organisms for intraspecific communication. In animals, the ability to detect and discriminate pheromones in a complex chemical environment contributes substantially to the species' survival. Insects primarily use specific chemical messages to attract mates, alarm conspecifics, or mark paths to rich food sources. These volatile molecules are generally detected through *sensillae*, specialized sensory neurons of the olfactory system located on the *antennae* [1]. Static *pheromone traps* have long been used to catch harmful insects in agricultural management. The team of very young researchers from BionikonLab & FABNAT14 (14–18 age) has experimented with ideas in the context of current developments in Precision Agriculture (PA). From the flying wing of *Zanonia macrocarpa's samara* analysis, inspired by some SEM ultrastructures of insect details and other bionic and biomimetic topics, this chapter proposed a new inflatable UAV concept at the service of resource optimization and sustainability management in agriculture. The goal is to reduce pesticides and other poisons that kill pollinating insects such as bees, to combat the chemical resistance of harmful insects that affect precious crops for humans. Unusual is the Tensairity®Solutions' technological approach to designing the flying body of *Pherodrone1.0*-UAV.

Keywords Bionics · Biomimetics · Environmental sustainability · Inflatable design · Precision agriculture · Pheromones · Samara · Sensillae · Tensairity solutions · UAV · Zanonia macrocarpa

M. Lumini (✉)
BionikonLab&FABNAT14, IIS "G.Asproni", Iglesias, SU, Italy
e-mail: m.lumini57@gmail.com

© The Author(s), under exclusive license to Springer Nature Singapore Pte Ltd. 2022   225
F. L. Palombini and S. S. Muthu (eds.), *Bionics and Sustainable Design*, Environmental
Footprints and Eco-design of Products and Processes,
https://doi.org/10.1007/978-981-19-1812-4_9

1 Homus Agriculus—Domina Agricola

Thousands of years ago, humanity emerged from the Paleolithic phase in which the hunting and gathering of occasional plant products allowed the sustenance of the first nomadic communities, discovered and developed farming techniques, and above all, agricultural cultivation. Starting from around 8,000 B.C., what has been called the *Neolithic Agricultural Revolution* developed. Thus was born the so-called *Homo Agriculus*. Probably favored by a milder climate than in the last *Ice Age*, agriculture started from the previous harvest of wild grains and was imposed when men carefully observed the various plants' birth and growth mechanisms, understanding the seeds. Through cultivation experiments, the settled human communities began to select seeds usable for food purposes, and the animals were domesticated and reared in the first rural villages of the world. Many scholars hypothesize that the advent of the *Agricultural Revolution* was accomplished by women, as men were engaged in hunting activities. In most ancient social structures, women have always been involved in the offspring [2].

Through centuries of slow trial and error processes, female communities were involved in foods preparation and conservation, from the treatment of meat, the tanning of hides and their derivatives, and the domestic management of the camps. *Domina Agricola* has probably developed deeper experimentation and knowledge of the spontaneous products of their territory through their acute sense of observation. Female hands have expertly collected, preserved, and processed many wild herbs, berries, and fruits. The ancient *Domina Agricola* has been able to map over time the food, nutritional and phytotherapeutic qualities of the natural flora, opening up to the discovery of future agriculture. The women of the *Neolithic* civilizations had to be very strong and resistant on a par with their companions, in order to be able to bear the strains of everyday life. Some recent anthropological researches have studied, with 3D laser scanning systems, hundreds of female bones, tibiae, and forearms, of women of the *Neolithic* (5300–4600 BC), of the *Bronze Age* (3200–1450 BC), of the *Age of Iron* (850 BC–100 AD) up to the *Middle Ages* (800–850 AD) in Central Europe. These anthropometric data were interpolated with data from leg and arm bone analyzes of female athletics, soccer, and rowing champions. The result of this comparison allowed the anthropologist Alison Macintosh who conducted this research at the University of Cambridge to affirm that, while the bones of the lower limbs had not changed particularly over time, the bones of the arms of Neolithic women were 15–30% stronger than those of modern athletes. This research suggests a specialization in differentiated manual labor (transporting weights, working the earth, grinding cereals) compared to men and further makes use of the hypothesis that the agricultural revolution was an all-female process [3].

In any case, agriculture, whether it is a product of male or female civilization, was a grandiose global process that imposed itself with the first fields of wheat, barley and spelled rationally cultivated by men in the world starting from the vast area of the so-called *Fertile Crescent* which included Mesopotamia with the Tigris and Euphrates valleys, Egypt with the Nile valley and the Anatolian plateau. In

Southeast Asia, agriculture was born with millet in Indochina and rice in China; in South America, Mexico, and Peru, with corn crops. Starting from these agricultural fulcrums, the knowledge and refinement of selection techniques, hybridization, and cultivation of seeds beneficial for human nutrition and livestock, will become the heritage of all humanity, evolving into increasingly specialized and intensive forms up to our days. Before modern chemistry was born and developed through the work and experimental research of Lavoisier, in ancient times, albeit unwittingly, *Homo Agriculus* was able to treat crops with the help of chemical molecules drawn from nature. The experience of direct contact and careful observation of natural facts pointed out that certain plants could defend themselves better than others against the assault and infestation of insects and parasites. From this natural wisdom, the idea of protecting plants and crops was born and developed over time with the help of natural pesticides. These protochemicals were created by treating leaves, seeds, flowers, and roots through decoctions, macerations, fermentation, and more. These practices are evidenced by ancient writings so much so that the *Rig Veda*, an ancient Sanskrit text of over 4000 years ago, reports the use in India of the margousier or neem tree (*Azadirachta indica*) as an effective natural pesticide as well as Ayurvedic medicine [4].

The term *neem* derives from the *Sanskrit Nimba*, which means *bearer of good health*, and current science has confirmed that more than a hundred phytosanitary molecules are contained in the leaves and seeds of this plant and, in particular, azadirachtin, a powerful insecticide and repulsive for insects. This substance has been shown to have an endocrine disruption power towards some insects that are particularly dangerous for bees from contemporary studies. Despite having a carcinogenic potential in large quantities, this fact has not prevented its use by *Homo Agriculus* for millennia. In India, farmers use it to fertilize the fields due to the maceration of bark and leaves as feed for animals. They plant it near their home because they claim that the breeze, passing through its branches, refreshes itself more, bringing refreshment and health to homes, freeing them from insects, worms, bacteria, and viruses. In ancient Mediterranean culture, Homer speaks of sulfur as a fungicide and Pliny the use of arsenic. Even the fundamental discovery of the rotation of crops, which is lost in the mists of time, is a cultivation technique that involves the alternation on the same plot of land of different agricultural species such as wheat, sunflower, clover, rapeseed, to rebalance the biological, chemical and physical properties of the soil, avoiding the drying up and sterilization caused by intensive mono cultivation. In any case, human agriculture has always had to fight against many negative factors that have affected its progress. The climatic factors such as sudden frosts, storms, aridity or are caused by nature, such as the heavy and invasive presence of parasites, plants, or animals. For example, the periodic famines caused by locusts have been of biblical proportions, so devastating as to be recorded and narrated in ancient memories. Furthermore, in recent times, with the impressive invasions of locusts in Africa, the introduction in Italy of *Xylella fastidiosa*, a lethal bacterium from Central America and carried by *Philaenus spumaris*, which since 2010 has been decimating centenary olive tree crops, devastating the economy of Italian olive oil without a possible effective cure, animal parasites have shown the enormous devastating impact on human labors. The aspect of the fight against animal parasites, mainly insects, through the use of competitions between species, is a rational and scientific intervention by the man on the natural balance that arises from observing the behavior of the various animal species in the victim-predator context within the food cycles that act in the various ecosystems. It historically takes the name of the *biological fight*, and in parallel with the recent discovery and intensive use of pesticides and poisons, with all the consequences they entail on human health and ecosystems, it has an intense development in recent decades. It seems that Erasmus Darwin, grandfather of the more famous Charles, was the first to propose in 1800 a series of objective methods of organized fight against aphids using hoverflies larvae and against cabbage caterpillars using predatory birds. To make these first experiments a real scientific practice with promising results was between 1880 and 1890, it was C. V. Riley, one of the first to understand the potential of this type of biological intervention in agriculture, importing predators and pests of species harmful to American crops into the United States. For example, he demonstrates the effectiveness of using the ladybug Rodolia cardinalis to combat the cochineal Iceryaparmisi. Parallel to the advent of this type of natural pesticide intervention, synthetic chemistry was introduced to strengthen the natural defenses of plants. For centuries man has resorted only to the contribution of organic fertilizers such as manure, guano, fish residues, animal blood, etc.) through which to convey the main elements of fertilization: Nitrogen (N), Phosphorus (P), and Potassium (K), defined macroelements of primary importance. As for phosphorus, often present in soils in an insoluble form (tricalcium phosphate) and therefore not usable by plants, the idea of making it soluble through a defined acidification process, was brought back to the observations conducted by the German chemist Justus von Liebig (1803–1873) who in 1830 noticed that bones treated with acid constituted a good vegetable nutrient. His *law of the minimum* is taught, a principle of agronomy developed by Carl Sprengel, considered the founder of modern agrochemistry, in 1882 but popularized by Liebig. This law claims that the plants' growth is controlled not so much by the total amount of natural resources available as by the availability of the scarcer ones. To illustrate it, we use the so-called Liebig-barrel example. Consider a barrel with the staves at different heights and filled with water. If each plank represents a chemical component of the soil, the height of the water, which identifies the plant's growth, contained in the barrel will be limited by the lowest plank and not by the highest or by the average height [5].

Applying this law makes it possible to improve crop growth by enriching the soil with minimal nutrients. In the past time, many herbal medicines of natural origin were used in the crops. Thanks to the development of chemistry, they have been tested for their molecular effectiveness. For example, nicotine extracted from decoctions of tobacco leaves, rotenone from Asian legumes, or pyrethrum from a sort of chrysanthemum. The official birth of synthetic agricultural chemistry can be considered in 1939 when the Swiss chemist Paul Herman Müller discovered the insecticidal properties of dichlorodiphenyltrichlorethane. This molecule belongs to the organochloride family and is better known by the acronym DDT in the laboratories of the Geigy® company in Bale. Subsequently, worldwide, the products born from synthetic chemistry will become the primary tool in agriculture for the fight against diseases caused by fungi, parasites, and harmful insects. These chemicals have replaced ancient natural practices as they are cheaper, initially more effective, and easier to use. For decades, synthetic chemistry has helped farmers reduce fatigue, labor, and economic losses caused by disease, infestation by plants, and parasitic insects, contributing to an initial increase in agricultural yield. However, starting from the 50 to 70 s, the coin's reverse was not slow to strike. Farmers who manipulate and use these substances get sick. The residues of pesticides and insecticides pass into fruits, seeds, and agricultural products, undermining the consumer's health; ecosystems suffer heavy repercussions, such as the death of bees and pollinators due to pesticides. The DDT that accompanied historical post-war reclamation campaigns to fight the Anophele mosquito, carrier of malaria, and other infestations of the territories, it turns out that it produces lethal effects on human and animal health. It turns out that organochlorines are not biodegradable and accumulate dangerously in ecosystems, undermining them at the base. This insecticide has been banned since the 1970s all over the world. The same fate will happen to atrazine, a substance that accumulates in groundwater and is highly carcinogenic, so much so that the EU will ban its use starting from 2003. Currently, for example, a molecule that is creating controversy worldwide is *glyphosate*. This molecule is an aminophosphoric analog of glycine and is known as a total herbicide. It was introduced into agriculture in the 1970s by the multinational Monsanto under the trade name of Roundup®. It has had a significant diffusion because some genetically modified crops can resist it: distributing it on the fields eliminates all weeds except GMOs. The combination of pesticides and genetically modified organisms have produced a total chemical dependence for intensive crops. It can also devitalize the hypogeal conservation organs of weeds, such as rhizomes and fleshy taproots. Which is no other way could be chemically devitalized. *Glyphosate* is used extensively throughout the world, especially for the cultivation of cereals and forage. For this reason, traces of it are found in meat, milk, and its derivatives, in feeds created from corn, rapeseed, and soy for the crops of which it is used abundantly to treat the land from grassy weeds. Extensive studies show that it causes genetic damage and oxidative stress in bacterial and organic cultures. However, in humans, its carcinogenicity has not yet been certified with absolute certainty, *probable carcinogens* [6].

All over the world, agriculture generates enormous profits, and the multinationals linked to synthetic chemistry are interested in controlling and monopolizing the system of fighting weeds and parasites. Over time, however, public opinion has become aware of the serious problems that these methods of treating soils and crops generate in ecosystems and human health. An intense review and search for alternatives to this synthetic chemical dictatorship are therefore underway. One of the exciting aspects, as we have seen above, is, therefore, the birth and development of the so-called *Biological Fight* or *Biological Control Program*, which makes use of living organisms such as viruses, bacteria, fungi, worms, and insects (which are predators of those they infest and destroy crops) [7].

This general aspect of research and expansion of knowledge related to possible more sustainable biological alternatives for natural ecosystems and the protection of

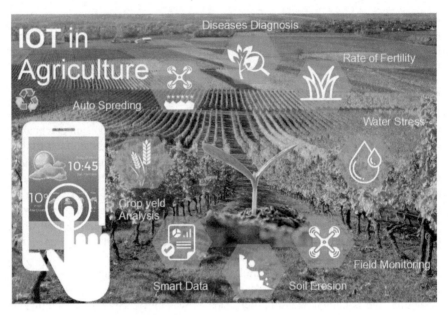

Fig. 1 IOT applications in smart agriculture or PA-precision agriculture

human health, combined with the potential of the development of digital technologies, has generated a vast area of research that goes under the name of *Agriculture Precision*. This is the specific area in which this research has moved (Fig. 1).

2   The Innovative Requests of the PA (Precision Agriculture)

Before addressing the topic, a premise is necessary to define the *BionikonLab & FABNAT14's* experience in the context of PA researches. In Sardinia, an important Italian Mediterranean island region, an interesting debate has been underway in recent years on the potential of digital technologies in the PA field. On the one hand, many companies and university research institutes are interested in seeking sustainable biological control solutions in the agricultural sector to counteract the harmful environmental effects created by the massive use of synthetic chemicals and GMOs. On the other hand, we are witnessing the birth of a research and development sector to test the potential benefits of introducing digital innovation in crop management assisted by typical PA systems. The participation in a Makeathon-Agrithon, between December 2020 and June 2021, dedicated to innovation in agriculture, inspired our interest in PA's issues. One of the possible themes proposed in this tech-tender concerned exactly the digital applications related to PA-Precision Agriculture, PF-Precision farming, Smart Agriculture or Agriculture 4.0 [8].

Thanks to the bionic and biomimetic *top-down* methodological research approach, during the contest, we have developed the first approaches to finding innovative solutions in the agricultural sector of pest control, mainly studying the feromone aspect in insects behavior. These aspects of the agricultural industry can make use of IoT technologies and, in general of those design aspects of PA [9].

Specifically, is possible to define the *top-down* approach as:

Defining a human needs or designing problem and looking to the ways other organisms or ecosystems solve this, termed here Design looking to biology (...) [10].

The steps of this design approach are not necessarily separate and unidirectional, as, during the research, the two flows can influence each other. Unlike design processes directly inspired and influenced by biology, a top-down design is initially generated by identifying a precise technological or design problem-setting. Following the drafting of a detailed checklist of required performance or design needs to which innovative solutions can be offered and a detailed analysis of the state of the art of the reference system, an attempt is made to find possible biological formal and/or functional analogies. The primary purpose of top-down research is to identify and extrapolate appropriate bio-inspired design principles, which can offer precise answers for optimizing materials, resources, and energy sources, in a logic of eco-sustainable approach to human technological future's challenges. This bio-scouting phase within the general research process pushes designers to enter the domains of biology, entomology, and natural sciences in general, making use of databases and scientific materials such as www.asknature.org di Biomimicry Institute. The problematic aspect of design-oriented bionic and biomimetic research is still that of access to scientific data useful for designers' hypothesis transfer processes. In 2004 Julian Vincent of the Center for Biomimetic and Natural Technologies Department of Mechanical Engineering at the University of Bath was developing a project to facilitate the technology transfer between Nature and Man [11].

The CBID-Center for Biologically Inspired Design at Georgia Tech-Atlanta - GA, is an institution that proposes itself as an interface for the development of bio-inspired design methodologies in the field of biomaterials, robotics, sensing, systems organization, and methodology & practice [12].

Once the biological sources of inspiration have been identified, we arrive at the definition of one or more problem-solving concepts through successive processes of abstraction and contextualization of the initial problem-setting. Progressive design developments through CAD/CAM processing, model development, material sampling, prototypes, and technological and functional tests follow one another until the engineered product is reached. The Biomimicry Institute has long developed and shared a methodology to guide the biomimetic process through its global educational network, specific flowcharts, various updated educational toolboxes, many online resources, and open-source materials to support the design. PA/PF and Agriculture 4.0 are modern farming management concept that uses digital techniques and the possibilities related to the IoT to monitor and optimize the different agricultural production processes. It is referred to as *precision* because, thanks to the technological tools and digital processing processes used and deployed, it is possible to carry out a range of monitoring and performance interventions calibrated in the suitable space and at the right time. The possibilities offered by digital applications make it possible to respond in a timely, optimized, customized manner to the specific demands of individual crops and individual areas of land with an outstanding level of precision. The principal advantages of PA/PF are:

- For farming enterprise, which can optimize efforts and resources, reduce consumption and waste, and boost land productivity. The work also becomes more profitable for farmers and contractors as the processes are managed more quickly and effectively, leading to a drop in hourly costs and reducing human resource fatigue;
- For the environment, given that there is a reduction in waste, fertilizers and herbicides, emissions, and soil compaction thanks to the more rational use of resources [13].

Thanks to the PA/PF methods, world agricultural activities enter a new 4.0 era, characterized by introducing and developing management and processing strategies. A considerable and complex amount of data is collected, stored, processed, analyzed, and combined with others information to guide business decisions according to the variability of space-time working conditions. The direct purposes are to improve and optimize the economic and functional efficiency in using energy, water, human and animal workforce, crop productivity, quality, profitability, and sustainability of world agricultural production. The remarkable development of this application sector of digital technologies and computerization was possible with the advent of GPS and GNSS technologies [14]. GPS-Global Positioning System is a satellite-based radio navigation system owned by the United States government and operated by the United States Space Force. GPS is one of the GNSS- Global Navigation Satellite Systems that provides geolocation and time information to a GPS receiver anywhere on or near the Earth. GPS operates independently of any telephonic or internet reception. The GPS provides critical positioning capability to military, civil, and commercial users around the world. Thanks to this system, the farmer or researcher can locate their precise position and that of their soils and land for cultivation and study to create smart maps in which to map a series of sensitive data such as e.g., crop yield, terrain features/topography, organic matter content, presence and quantity of water, moisture levels, nitrogen levels, pH, EC, Mg, K, and others. Through sensors and multispectral imagery, it is then possible, for example, to monitor the state of water in cultivated vegetation by mapping the levels of chlorophyll present in the foliage. An aspect significantly under development is the introduction of specific aircraft such as remotely guided drones and equipped with unique multispectral cameras or RGB cameras used for mapping land and crops to assess their state of health. The drones are then used to spread water, fertilizers, pesticides, and growth regulators through variable-rate applications (*source Wiki*).

From our point of view, once we have identified the design application sector, we found it particularly interesting to orient biomimetic research in the direction of using IoT technologies, sensors, and drones with particular attention to the problem of combating parasites. Inspired by the analysis of the flight performance of the *Zanonia*

Macrocarpa seed and by the *sensillae* of insects, during the research process, we tried to explore possible alternatives of *softdronics*, moving away from some technological and functional aspects of the typical design of the products that constitute the current state of the art of the PA-drones technological market. The idea was also suggested to us by the exciting developments in the evolution of robotics. In recent years, through *soft robotics*, new technological concepts have become more friendly, functionally flexible, and sustainable in impact and performance towards man and the environment.

3   The Flight

In 1640 the Italian scientist Evangelista Torricelli, during his experiments, discovered that the air exerted pressure on a slender little column of glass with mercury: *We don't live on top of the earth,* claimed the physicist, *but on the bottom of an immense ocean of air.* In fact, despite its apparent lightness, if multiplied by the height of the earth's atmosphere, the air is a physical substance with considerable weight. On average, the mass of an invisible high air column weighing about 3000 kg gravitates on the shoulders of a human being. This immense weight force is not perceived by our body, as it is opposed by the internal pneumatic pressure system, the blood, and the lungs, which try to balance the atmospheric and gravitational load to allow our survival to the conditions dictated by our planet. Air is a mixture of different gases (oxygen, carbon dioxide, and nitrogen) and therefore consists of a mass of molecules that are in constant motion. Moving molecules create air pressure. All things that fly need air. Both living organisms, which during the millions of years of biological evolution, have transformed their anatomy and physiology to lighten themselves and overcome the attraction of the earth, and the inventions that have led man to conquer the aerial domains, to be able to fly, they had to deal with the physical and geometric constraints imposed by the rules of the weight of the earth. The force that allows a bird or a technological aircraft to lift off the ground, overcoming the force of earth's gravity, in addition to the muscular force or the thrust of a propeller or a reactor, is called lift. In summary, it is the thrust, perpendicular to the direction of motion, produced by the effect of the air flow that beats and splits as it flows around the wing of the aircraft. For this phenomenon, the geometry of the aerodynamic profile is of fundamental importance, which must allow the correct flow of the air current. Lift depends on the wing profile and increases with speed. This is the reason why birds and airplanes do not take off while stationary but must accelerate for a while until the lift becomes equal to or greater than the weight of the airplane. This flow of air that meets the wing is, in fact, divided and *forced* to pass part above and part below the wing itself: given the shape (profile) of the latter, the two air's *veils*, dividing, they move at different speeds (faster above, slower below). This condition means that on the upper part of the wing (back) the air pressure is lower than the lower one (belly): the resulting force, therefore, creates a *sucking effect* upwards, which—exceeds the intensity of gravity—allows the aircraft to sustain itself in flight (Fig. 2).

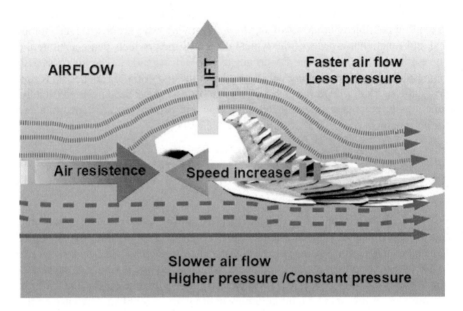

Fig. 2 The *sucking effect* in the lift

Basically, there is a kind of tug-of-war between two forces on a flying plane. It is the *weight* that brings it back to the ground, which must be able to counteract another aerodynamic support force, precisely the *lift*. Lift comes into play in every moment of the flight and depends on the relative speed and configuration of the wing: in particular on the surface and the inclination, which are modified through the movement of the mobile elements like *flaps*. Ailerons, on the other hand, are moving parts, fins placed at the ends of the wing on the trailing edge that make the aircraft turn left or right according to their movement through the flight controls placed in the cabin. As lift increases, the wing's curvature increases while maintaining constant airspeed. However, another essential factor is the density of the air, which decreases as the flight altitude increases. In fact, at the cruising altitude of airliners, around 10 thousand meters, the thinner air offers less resistance: on the one hand, this increases the speed of the aircraft, with the same thrust of the engines; on the other hand, it forces them to maintain a higher pace to maintain lift. If the speed is reduced excessively, the underlying thrust becomes insufficient and the aircraft descends in altitude. At a certain minimum detection speed that each aircraft has and/or due to too high an angle of attack, the airflow over the wing is interrupted, breaking away from the wing surface and creating the stall effect. The other essential parts are the aircraft and the tail rocker, the fin, and the tail rudder.

Due to their morphology and arrangement, the feathers in birds' wings are divided into primary and secondary. In millions of years of evolution, each of the feathers has specialized in performing certain functions. In addition to mechanically protecting, isolating, and maintaining the body's temperature, waterproofing, and identifying the sex and species of each bird, the feathers perform the primary task of determining the aerodynamic profile of the wing. The bird controls them by modifying their attitude to act on the fundamental parameters that determine the possibility of raising in flight, navigation that exploits updrafts, and maneuvers related to attack or defense or other functions such as courtship and others. The analysis of the flight of birds introduces elements of extreme complexity that go beyond the scope of this article; the study of the main parameters of aerodynamics, however, allowed the teamwork to master the basic knowledge necessary to master the first design hypotheses.

3.1 The UAV-Unmanned Aerial Vehicles Use in PA

A remotely piloted aircraft, commonly known as a *drone*, is a flying device characterized by the absence of the pilot on board. The aircraft's flying is guaranteed by a digital technology that provides a computer placed on board or via remote control systems of a navigator or pilot, positioned to the ground or in other positions, including in-flight positions. The possible range depends on the power and sophistication of the digital control system and can vary from a few hundred meters to hundreds if not thousands of kilometers in the most advanced military and civil models. The drone technology was developed in the military and war fields, but in recent years, the development of systems for civil applications such as [15]:

- Territorial monitoring for the prevention and emergency intervention in emergency fires or environmental disasters;
- Territorial monitoring and search for missing persons in situations of environmental disasters;
- Industrial and public utility safety use;
- Video surveillance of oil pipelines and industrial equipment for remote
- sensing and research purposes;
- Archaeological missions and urban and architectural documentation;
- Cinematographic and documentary video shooting;
- In general for dull, dirty, and dangerous missions, with competitive human and economic costs;
- Interventions in the Precision Agriculture sector.

The current exponential development of UAV technology has brought the design and construction of vehicles to such a state of the art that the market offers highly sophisticated flying models at an affordable cost. Drones, particularly the type of modern quadcopters, are offered to the public starting from toy models costing a few hundred euros. However, the first attempts to create a radio-controlled flying object can be traced back to some experiments conducted towards the end of the nineteenth century, precisely in 1849 when the Austrian army attacked the Italian city of Venice through the first actual aerial bombardment in history. The Austrians, led by Colonel Benno Uchatius, tried to use a chain of a dozen hot-air balloons, a real flying spider's web, loaded with explosives, launched from the Vulcano ship off the

Venetian lagoon. It so happened that some of these aircraft carried out the mission by unloading explosives on the city, but due to the wind, many of them ended up hitting the Austrian attack lines placed on the ground. The first flying quadcopter in history, the ancestor of modern APRs, called *Gyrolpane*, was built in 1907 by brothers Jacques and Louis Bréguet. It was a hefty and bulky model that flew just sixty centimeters off the ground, but it possessed the concept that led to the idea of the current four-prop drone. In 1917, during the First World War, the Ruston & Co. company designed the *Aerial Target*, an aircraft controlled through remote radio-control techniques developed by Nikola Tesla. Despite promising experimental demonstrations, it was designed to be a flying bomb, found no practical application for war purposes. Thanks to the development of innovative uses of the gyroscope by the American inventor Elmer Sperry, who in 1911 installed it for the first time aboard the American battleship Delaware, the use of the gyroscope was extended to guide torpedoes to gyro pilots for the government of ships and to stabilize aircraft. In 1916, the Hewitt-Sperry automatic airplane, known as the flying bomb, was the first example of a gyroscope-controlled aircraft. The first drones were produced on a large scale during World War II thanks to the ideas of Reginald Denny, who was a famous American film actor of the time, and the inventor Walter Righter, who developed the so-called *OQ2-Radioplane* drone. The demonstration of the potential of radio-controlled flight that Denny did with the prototype of one of his *Dennyplanes,* amazed the American military leaders so much that he was able to persuade them to grant the funds for a large-scale production. Later, during the Cold War and the Vietnam War, technological development made it possible to reach a high technological level, bringing to the market solutions of ever-smaller aircraft and with characteristics of control of their flight, such as to be able to be used in a different range of operational scenarios. In 1943 the Nazi army created the *FrtitzX 1400*, the forerunner of modern remotely guided anti-ship missiles. Throughout the 1960s, different radio-controlled aircraft models spread thanks to the development of electronic transistor technology, making it possible to produce miniaturized and low-cost components. Also, in Italy, starting from the Second World War, and especially during the Cold War, various models of APR were invented, such as the *CL-89* or *AN USD 51*, produced by Canadair and, more recently, other types of aircraft equipped with on-board camera with a range of about 120 km. The popularity of radio-controlled aircraft exploded in America and Italy, where the Olivetti industry of Ivrea, in collaboration with Telecom, developed some control cards still contained in most crewless aircraft. The Federal Aviation Administration (FAA) from 2006 will issue the first permits for the marketing of drones, opening up to the development and diffusion of these aircraft for various civil applications from aerial photography to construction and control of industrial structures. In 2019, the French company Parrot launched the *Parrot AR Drone* model, which was controlled via WIFI technology simply from a smartphone. More than half a million units of this model that have changed the rules of the sector market, awarded CES Innovations 2010, are sold worldwide. The first-person view (FPV) technology is then developed, which allows the construction of drones that allow you to view the drone's point of view in real-time. This technological expansion that allows drones to easily control and spy on vast environmental areas subtly and remotely produces a strong media impact. Numerous controversies, debates, reflections, and legislative promulgations develop to regulate their use in public and private spaces. Institutional bodies such as ENAC will draw up increasingly stringent safety rules to regulate the flight and use field. Currently, drone technology is in continuous development both in technology and civil and industrial uses, including uses in Precision Agriculture, the application sector that affects this project [16].

To date, Amazon has obtained, for example, permission in America for deliveries through a system managed with specific drones. Technologies such as Intel RealSense allow aircraft to recognize and manage obstacles independently. However, drones are involved in developing military technologies for the development of lethal autonomous weapons systems. Vehicles of this type, such as the *Krugu* attack drone, developed by the Turkish company STM, can hit human targets in complete autonomy, excluding remote intervention. This type of AI development applied to drones generates disturbing scenarios for the social and ethical implications involved [17] In general, drones are divided into these categories (Fig. 3):

- **Fixed-wing**
- **Multirotor**
- **VTOL** (Vertical Take-Off and Landing) /STOL (Short Take-Off/Landing)*.

*(The VTOL/STOL drones are hybrid technologies that combine the fixed-wing design with that type of rotor, arranging a series of propellers arranged in various ways. This type was developed to optimize the distinct advantages of a fixed-wing to those of a multirotor).

From an aerodynamic point of view, fixed-wing drones work like standard aircraft, generating lift with their more or less rigid wings. Compared to multirotor models, they take advantage of updrafts and lift themselves; they do not consume large amounts of energy to stay in the air. As a result, they usually have a more extended flight range, up to two hours, which makes them suitable for use in activities that require coverage of vast land areas such as pipeline inspections or agricultural investigations (up to 1 km^2 with a single flight). Due to their physical flight characteristics, they are more stable in strong winds and can carry heavier weights using less power. The disadvantages of fixed-wing drones are related to their higher cost, size, and difficulty of transport and configuration on-site [18].

They are also more challenging to maneuver, and they cannot hover over a target, and use in complex or confined areas can be dangerous due to their poor ability to interact with navigation obstacles. Depending on their size, they may need a launch and landing runway. Multirotors are aircraft with flight characteristics similar to helicopters, characterized by a frame equipped with two or more rotors (up to 8) that move propellers. The drone model certainly most popular today is the four-helix, which produces the typical hum reminiscent of a bee or a hornet (in the Anglo-Saxon language, the term drone indicates the male of the bee). These drones are extremely unstable, and for this reason, they require a series of integrated systems to ensure static and dynamic stability. Each rotor produces a thrust but also a torque that is formed in its center of rotation. In the same way, it also produces a strong

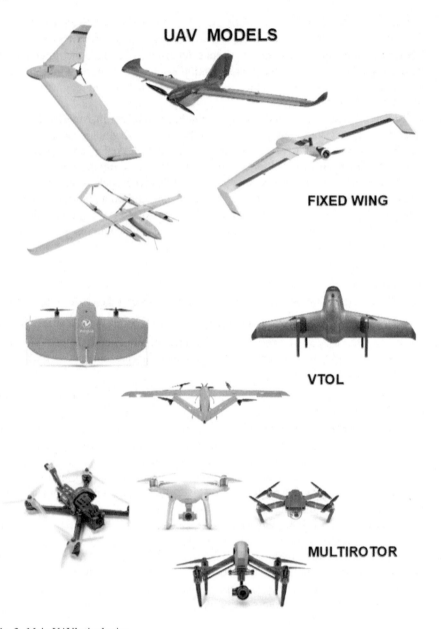

Fig. 3 Main UAV's tipologies

resistance generated on the entire structure of the multirotor, the arms, and the various components exposed to the air flows from the propellers that turn into turbulence. The balance of a multirotor is obtained when the lift is equal to the weight, the thrust equal to the resistance, and no other moment or force must act on the flying machine. Two rotors will have a clockwise revolution in a four-helix medium, while the others will have an anti-clockwise revolution. This mechanism generates reaction torques on all four rotors, which, however, will be mutually canceled. This condition is complicated to achieve due to the numerous physicomechanical variables in action. Thanks to the provision of a Flight Controller (FC), the autopilot, through unique algorithms and gyroscopic-accelerometer sensors, manages the rotation speed of each rotor in a dynamic relationship between them and with the flight variables. Furthermore, the rotors in action generate a series of complex turbulences that act on all the physical and mechanical components of the drone itself, helping to create a condition of complex management of its attitude during navigation. A typical flight mission of a drone starts from the vertical elevation along the z-axis to reach the hovering state at a certain height, where the vehicle stands in equilibrium, at zero speed and constant altitude, without moving along the other axes. By analogy, in nature, there is a flight technique typical of birds of prey, kingfisher, and hummingbirds called Holy Spirit. With small wing movements, the animal can maintain a stall position at a point in space, even for many minutes. The primary technologies of flight found in birds and insects will be explored in the specific biomechanical section on the flight in nature. This condition is possible because the wing parts are held in place while the distal parts are rotated quickly. These flight dynamics allow the bird to remain motionless in the air, assuming a figure that recalls the Holy Spirit's typical representation in the dove's classic iconography with open wings. Briefly, the navigation dynamics controlled by a radio control of a four-rotor drone, concerning the system of spatial axes X–Y–Z, provides for the following states of movement (Pitch, Roll, Yaw, and Throttle):

- The pitch is the rotation around the transverse axis; by moving the pitch stick forward, the drone will lower the nose and raise the tail as it moves forward; the opposite will happen by moving the lever backward;
- The roll describes the rotation of the drone around the longitudinal x-axis. By moving the roll stick to the right, the drone will lower the corresponding side and vice-versa;
- The yaw describes the rotation around the vertical axis and occurs by moving the stitch to the right or left;
- With the throttle control, which corresponds to acceleration, the upward movement of the stick increases the revs of all the motors simultaneously, making the drone rise along the vertical axis; the descent is obtained with the reverse movement.

3.2 Main Components of a UAV

3.2.1 Frame

The frame is the supporting structure on which mounts the various components. The frame's shape varies according to the number of rotors which can range from two to six propellers or even more. The materials for the construction of the frames are wood, plastic, aluminum, carbon fiber.

The choice of construction material determines some critical factors of the aircraft:

- Characteristics of the flight
- Weight
- Range in flight
- Impact resistance
- Safety of use.

3.2.2 Drone Transmitter and Receiver, FPV (First Person View) Radio Control System

A drone is a flying object guided by a radio control system that is composed of two elements: the transmitter which is held in the hands of the pilot on the ground and the FPV receiver that is inside the UAV (Fig. 4). This technology allows sending the live image of the camera to the ground, thus allowing the pilot to fly the aircraft while looking at the screen as if he were flying for real. The transmitter reads the joystick inputs and sends them through the air to the receiver in near real-time. Once the receiver has these pieces of information, it passes it to the drone's flight controller, which causes it to move accordingly. A radio can generally have four separate channels for each direction on the stick along with some extras for any auxiliary switches it might have.

3.2.3 VTX Video Transmitter

Some models integrate an integrated all-in-one video transmitter.

3.2.4 Power Distribution Board (PBD)

PBD is a printed circuit board that is used to distribute the power from the flight battery to all different components of the multirotor.

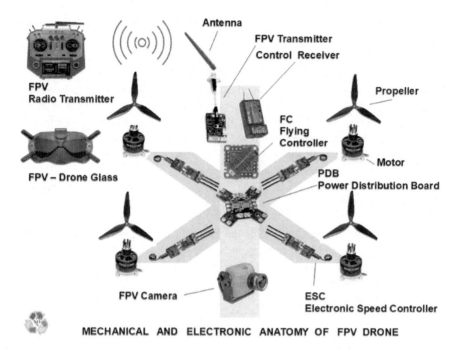

Fig. 4 Mechanical and electronic components of an FPV drone

3.2.5 Flight Controller FC

The FC is the drone's brain that receives remote commands from the pilot and, thanks to the built-in sensors, manages the various rotors allowing them to calibrate their operation. The most common models of microprocessors used are the STM32F1, F2, F3, F4, and F7 chips. Currently, the available processors are:

F1 72 MHz 2 UART.
F2 72 MHz 3–5 UART.
F3 168 MHz 3–6 UART.
F4 216 MHz 7 + UART.

Each UART (Universal Asynchronous Receiver/Transmitter) has two TX pins for transmitting data and RX for receiving.

3.2.6 Gyroscope (IMU)

The gyroscope detects the angular velocity or the speed with which a quadcopter rotates in the roll, pitch, and yaw axis. This is the only sensor needed to fly the drone in ACRO mode. The control sticks are only for making acceptable corrections to make the flight more precise.

3.2.7   SPI and i2c

These are the types of BUS or communication protocol between the IMU (Inertial Measurement Unit) sensor and the processor. An IMU unit is a card that determines if an object is moving, translating, rotating, or tilting the horizon. An IMU system for navigation and control is an avionics system, which has 9 degrees of freedom (9DOF) in the perception of motion; with these units, the angular velocities, the accelerations, and the magnetic field are measured on the three orthogonal axes XYZ through accelerometers and rate-gyro sensors of the solid-state type.

3.2.8   Accelerometer

The sensor detects the acceleration of the drone in the roll, pitch, and yaw axis. The roll indicates the right-left movement concerning the longitudinal axis of reference passing through the center of gravity of a body. The pitch indicates the rotational movement for the transverse forward–backward axis, while yaw indicates the rotational movement to the vertical z-axis.

3.2.9   GPS and Barometer

A GPS antenna allows the drone to identify its position and autonomously return to a set base point. The GPS data is displayed on display (OSD One Screen Display), while the barometer provides altitude data to allow the HR to maintain a constant altitude during the flight.

3.2.10   Current Sensor

The current sensor allows the flight controller to calculate and display the instantaneous current draw and battery consumption on the OSD.

3.2.11   Obstacle Avoidance Sensors

Almost all drones have an *obstacle avoidance system* enabled by technologies such as infrared sensors, stereo vision sensors, ultrasonic sensors, and GPS. These sensors work together to make sure the drone detects and avoids obstacles in the flight path to prevent crashes.

3.2.12   OSD (On Screen Display)

Allows telemetry data to be displayed on the rider's LCD screen and other operational controls (PID, speed, VTX channel change).

3.2.13   Black-Box

The black-box allows the recording of all flight data on an SD card.

3.2.14   Engines and Electronic Speed Control (ESC)

Drones can have several engines; generally, the electric motors mounted are brushless, without brushes, which do not require sliding electrical contacts on the motor shaft. This avoids lower mechanical resistance, absence of sparks, and lower weight. Thanks to Electronic Speed Control, the brushless motors are connected to the Fly Controller. The ESC receives the command from the FC and transfers it to the engines and propellers, allowing the correct synchronism and balancing of the flight parameters necessary for the aircraft's stability in flight.

3.2.15   Propeller

They are permanently mounted on the motors and rotate according to the power delivered through the remote control. The materials used are plastics or carbon fibers, which undergo minor deformation during flight.

3.2.16   Propeller Guard

They are super-light plastic fairings mounted on the drone that can protect the drone's body and, above all, the propellers from accidental impacts.

Exciting models of current drones have been studied starting from Gimball, a prototype presented by the Swiss team Flyability, which, by caging the drone in a polyhedral grid-shell made with carbon fiber elements, won the UAE Drones for Good Award held in 2015. in Dubai. This competition is dedicated to using drone technology in the social sphere rather than in the war.

3.3   Living Flying: Birds, Fishes, & Other Aeromobiles Creatures

This section introduces the general elements relating to the main characteristics of the evolution of flight in nature. There are about 11,000 known species of birds that are the flying animals par excellence, even if for efficiency, originality, and differentiation of technologies adopted, the world of insects offers extraordinary applications of aerodynamic engineering, probably less known as they are less conspicuous. Over millions of years of biological evolution, birds have experienced extraordinary morphological, anatomical, and functional adaptations of their whole bodies in search of more extraordinary lightness, power, control, and optimization of their ability to overcome the constraints of the earth's gravity and to be able to soar in the air and conquer the sky. This process seems to have started with *pterosaurs* (from the Greek lizards with wings), an order of flying reptiles that lived throughout the *Mesozoic*, from the upper *Triassic* to the end of the *Cretaceous*, about 230-65 million years ago. *Pterosaurs* were the first vertebrates known to have evolved and adapted to flapping flight. Today about 130 species of *pterosaurs* are known. The first fossil of a *pterosaur* was described by the Italian naturalist Cosimo Alessandro Cellini who in 1784 interpreted the specimen as marine, confusing the long wings like fins. In 1801, Georges Cuvier proclaimed it as a flying creature, and in 1809 he coined the term *ptero-dactyl* (winged finger) concerning a remnant recovered in Germany. The study and classification of *pterosaurs* and *pterodactyls* are still incomplete today, and their evolutionary genealogy is a work in progress that is updated by continuous findings. *Pterosaurs* include the largest flying animals to ever live on earth. The specimens of *Quetzalcoatlus northropi* and *Arambourgiania philadelphiae* are estimated to be around 5.5 m tall (like a current giraffe), 6 m long, and about 250 kg heavy, and have an estimated wingspan of around 11/13 m. The flight dynamics of these winged beings are not yet evident. Over time, various biomechanical hypotheses have been proposed that presuppose relationships with the state of the earth's atmosphere at the time. In the past, the flight capacity of these gigantic animals was attributed to the dense and warm atmosphere of the upper *Cretaceous*. Today's theories generally agree that the larger models could keep in flight thanks to the air pockets present in the wing membranes and the powerful muscles of the arms that on the ground allowed the pterosaurs, up to 15 m high) both to take off quickly and to walk in a quadrupedal position like modern bats. Two thousand nine studies show that these beings possessed an air sac system and a precisely controlled skeletal respiration pump that supported a pulmonary ventilation system similar to that of modern birds. Subcutaneous aeration systems found in some pterodactyloids support the idea of lightning systems for animals in flight, searching for updrafts to sail even in gliding flight. Their bones were hollow as in modern birds, and their wings were membranous, similar to those of modern bats. However, it was not a superficial, leathery epidermal tissue but a real organ evolved for flight, consisting of a highly specialized structure. The external wings from the tip to the elbow were strengthened by fibers called actinofibrils of keratinous origin, which included three layers overlapping each other in the directions of the fibers to create a crossed canvas structure that supported the wing. The wing membrane also featured a thin muscular structure with fibrous parts, elastic structures, and a complex blood circulatory system. X-ray analysis of their brain cavities revealed a massive presence of floccules, regions of the brain delegated to the integration of signals from the joints, muscles, skin, and organs of balance. The flocs present in the skull of *pterosaurs* occupy about 7.5% of the entire brain mass, a more significant amount than in any other vertebrate. Modern birds have brain flocs, unusually large but which occupy a percentage of 1–2% of the entire total mass of the brain. It is estimated that in flight, the pterosaurs, to which order also much smaller individuals belonged, could reach the speed of over 100 km/h for a few minutes and glide at a cruising speed of around 90 km/h. *Pterosaurs* and *pterodactyls* belong to a clade different from dinosaurs and should not be confused with *Archeopteryx*, a theropod of the saurischi group, which marked the transition between reptiles and birds, and in which characteristic features of modern birds began to appear. The evolution of modern birds is thought to have begun in the *Jurassic*, with the first *theropod* beings derived from dinosaurs such as the Archeopteryx mentioned above [19].

The evolutionary line of the dinosaurs that turn into birds has been recently drawn. It foresees the passage from the *neoteropods* (220 million years ago) through the *tetanurs* (about 200 million) passing through *celosaurs* (175 million), *paravians* (165 million) up to *Archeopteryx* (150 million). This rapid passage, four times faster than the times of other evolutionary branches, has led to a drastic reduction in the size and weight of the animals, which, starting from 160 kg of the tetanurs, reached 800 g *Archeopteryx* in about thirty million years. By shrinking, the dinosaurs/birds have further specialized the feathered body; they have acquired a shorter snout, a proportionately larger brain, large eyes, small teeth, a more agile and faster body, a stiffer and shorter tail. The environment had changed and prompted the change, and from animals of extensive grasslands, birds became a species more suited to the ecosystem of the undergrowth. They began to climb trees to escape their enemies and to find new sources of food. The trees most likely began to take their first flights, inventing the actual beaten flight to become the masters of the air finally. The shrinking in size probably allowed the birds to survive the tremendous *Cretaceous* mass extinction that killed most of the dinosaurs of yore. Modern birds then evolved from *therapods*, a line of feathered dinosaurs. For a long time, paleontologists have been wondering about the evolutionary nature of feathers as it now appears confirmed, by careful investigations and analyses on available fossils, that dinosaurs were covered in brightly colored feathers. To understand the usefulness of a feathered body, scholars have established that as short, fluffy feathers help modern birds stay warm, and brightly colored feathers can help birds attract a partner, feathers likely have evolved for different functional reasons and not just for the flight. Birds are homeotherms and therefore need to keep their body temperature constant as environmental conditions vary. Plumage works the same way as fur, protecting the body from the sun, cold, wind, and rain and protecting against injury. The plumage gives shape to the bird's profile, which is an essential element in flight. Only much later did evolution lead to the development of more potent and longer feathers on the chestnut bodywork that build a wing suitable for the flight typical of almost all modern birds. In order to fly, birds have had to modify many crucial aspects of their anatomy and physiology. They lightened their bones through trabeculae that create internal scaffolding that takes away weight and offers the skeleton greater strength and elasticity. They changed their respiratory system, skull, lower and upper joints. They are equipped with highly sophisticated vision systems and electronic navigation technologies that have led them to explore the whole planet's skies and undertake extraordinary routes around it, exploring all of its continents. In a certain sense, the admiration and envy for their ability to fly was the driving force that allowed humanity to venture into a territory forbidden to it: the air, the sky, and the excellent atmosphere. Leonardo da Vinci was one of the first and most attentive observers of the flight of birds, obsessed as he was with the desire to put the wings on man. In particular, his observations focused on birds of prey, especially the kite, which we can probably consider among the most sophisticated and elegant flying objects in the natural world [20].

The large birds of prey that lived in mountainous or desert regions, the harshness of the places, and the scarcity of food have required prolonged stays in flight at high altitudes to discover and capture possible losses. Between the various hunting phases, many hours could elapse, and remaining in flight for a long time represented the key to success for survival. These needs have led to the development of complex evolutionary expedients to allow birds to equip themselves with efficient winged and feathered systems capable of exploiting the updrafts for a gliding flight with low energy consumption. It was also vital to lift off the ground quickly and by one's means. This need has led to a progressive reduction in the size of all organisms suitable for flight. Their body evolved towards forms of high aerodynamic efficiency and flight techniques towards minimum energy expenditure. Gaining altitude without flapping was a discriminating factor in the selection process, but differences were created between seabirds and birds of prey. The former was able to take advantage of the winds and breezes sustained on the surface of the water and at altitude, while the latter, in order to climb, had to adapt to take advantage of the rarer and weaker climbs. The winning formula involved sailing by extracting energy from vertical currents to gain altitude and then using the energy thus accumulated most productively during the glide. To obtain an effective glide, the wings of the great planers have been lengthened, reducing their planer surface proportionally. It is necessary to go up with a slow circular motion to sail upward, obtainable only with large wing surfaces or low weight. The variation of the wing geometry to have a large surface in the take-off or soaring phases, then reducing it, and increasing the elongation, in gliding, has led their evolution to give them the ability to fold more or less behind the outer part of the elbow in order to overlap a good portion of the flight feathers. As for elongation, the requirements for take-off and upward flight prevailed, so the wing remained less slender than marine planers. However, at low flight speeds, this would have produced a dangerous multiplication of eddies at the wingtips that would have added to the friction with the air. Natural technology has developed a particular sectional wing profile in birds of prey: the front edge is curved and thinned. On the lower face, it is immediately followed by a sort of step formed by the bones of the arm and the extensor muscle that makes the flow of air uniform and controllable. This source of turbulence created at a specific point prevents the development of dangerous random turbulence and cancels the harmful effects of friction. To counteract the eddies that arise at the wingtips, one-third of the wing, on the outside, is made up of separate feathers; the front three or four have a very elastic rachis and, in flight, an energetic flow acts on their extremity which, when the wing is stretched, inflects them upwards in a pronounced way, managing to decrease the creation of vortices and therefore of friction. Thus optimized in design, the eagle, the hawk, and their similars can travel over fifteen meters per meter of altitude lost, which brings them not very far from the efficiency of large seabirds, which, however, cannot boast equal agility and take-offs powerful. Perfect flight control is possible thanks to the width of the wing and the small displacements of the load-bearing canopy concerning the body's center of gravity to balance and direct the body with extreme precision and effectiveness. The function of the tail is to implement minimal corrective adjustments, and for almost the entire duration of the flight, it is kept well together and stretched back to increase penetration into the air. To produce highly rapid flight, the wings fold back, like those of swallows or swifts; when instead the flight speed is not very important, but the need for a long flight with a lot of thrusts and little resistance prevails, long and narrow wings are needed, like those of albatrosses or gulls. The flapping of the wings produces the forward thrust, and it is the means that allows lifting and support in the air: the first function is performed by the distal portion of the wing (the "hand" of the bird), the second by the proximal portion wing (the bird's arm). Behind the rear edge of the wings of a flying bird, turbulence of the air is created, swirling upwards. Some birds benefit from this by flying in formation, flapping their wings in unison, as do pelicans, for example. In-flight formations, ducks, and wild geese in their very long and tiring migratory flights use a V-shaped formation. This formation allows each bird to keep one tip of its wings in the swirling wake caused by the bird's flight in front. In this way, they can exploit the vortex motions to sustain themselves by decreasing fatigue and energy expenditure. Since the pilot bird, at the head of the formation, cannot take advantage of this support, from time to time, it exchanges its position with others, assuming a less tiring one. The behaviors and technical performance of bird flight are extraordinary. The peregrine falcon (*Falco peregrinus*), which is perhaps the fastest flying bird, has been accurately timed, and it has been found to reach 280 km/h. A mountain passerine, black-headed tit (*Parus atricapillus*), if startled in flight, can change direction in 0.3 s [21].

The alpine croaking (*Coracia graculus*) has been reported at more than 8000 m high, near Mount Everest. A sea swallow (*Sterna fuscata*) remains in the sea for some months, coming ashore only during the breeding periods; it is known that this bird cannot rest on the water for long, to avoid damaging the feathers. It must therefore be concluded that during the period of life at sea it flies almost continuously. In one year it can cover a distance of 40,000 km. A swift that has to provide for its nourishment and two young ones, covers an average distance of 1000 km per day, flying incessantly even for 14 h a day and reaching 145 km/h. On the hottest days, the swifts can take to great heights and, after spending the whole day in flight, they spend the night at high altitude, gliding in the air with hovering flight, probably suspending many brain functions as to fall into a sort of sleep. Recently, a group of researchers from the Royal Veterinary College of London developed an original experiment to deepen their knowledge on the flight of birds of prey. Through particular video recording techniques through a soap bubble chamber, it was possible to observe with extreme precision the use of the tail to get rid of a part of the resistance in the air and to be able to glide more easily [22].

A trick that these birds employ to reach prey quickly but which, from a technological point of view, could provide a helpful idea to revolutionize the design—to begin with—of small airplanes and drones. Compared with these performances, even the most efficient products of human technology fail in some respects to compete in terms of efficiency, energy-saving, and resource optimization. For this series of reasons, the bionic and biomimetic study of the flight of birds and all flying organisms covers an important area of development of bionic and biomimetic studies [23].

In its unstoppable evolutionary process, studded with extraordinary biological designs, nature has also allowed the inhabitants of the waters the opportunity to experience flight. *Potanichthys xingyiensi* is now the oldest flying fish in the world lived 240 million years ago, and were able to leap by flapping two pairs of fins like wings and using the tail as a propeller. *Exocoetus volitans*, also called flying fish and *Exocoetus obtusirostris*, swallow-fish, are current modern flying-fish; in order to take flight, they must first reach a speed of 60 km/h underwater, then take off. Their flight can reach 50 km/h before breaking the planar surface over distances of 200–400 m. In these fish, the pectoral fins have turned into real wings that allow the fish to glide, like gliders, on the surface of the sea aided by a very tapered and aerodynamic body structure. Thanks to the frenetic movement of the tail that beats up to 70 times per second, the fish manages to make the leap, get out of the water, take flight, and reach heights of a few meters. The winged fins open, and as if by magic, the fish begins to circle over the water and exploit the strength of the air currents to get carried away and increase the distance to travel. Of the genus *Exocoetidae* there are about 70 species grouped in 9 species; some have individuals with large pelvic fins and are commonly called *four-winged flying fish*; in practice, a quadcopter *bio-drone* model. From an evolutionary point of view, Ichthyo-paleontologists argue that the ability to fly of these fish may be a technique born and developed as a solution to avoid the attacks of their predators. Coming back to the mainland, we meet a whole series of animals more or less adapted to face the flight. Bats are the only mammals capable of natural flight, with entirely different biotechnological solutions from birds. They belong to the order of the *Chiroptera*, and currently, there are about 1380 species; they are the second order of mammals by number and represent a fifth of all mammal species on Earth. The dimensions of the bats are very variable: the most notable species is that of the flying foxes, which can reach a wingspan of about 2 m and a body weight of 1.5 kg; the smallest species is instead the bumblebee bat that with its weight of only 2 g it is the smallest mammal in the world. Their numerous presence of species, the considerable morphological differences within them, and their widespread presence in all continents and environmental typologies (except for the most remote poles and islands) confirm that the invention of the bats is, among mammals, a remarkable evolutionary success. The keys to this success certainly include the evolution of active flight, a unique case among mammals, and the ability to orient themselves through echolocation. Reconstructing the evolutionary history of bats is not easy since there are few intact finds available due to the difficulty of fossilization of these often small animals, thus making it difficult to compare the past species with the current ones. The oldest bats currently date back to at least 50 million years ago (*Eocene*) and strongly resemble the current forms. Recently, the rest of a rather bizarre creature was discovered, a hybrid between dinosaur and bat, named *Ambopteryx longibrachium* [24]

The discovery of *Ambopteryx* shows that *Yi* was not alone and raises the possibility that *Scansoriopterygidae*, in general, may have had bat-like wings in addition to feathers. In the universe of terrestrial living beings, many sporadic animals use gliding flight to move above all from the branches of trees, such as flying squirrels, flying lizards, and even snakes. However, the most extraordinary family that started millions of years ago to build a career as a pilot is undoubtedly that of insects.

3.4 Insects

While the flight of birds has attracted curiosity and the spirit of human imitation since ancient times, the most secret and mysterious of other organisms such as insects and moths, we only begin to know in detail in the last few decades. This evolution in knowledge is possible above all thanks to the development of slow-motion video shooting technologies and other sophisticated digital microscopy and imaging techniques. Insects a few millimeters long have revealed extraordinary flight capabilities with technological performance, if compared to our technological parameters, at the possible limit. At the Max-Planck Institute in Martinsried, it was discovered that the gene, called Salm, has been preserved over 280 million years of evolution and controls the development of individual muscle fibers, different from all the others present in the insect body. They are highly specialized muscle fibers similar to the human heart, used in flight as "indirect" muscles. Thanks to them, insects can swing their wings at high frequencies, over 1,000 Hz, and are capable of developing a mechanical force of 80 Watts per kilogram. To understand the factors that determine the formation of these types of muscles during the development of the individual, the researchers, led by Frank Schnorrer, analyzed all the factors involved in the morphogenesis of the muscles of the fruit fly, Drosophila melanogaster, verifying the importance of the Salm gene in this process. Researchers hypothesize that this gene, capable of activating the mechanisms for the development of the special muscle fibers of flight, has some equivalent also in mammals, in the mechanisms of heart muscle development [25].

Much interest in robotics and avionics research centers are dedicating essential investments in the systematic study of the flight systems characteristic of insects. Studies have been developed on the ability of many organisms to fold and pack their wings through exciting Bio-origami techniques. Even NASA has produced a series of biomimetic departmental research to uncover the secrets of these natural behaviors. Control of birds and insects is an EU-funded initiative investigating how birds and insects use sensory information to control the beat and shapes of their wings to direct their flight. The flight dynamics and control of birds and insects (DCBIF) project investigated how birds and insects use vision to stabilize and control flight [26].

This ability trained birds of prey and virtual reality with a flight simulator for insects, which measured the tiny forces and torques exerted by the insects while in flight. The results showed that the insects responded to simulated rotations of their field of view by producing controlled pairs to help them stabilize and curve their flight. The team also found that by turning their heads, the insects could see a much faster range of motion than they could otherwise have done. Field experiments with trained eagles and falcons have shown that birds move in similar ways. After observing themselves as they move through the air, birds and insects must then use this information. Therefore, DCBIF studied how birds and insects control the flapping of their wings, particularly how they change the shape of their wings to control flight. The issue was investigated using high-speed cameras to measure the movement and deformation of eagles and a variety of insect wings. Computational techniques were also used to predict the aerodynamic forces involved. It was found that the deformation of the wing improved the aerodynamic lift produced by an insect by 70%. The internal dynamics of the insect flight engine have been studied using a new technique that exposes the insect to X-ray illumination while it is in flight. The insect was spun to allow to see the flight engine from all angles and push it to turn by varying the beat of its wings. By putting together X-ray images taken from different angles in the same wing-flapping phase, the researchers could 3D reconstruct the musculoskeletal movements that provide force and control wing flapping. The insights gained from DCBIF are now being applied to control the next generation of small unmanned devices capable of flying.

3.4.1   Dragonflies

The order of dragonflies, the *Odonata* that is with teeth, is one of the oldest. Insects such as *meganeura* (*Meganeura monyi*) and *Meganeuropsis permiana* of which we have very well preserved fossils, were gigantic prehistoric insects that lived in the *Carboniferous*, about 300 million years ago and in the *Early Permian* of North America (Fig. 5).

Their appearance is very reminiscent of a current dragonfly but gigantic: with a wingspan of 75 cm and a length of 50, authentic natural drones. Science has not yet solved the enigma of the greatness of these creatures, and there are many theories about it. Some consider the composition of the archaic atmosphere, perhaps richer in oxygen than today's, may have influenced their size, but in fact, a specific and irrefutable theory that explains the phenomenon has not yet been hypothesized. Dragonflies are, therefore, very ancient insects, and still today, their flight mechanism is defined as *primitive*, but it is one of the most efficient of all those that insects use.

The wing movement is made fluid by the inserted *direct muscles of flight* directly at the base of the wings. When the muscle contracts, the wing lowers, and when the muscle relaxes, the wing rises. The muscles moved by a discharge of nerve

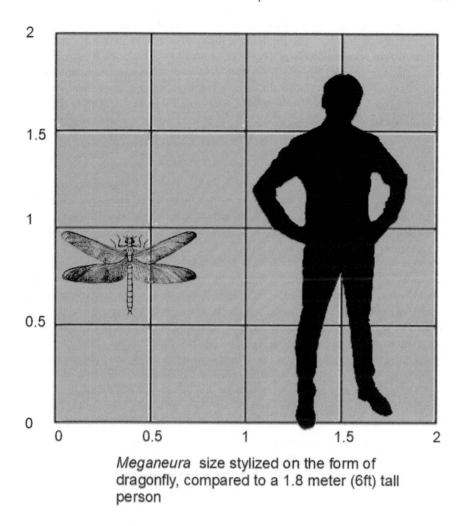

Meganeura size stylized on the form of dragonfly, compared to a 1.8 meter (6ft) tall person

Fig. 5 *Meganeura* size compared to human ones

impulses, and, surprisingly, the two pairs of wings of the dragonflies are independent, thus allowing them to fly backward and glide. They are equipped with large composite eyes, and their sight is exceptional, thanks to the movable neck that allows a panoramic vision. The double wings are very elongated, very thin, and sprinkled with a dense network of capillary canals in which hemolymph circulates. A particular feature of the wings is a thickening called pterostigma, which functions as a flight stabilizer, as it dampens the oscillations induced by air turbulence and strengthens the wings at the points of most dangerous stress taken by vibrations. The dragonfly can reach a speed of about 50 km/h, flapping its wings about twenty times per second, managing to stay suspended in midair, and even fly backward. In particular, the anchoring of the four wings to the insect's body is highly flexible and independent from each other. The flight capabilities of flexible-winged insects are the subject of Professor Haibo Dong's research in the Flow Simulation Research Group at the University of Virginia. His team uses high-speed video cameras and advanced computational systems to build 3-dimensional models of flying insects and the vortex structure of airflows created by their flight. The project's goal is to improve the design of small flying robots [27].

3.4.2 Bumblebee

Antoine Magnan, a zoologist and aeronautical engineer, was a French zoologist and aeronautical engineer. They studied the flight of insects and birds in an early bionic way to apply to powered flight. He is best known for a remark in his 1934 book Le Vol des Insectes (*Insect Flight*) that insect flight was impossible and taught that *the wing structure of the bumblebee, concerning its weight, is not suitable for flying, but he does not know and flies anyway.* With the distance of time, contemporary research that uses macro photographs, sophisticated slow-motion video footage, and high-resolution digital microscopy has discovered that the secret of the bumblebee's flight lay precisely in the wings of these fantastic insects. The mistake was in considering the smooth wings and vertical movement. Microscopic images show how micro ripples are present on the bumblebee's wings, which even escaped the microscopes of the past. These, combined with the speed at which they are thrown into the air (which exceeds that of the hummingbird by five times), allow the creation of small vortices, air pockets around the central core. This way, they can increase lift (the force that keeps them flying) much more than they could if the wings were smooth.

> The fluid dynamics behind bumblebees' flight are different from those that allow a plane to fly. An airplane's wing forces air down, which in turn pushes the wing (and the plane it's attached to) upward. For bugs, it isn't so simple. The wing sweeping is a bit like a partial spin of a "somewhat crappy" helicopter propeller, Dickinson said, but the angle to the also creates vortices in the airlike small hurricanes. The eyes of those mini-hurricanes have lower pressure than the surrounding air, so, keeping those eddies of air above its wings helps the bee stay aloft. Other studies have confirmed that bees can flying 2001, a Chinese research team led by Lijang Zeng of Tsinghua University glued small pieces of glass to bees and then tracked reflected light as they flew around in a laser array. But now, Dickinson says, researchers are more interested in the finer points of how insects control themselves once they're in the air. Those studies will be especially important for a fleet of robotic insects in development, including robobees created by a team at Harvard University [28].

3.4.3 Flies

Flies are insects of the order *Diptera*, the name is derived from the Greek δι-di-*two*, and πτερόν-pteron *wing*. Insects of this order use only a single pair of wings to fly. The hindwings have evolved into advanced mechanosensory organs known as *halteres*, which act as high-speed sensors of rotational movement and allow dipterans to perform advanced aerobatics. Halteres function as *gyros* that help flies orient themselves in flight. They measure torque and angular momentum around the body as a balance and guidance system. If you observe a fly in routine flight, you will notice the white rod-shaped halters that flutter up and down in opposition (antiphase) to the front wings. Diptera is a large order containing an estimated 1,000,000 species, including *horse-flies, crane-flies, hoverflies,* and others, although only about 125,000 species have been described. Flies are the best aerial of all insects; they can hover, move vertically, and fly backward. Most flies flap their wings at over 200 Hz or 200 cycles per second. *Drosophila melanogaster* flaps its wings once every four milliseconds. To achieve this performing flight, the flies have a *gear* that helps them turn their wings. Flies can measure pitch and yaw and recalculate on the fly to correct their flight. It is a model that has been tested for millions of years of evolution, and now this knowledge has been employed for bionic inspiration in robotic and dronic technologies. In this regard, it is interesting to analyze how the mechanical detail of the gears of one of the most sophisticated two-propeller drone models on the market (V-Coptr Falcon produced by Zero Zero Robotics). Its design, which deviates considerably from the standard four-propeller, revolutionizes the engineering approach to drone design. This flying model is less noisy than other 4-rotor models; they consume less and therefore has greater flight autonomy for the exact weight of the power supply battery and has a particular joint of the two rotors that make it act in a similar way to the muscular system seen on dragonflies and flies. Flies possess a sophisticated system of connection and coordination between the wings and the anthers. Mechanical mating explains how flies keep their sense of rhythm. When the front wings rise, the halteres descend. When the front wings accelerate, so do the halteres. Flies have highly specialized flying techniques and technologies: sometimes, they flap only one wing at a time, or a wing has to tighten faster for a tight turn. To control all this, it seems that they possess a sort of reduced gear in analogy with the mechanics of human means. We can find an exciting and comprehensive demonstration of the dynamics of fly flies in Michael Dickinsons' TED documentary: *How a fly flies*. The researcher shows how flies know how to make lightning-fast decisions in-flight to overcome obstacles, hunt, and escape danger [29].

This happens while they flap their wings 220 times per second. A superfine brain and nervous system seem to guarantee the capacity for excellent coordination. But Dickinson's research also carried out by creating simulator robots that move in oil chambers to study the turbulence created, has shown the existence of a crucial technological factor in the flapping of wings. Flies move the wings in a way capable of generating a turbulent movement at their ends, a leading vortex edge, which allows the insect's wings to support it in flight. A bit like we have seen happening at the ends of the feathered wings of birds that in flocks in-flight approach each other to support each other as a whole. Another fascinating aspect is the analysis of the muscles involved in the power to be imparted to the movement of the wings. Flies possess two types of flight muscle; the first or power muscle, activated by stretching, which is triggered automatically with a system of contractions by the nervous system. They are specialized in imparting the great necessary flight power and occupy most of the insect's body. These muscles are attached to the base of the wings, and there are a whole series of small control muscles; they are not particularly powerful but fast and can reconfigure the wing hinge very quickly, with one hit after another. This biological mechanism underlies the flight dynamics of flies that can manage the movements of the individual wings, generating changes in the game of aerodynamic forces. Their nervous system's role is to coordinate all of this through an impressive array of bio-sensors. Flies have antennas and sensors capable of perceiving odors and detecting the direction and intensity of winds. They have a complex eye, which translates into one of the fastest visual systems in the world of living beings. They have other ocelli in their heads whose function is still unknown; there are a series of sensors in the edges of the wings and on the entire surface of the same, which informs the central system of the deformations of the wings in flight. Halteres are one of their most sophisticated sensors that have been seen to act as gyroscopes that oscillate at 200 Hz, and the animal can use them to control the rotation of its body and activate corrective maneuvers very quickly. A brain of approximately 100,000 neurons controls these complex flight information management systems. Interesting is the evaluation that Dickinson shows in comparing the human brain of a mouse to that of a fly. It shows that this brain which is 400 million years old, has fascinating characteristics of miniaturizations and, in particular, refers to one of the most miniature models of wasp: *Megaphragma* (which with *Megaphragma caribea* and *Megaphragma mymaripenne* represents two of the smallest insects in the world whose dimensions are comparable to those of an amoeba or a paramecium). This tiny wasp has 7000 neurons; how modest brains can control sophisticated systems of flight and movement is still an open research threshold. The intervention of a series of neuromodulators such as *octopamine* seems to respond preponderantly [30].

Overall, we can only hint at the complex neurophysiological implications of the multitasking brain-behavior of insects' brain and nervous system that have developed an excellent optimization by making a minimal number of neurons perform interlaced actions of considerable complexity. Instead, the aspect that we will try to identify for the purposes of our project is that relating to the sensor components that accompany the body of flying insects.

4   Design Inspirations

This section considers the sources that inspired and guided the process of defining the concept of a *Pherodrone1.0*, a technology able to help farmers in pest capture and sexual disorientation as the chemical detection of pheromones by males takes the place of insects harmful to crops. We set out to define a concept of light flying aircraft, partially inflatable with lifting gas, safe and harmless to the user, low consumption, and more versatile and maneuverable in targeted uses for the fight against parasites, to be used for specific purposes in the PA (Fig. 6).

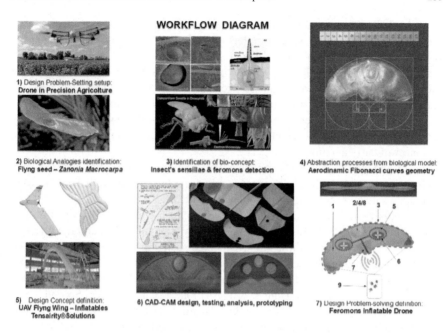

Fig. 6 Workflow steps of *top-down* research in *BionikonLab&FABNAT14*

4.1 Design Inspiration 01: Flying Seeds—Zanonia Macrocarpa *(Venustas)*

Man has nourished the desire to conquer the dominions of the high skies inspired by the observation of the flight of birds and other animals such as Lepidoptera, Insects, and last but not least by the extraordinary specimens of flying seeds present in anemophilous plant species (anemòcora). Wind dissemination appears to be the oldest natural invention to ensure the possibility of plant germination. The biology of flying seeds is not much studied, and from a biomimetic and bionic point of view, there are still many aspects to be discovered. Perhaps the most famous case in this sense is linking some flying inventions to the great seed of the *Alsomitra macrocarpa (Zanonia)*. This plant was first described in 1825 by Carl Ludwig Blume, who baptized it *Zanonia* macrocarpa, starting from fruiting material collected from Mount Parang, on the island of Java. In 1843 Max Joseph Roemer published an article about the same plant, which he called *Alsomitra macrocarpa*. Our research draws its inspiration from this famous flying seed, whose study inspired some attractive engineering solutions at the dawn of aeronautics. Prof. Friedrich Ahlborn had published an article in 1897 in which he had described the characteristics of the flight of the *Zanonia* seed. The winged seed of the exotic Java cucumber, then known to botanists as *Zanonia macrocarpa* (although today it is classified as Alsomitra), indeed possesses an extraordinary property. The seed bracts determine a crescent-shaped wing, with the seed providing the weight needed for stable flight. The wing is slightly rolled up at the back. This fact creates a washout that reduces the angle of attack concerning the tips. As the tips curve back from the midsection, this washout provides stability in the step. The seeds disperse by sliding off the parent tree. The German investigator studied this naturally stable wing and recognized its potential in airplanes. Studying him inspired many early aeronauts, especially those who recognized the need for safe and manageable flight characteristics. He soon realized that more or less the exact reversal could be seen in the wings of the common pigeons and the crows they saw fluttering. The relative stability of the seed in pitch and roll inspired Igo Etrich, a pioneer of early aviation.

Etrich and his collaborator Franz Xaver Wels in 1904 designed the Taube airplane, with a wing planform resembling the *Zanonia macrocarpa* shape and demonstrating extremely safe and forgiving flying qualities [31].

Attempts to add an engine failed, but a successful crewed glider was flown in 1906. Among these were José Weiss and Handley Page in the UK and Igo Etrich in Germany. However, when these pioneers copied such wing shapes, with or without a tailplane, they found it necessary to add a tail fin before their planes flew straight. Weiss returned to painting, Etrich to a conventional tail, and Page to a conventional straight wing plus a tail. J.W. Dunne, a British pioneer of aeronautical Second World War engineering, also studied the seed but discarded it as inspiration because it was not directionally stable. At the slightest wind, the seeds zigzagged madly all over the place as they flew. This was undoubtedly good for wider seed dispersal for a bird moving in and out of trees but hopeless for a stable airplane. A better solution could be found in seagulls. These birds could soar for significant periods, making no perceptible movement of their wings, while their small tails were not needed, closed up and tucked away to minimize drag. Yet they could still maneuver adroitly when they wanted to. Dunne observed the gulls more closely than most, for his sister May would draw them in by feeding them, allowing him to observe their maneuverings from close quarters. This close-up observation revealed that they banked by dipping down the leading edge of an outer wing. This had the same effect on the lift as raising a trailing aileron. Indeed, the leading edge there tended to have a permanent slight droop. This helped impart the same washout as seen on the Zanonia, but its effect on drag and directional yaw was quite different. Dunne found that, crucially, provided the wing was swept, turning down the leading edge towards the tip made it stable in pitch and yaw [32].

The seed or samara of this species is unusual in having two flat bracts extending either side of the seed to form a wing-like shape with the seed embedded along one long edge and the wings angled slightly back from it. As the seed ripens the wings dry and the long edge furthest from the seed curls slightly upwards. When ripe, the seed drops off and its aerodynamic form allows it to glide away from the tree. The wingspans some 13–15 cm. and can glide for great distances. The seed moves through the air like a butterfly in flight, it gains height, stalls, dips, and accelerates, once again producing lift, a process termed phugoid oscillation.

Unlike many seeds that make a gliding flight using auto-rotation, the seed of the Javan cucumber vine exhibits a stable gliding flight with its paper-thin wings. The seed's design is efficient enough to achieve a low descent angle of only 12 degrees and therefore it is able to achieve a slower rate of descent (0.41 meter per second) compared to that of rotating winged seeds (1 meter per second). This aerodynamic advantage allows the seed to be easily carried by the wind. The construction of the seed and wing gives it this advantage. The seed itself is thin, about 1 millimeter in thickness, and positioned almost exactly at the structure's center of gravity to give it balance. The wings are even thinner, about a few micrometers to some 10 micrometers. Because the wings are so thin, as the samara is angled up or down, the center of pressure from the wind will shift to reduce that angle. This effect stabilizes the seed and also prevents it from diving. When viewed from above, the wings are angled behind the center of the seed to give it more stability and are slightly tapered toward the tip to make it lighter with less drag. When viewed from the front, the wings are angled upward which helps it fly in a straighter path and prevents spiral instability. The wings also have a sharp leading edge and an aspect ratio (AR = 3–4) that results in an appropriate lift-to-drag ratio (L/D = 3–4) to support their gliding flight. The form of the samara allows it to travel long distances in the wind. It is possible for the seeds to glide up to hundreds of meters, ensuring that they spread far from each other as well as the parent pod. This wide dispersal prevents the seeds from competing for resources once they fall to the ground and begin growing [33].

We're all familiar with such natural flying machines as spinning maple seeds and airborne dandelion fuzz. But few people in this part of the world have heard of the flying zanonia seed of tropical Asia. Long before man even thought of flying, the seeds of the climbing zanonia vine were gliding through the jungle on a very "modern" swept-back wing. Nature shaped and balanced these winged seeds so efficiently that they drop from the vine and glide to the ground to begin their life cycle at a great distance from the mother plant. The late Dr. A. M. Lippisch, the inventor of the delta wing, pioneered the design of all kinds of tailless aircraft. For many years, he kept a large zanonia seed in his laboratory. Perhaps it was a symbol to him-hinting that nature still held undiscovered solutions to the mystery of flight. (Model Aviation) [34].

In an issue of Model Aviation magazine edited in 1983, retrieved on the web, we found an article where is proposed the construction of a flying model inspired by the morphology of the *Zanonia* seed. We contacted the publisher office in America, who kindly sent us scans of the original pages. The article written by Paul McIlrath introduces the description of the aerodynamic characteristics of the suit and proposes a template for the construction of a foam model. The success of the flight performance of this model foam is due to the placement of a correct center of gravity weight. In the natural flying wing, this aerodynamic aspect is resolved by the wild seed's lenticular body, which guarantees the right lift for the whole biological design. Thanks to this template, intrigued and eager to play, we have created some straightforward cardboard models to experiment at first glance with the flight characteristics and the possible alternatives to the proposed airfoil. We analyzed the plan profile of a Zanonia seed, and through an interpretation of the possible generative geometric scheme, we stylized a profile to be used in the design phase of the wing-body. Based on these indications, we have made very *naive* early cardboard models, to begin experimenting with the flight of the profile of this and its some simple variants (Fig. 7).

Subsequently, starting from some photographs, we analyzed the shape of the *samara*, applying a theoretical geometric and constructive scheme, to arrive at defining an abstraction from the initial biological model. This result was used to define the morphology of the wing for the subsequent CAD steps (Figs. 8, 9 and 10).

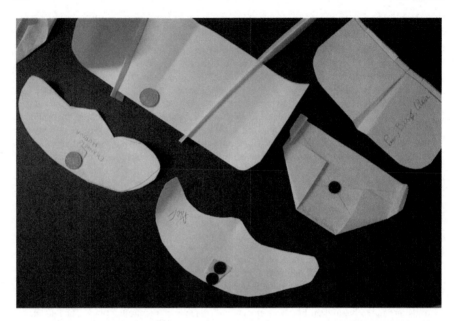

Fig. 7 Very *naive* early cardboard models of *samara* and some profile variations (From BionikonLab's photo archive)

4.2 Design Inspiration 02: Pheromones Insect's Detection (Propinquitas)

4.2.1 The Insects' Smell Universe

As we have learned from Dr. Dickinson's research and a whole host of other scientific contributions, insects' and moths' bodies are equipped with a series of *antennae* and *sensillae* that perform sophisticated and vital functions for flight, spatial orientation, communication, and environmental scanning. They are evolved with surprising technological refinements for the survival of organisms under the conditions and characteristics of their vital ecosystem wing. The impressive development that took place in science and technology from the Second World War to the present day of cybernetics, electronics, nanotechnology, and neuroscience, combined with the enormous possibilities of exploration of the microcosm by SEM and digital imaging, have opened unexplored human eyes domains of the so-called *Invisible Factories* [35].

In parallel, IoT technology has led humanity to create an artificial world in which microsensors and actuators govern, through sophisticated and complex digital and electronic interaction and control activities, our physical, prossemic, and private home universe. WIFI systems and telecommunications developments are wirings, installing millions of antennas and detectors all over the planet, enveloping it in a cloud of electromagnetic waves and micro detectors that convey information, words, sounds, images, and codes connecting billions of humans to billions of apparatuses.

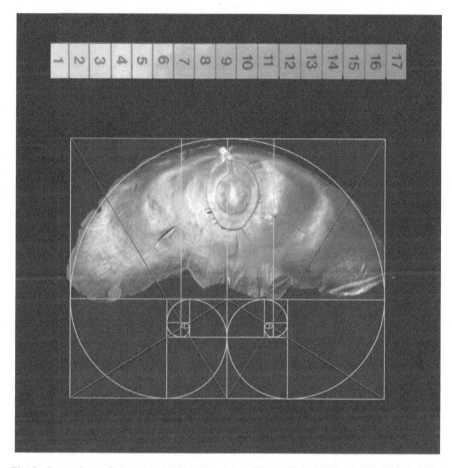

Fig. 8 *Samara*'s profile geometrization, based on Fibonacci Golden Spiral (S. Musa CAD rendering)

In this behavior, humanity seems to take insect's *swarm intelligence* as a model, a term coined for the first time in 1988 by Gerardo Beni, Susan Hackwood, and Jing Wang following a project inspired by robotic systems [36].

It considers the study of self-organizing systems, in which a complex action derives from collective intelligence, as happens in nature in the case of colonies of insects or flocks of birds, or schools of fish, or herds of mammals. Humans are highly visual beings; the activity attributable to the functions of sight occupies more than 70% of the total brain: we see the world around us, and the sense of sight guides us for most of the primary activities. Hearing, touch, smell, and taste are splendid sensorial corollaries of the uncontested domain of the eyes. This is not the case; they rely mainly on a chemical sense that we call olfaction. This sense is used to identify the nutrients that provide energy and the raw material for their survival and growth. Through chemical scanning of the environment in which they are

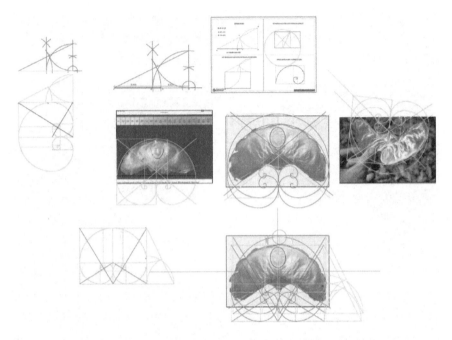

Fig. 9 Geometric CAD abstraction of *samara*'s profile (S. Musa CAD rendering)

inserted, they identify and avoid the dangers posed by predators and other threats. In particular, through the sense of smell, they trace those molecules in the environment that promise a possibility of drafting a female and being able to mate. For this reason, the chemical senses were the first to evolve, and for a vast number of living creatures, it represents the keystone for building an image of their living in the physical earth world. There are sip animals, blind but beings who cannot decode molecular-chemical sources are unknown. This behavior is called chemotaxis in bacteria, which pushes them towards food sources and removes them from toxins and poisons. Particular protein molecules are positioned in the bacterium's membrane in the manner of various chemical receptors. These devices can be conceived as a series of key locks. A particular chemical key substance can enter a specific receptor lock capable of sending an impulse to the motor system of the flagellum or cilia. If the key-lock has identified a sugary substance, it pushes the action of the bacterium towards that source of chemical impulses. Conversely, detecting a toxin causes the bacterium to move away from it. This key-lock system is the basis of all animal chemistry detection. Even in the case of the release of pheromones linked to sexual attraction, a similar process takes place in a much more complex and intricate dimension. The evolutionary advantage of smell oversight is that it works well even in the dark. Many mammals have developed the sense of smell by delegating the nose organ the most suitable biotechnology for scanning and decoding olfactory information from the external environment. The question takes some interesting differentiation

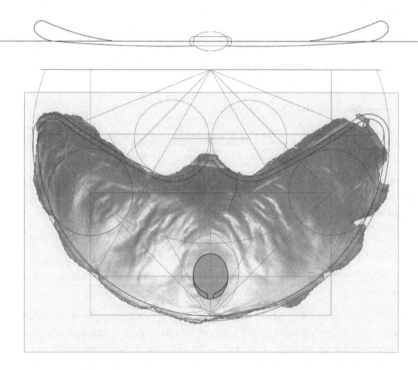

Fig. 10 Geometric CAD abstraction of *samara*'s profile (S. Musa CAD rendering)

in insects as they do not have a natural nose similar to ours. In this whole universe of olfactory information, pheromones are of considerable importance. *Pheromones* are chemical compounds that have a lot in common with *hormones*. The most interesting peculiarity of *pheromones* is that of triggering, through the smell, a whole complex series of behavioral reactions generally related to the sphere of sexual or social attraction in a broad sense, such as recognition between species or parental [37].

4.2.2 Pheromones: The PiedPiper-Chemical Effect

From BugInfo, *Pheromons in Insects, Smithsonian Library*, we take a definition of *pheromons'* main characteristics:

Pheromones are chemicals produced as messengers that affect the behavior of other individuals of insects or other animals. They are usually wind borne but may be placed on soil, vegetation or various items. Tom Eisner, a foremost authority in the science of chemical use by insects, claims that each species of insect relies on some one hundred chemicals in its life, to engage in such routine activities as finding food and mates, aggregating to take advantage of food resources, protecting sites of oviposition, and escaping predation. It has been found that pheromones may convey different signals when presented in combinations or concentrations. Pheromones differ from sight or sound signals in a number of ways. They travel slowly, do not fade quickly, and are effective over a long range. Sound and sight receptors are not needed for pheromone detection, and pheromone direction is not limited to straight lines. Pheromones have long been known to be important to the lives of insects in mating, as witnessed, for example, in some of the larger silkworm family moths, where males are noted to travel nearly 30 miles to a female, following a pheromone trail in the air. [38]

4.2.3   Insect's *Antennae* and *Sensillae*

The insects are receptors capable of transforming a stimulus from outside or inside into a nerve impulse. Three main types of *stimuli* are able to act on insect receptors:

- **vibration or pressure**
- **activity of chemicals**
- **light and heat**.

The receptors sensitive to vibration or pressure are referred to as mechanoreceptor *sensillae*. They include:

- **s. tactile** distributed throughout the body, especially on the antennae and tarsi;
- **s. auditory** (or phonoreceptors).

The auditory *sensillae* (or phonoreceptors) range from simple auditory bristles identical to tactile ones up to the more complicated type located in the antennae through which the sensation is initiated to the antennal nerve from this to the *deuto-cerebrum*. The tympanal organs of the *Noctuid Ledidoptera* can pick up the ultra-sounds emitted by the bats and, in the case of the *Malay Laodamia*, also emit them in turn, to confuse the predator. The receptors sensitive to the activity of chemical substances are referred to as chemoreceptors *sensillae*. The taste senses are located on the tarsi, palps, and in the oral cavity and must directly contact the chemical. The smell *sensillae* is located on the *antennae* and the genital appendages (Fig. 11).

They are sensitive to substances volatile like pheromones in shallow doses. Production of insect sex pheromones is usually by the female of the species and involves either specialized glands secreting to the surface cuticle or direct release to the air, e.g. by pneumatic eversion of the gland. *Pheromones'* proprieties have inspired ecological solutions in the fight against parasites (Fig. 12).

The *pheromones* used in agriculture and gardening are chemical compounds that reproduce the natural chemical communication between individuals of the same species emitted by female insects during the mating season. These substances have long been used in traps, in the form of dispensers, to monitor the presence of insects, and in diffusers to prevent mating. To ensure the certainty of reproduction, insects use silent remote communication based on *semiochemicals*. These molecules regulate many aspects of insects' life, such as the choice of host plants, the places in which to lay, the location of the prey or host, the search and option of a partner, the organization of social activity. There are different types of *semiochemicals: allomones, kairomones, sinomones*, and, finally, *pheromones*, when the emitter and the recipient concern the same insect species (Fig. 13).

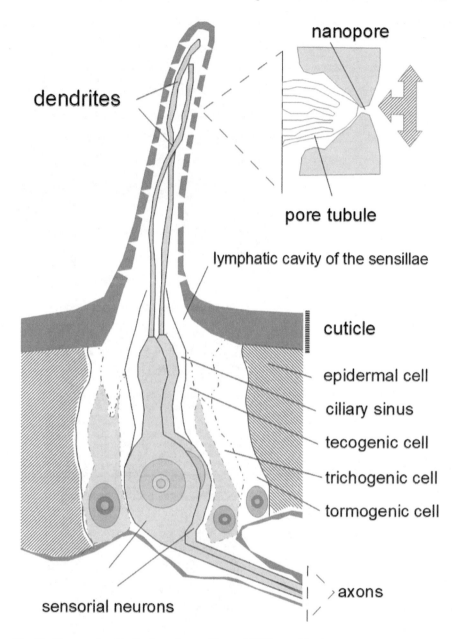

Fig. 11 Cross-section idealization of a *sensillae* and SEM magnification

Fig. 12 PA tools for insects'
pest fight with *pheromones*
chemicals

PHEROMONES AEREAL DISPENSER

PHEROMONES TRAPS

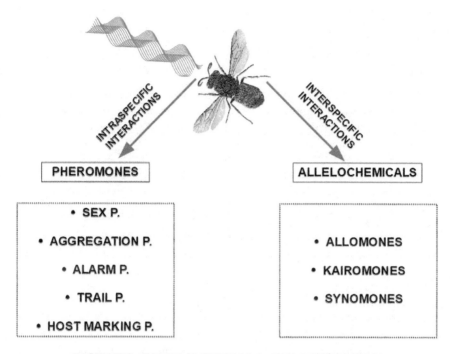

INSECTS SEMIOCHEMICALS CLASSIFICATION

Fig. 13 Semiochemicals classification

According to the transmitted signal, *pheromones* are divided into aggregation, oviposition, sexual, alarm, arrest, and trace chemicals. The sexual *pheromones* of insects are volatile substances, natural chemical messengers released by females through the exocrine glands, removed and transported by winds and currents, to attract males of the same species. These substances are detected by male individuals through the olfactory sensors present on the *antennae* and cause a specific reaction such as behavior, physiological change, or development. Studies on *pheromones* were initiated by the French entomologist and naturalist Jean Henry Fabre (1823–1915), father of modern entomology, who began a deepen study of insect behavior. About 100 insect species are currently covered, and there are innovative solutions for both capture and disorientation of insects. Possible applications of *pheromones* are monitoring, reporting male flights, and the study of population density, with the consequent estimation of the risk of damage before they occur, to have the necessary data to carry out a rational integrated or biological fight. Other proceedings include the use of specific traps for the direct control of a harmful species with the capture and removal of most males in the controlled area and consequent reduction of mating. These techniques are used in the forestry field to conserve wood, foodstuffs, and fruit against aggression by specific parasites. In general, the following types of procedures alternatives to chemical irrorations are known:

- sexual **distraction**;
- sexual **disorientation**, and sexual confusion;
- direct defense systems to inhibit mating employing **trap-killing** treatments in certain agronomic situations.

The disorientation, consists of creating numerous artificial traces predominant on the females, and it requires 20–30 g of pheromone per hectare and about 2,000 diffusers per hectare; monitoring with sexual traps within the plot is possible. Instead, the confusion is the shielding of the females' call through the saturation of the environment; it requires 170–200 g of pheromone per hectare, 300–800 diffusers per hectare; in this case, monitoring with traps is more complicated.

Allomone is a volatile chemical substance similar to pheromones, which acts as a chemical messenger between organisms of different species and intra animal and vegetal domain. Kairomones is a chemical substance emitted by one species and especially an insect or plant that as an adaptive benefit, such as a stimulus for oviposition, to another species. Synomones is a semiochemical that is beneficial to both interacting organisms, the emitter, and the receiver.

Cydia splendana is a moth that lays 100–300 eggs on chestnut leaves. The *larvae* penetrate the chestnut hedgehog by digging long, deep tunnels that soon fill with excrement. Once it reaches maturity, the larva comes out of the hedgehog and falls to the ground cocoons to overwinter. In Italy, the *carpocapsa* is present in all chestnut areas where, depending on the severity of the infestation, it can lead to the loss of 50% of the harvest. The affected hedgehogs can fall to the ground prematurely, while the ripe chestnuts attacked by the larvae are unsaleable. Sexual confusion has proven effective in combating this pest, but positioning the diffusers is tricky, especially for larger plants. Operators must climb the canopy to place the diffuser at the top of the tree; a long and dangerous job. To meet the needs of a PA technological intervention, in 2016 BioPose© French company, has developed a drone that can perfectly replace a human operator to place the traps (Fig. 14).

A first drone flies over the chestnut wood, identifying the plants to which the diffuser is applied thanks to a GPS. A second drone, equipped with a particular release system, takes off and follows the route plotted thanks to the surveys of the first uncrewed aircraft. Once it reaches the top of the desired plant, the drone drops the diffuser in the shape of a ring, which becomes entangled at the top of the tree. With just one flight is possible to protect one hectare of the chestnut grove [39].

Many of the examples of the application of drones in the fight against pests through *pheromones* are now available on the market and this technology is rapidly establishing itself as an exciting solution for reducing the use of pesticides and chemicals that have now been shown to be harmful to both healthy humans and ecosystems life. The current development project of a *Pherodrone1.0* has been oriented towards these development hypotheses.

Fig. 14 The drone *BioPose.* © while placing *pheromenes* traps on trees

5 *Pherodrone1.0*'s Design Concept

5.1 *Premise*

BionikonLab&FABNAT14's teamwork developed the *Pherodrone1.0* design concept through a top-down biomimetic process that tries to work by simulating the chemical behavior of a female insect. The concept we have outlined fits within the methods of using the *pheromones* analyzed. The *pheromone* drone is designed above all to increase the effectiveness of hunting campaigns and disorientation of insect colonies harmful to specific crops. Unlike drones currently produced for agricultural uses, which are generally four-propellers, very heavy as they are delegated to carry out fatigue work or environmental and crop monitoring, the *soft-Pherodrone1.0* is a relatively small aircraft, from the design of the profile directly inspired by the morphology of the fixed-wing of the *Zanonia macrocarpa* seed (*samara*), previously studied for aerodynamic research purposes. The drone is then equipped with a series of diffusers of *pheromonic substance*, the design of which drew inspiration from the study and observation of the functional morphology of *antennas* and *sensillae* insect sensors. Finally, the drone has onboard, an apparatus that acts as a trap. Unlike the models we have analyzed on the customer market dedicated to PA-Precision Agriculture, it actively seeks to intercept through the artificial digital micro-sensillae set arranged in the margins of the wings, the *pheromones* flow of chemicals traces released by female individuals of a specific parasite. Once a *pheromones'* plume target has been identified, the drone tries to work alongside them to release the appropriate amount of substances from its tank, acting according to different possible action scenarios. These intervention technologies already exist in the sustainable agriculture tecnologies, but they generally work through special dispensers placed in fixed positions, activating and spreading their chem-plume according to the prevailing winds diffusers. Their design resembles exactly like the replicas of small bags or *pochettes* used to hang in cabinets as a mothproof or the home's fragrance diffuser [40].

This intervention modality seemed to limit and passive, and this observation stimulated to imagine an apparatus that, by imitating the behavior of insects, literally becomes an antagonist that creates real *hormonal sabotages* in the process of hunting and reproduction, in creating sexual confusion in the structure of the colonies of parasites. However, it must be emphasized that the scope of this research falls within a school training process that engages students aged 14–16 to the bionic and biomimetic design approach. The path we have developed with the teachers and designers involved in the work aims to train the team to think of solutions with high technological, functional, and economic feasibility coefficients. However, we did not set as working conditions the development of a working prototype and an electronic and constructive engineering process, at least in terms of the development of the current concept. It is not unthinkable that this idea will reach, through successive research phases, the precise development potential of functional prototyping, shared with the experts of the R&D centers and Technological Departments of the International University network with which *BionikonLab & FABNAT14* have been collaborating regularly for several years (see *Conclusions*).

5.2   Functional Layout *(*Utilitas *and* Firmitas*)*

A utility layout of our *Pherodrone1.0* is briefly illustrated in the graphic scheme of Fig. 15. We can divide the system into nine *functional blocks*:

A.   *Zanonia macrocarpa* inflatable body-wings case, 100% natural rubber latex
B.   FC + (see previous *main components of a drone* section)
C.   Semirigid boundary edges equipped with wired rubbery *sensillae-pads* for chemical detection of female pheromones
D.   Flight control cabin (recycled 100% plastic and regenerated vegetable hemp-fibers 3D-printed)
E.   Ringed rotor in super-light aluminum body foam or recycled 100% plastic and regenerated vegetable hemp-fibers 3D-printed

Fig. 15 *Pherodrone1.0* UAV layout

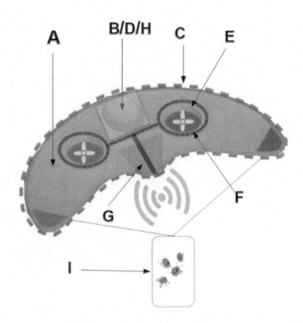

F. Electrical wiring and battery side, via an external interlocking in super-light aluminum body foam platform or regenerated vegetable hemp-fibers 3D-printed
G. Cartridge-spry and pheromone's nozzle dispenser
H. Arduino & Raspberry interaction drone's sensory hub
I. Optional Trap system.

5.2.1 Block "A"

A structural feature of the *Zanonia-wing* that constitutes the constructive body of our concept-drone is that of being inflatable with a gas mixture slightly lighter than air. We had this intuition to overcome the bulky, massive and tiring weight of many *slave-drones* that, with dangerous propellers, action, and 10–45 kg of weight in flight, are often hard machines, of difficult flight control, potentially dangerous if not deadly. We got inspired by looking at hundreds of inflatable games and other pneumatic items explored in the current market. We then found traces of an idea for an inflatable drone in a video that features a quadcopter developed in China [41].

The still few visualizations recorded comforted us on our intuition, and we did not feel frustrated by this fact. We imagine that it is possible to make, at affordable costs, our single wing, which measures about 50 × 25 × 2 cm using an elastic material that is 100% natural and easily recyclable. The choice could fall on a rubber-based latex material that we have seen used by an Italian factory that produces fully compostable balloons for parties [42].

We could not get samples of this material in time, but we have assessed that if it works for these inflatables, it could also work well for our drone frame with the necessary technologies and adaptations. To find out about the current processes available on the market for inflatables, we viewed dozens of videos, primarily promotional material from Chinese companies, which showed the main processes through which an inflatable promotional toy or accessory is made, mainly in PVC or similar polymers (Fig. 16).

Fig. 16 Making inflatable toys

Latex is also basically treated and assembled with similar processes. The wing has different ripples and embossing that make the surface texturized, according to the bionic observations carried out on the seed of *Zanonia*. This surface treatment seems to guarantee the seed an optimized aerodynamic profile. The body is inflated in various communicable sections, which generate a total stiffening of the wing. The mechanical components of flight, flight control, and environmental and sensor interaction control are inserted into this body.

These are the typical factory's steps for to carry out inflatable toys:

1. **CAD-CAM design of the model**
2. **Cutting and cropping with a Vinyl plotter CNC manufacturing**
3. **Heat sealing assemblage of parts**
4. **Machine sewing assemblage of parts**
5. **Pneumatic test**.

During the development of the *Pherodrone1.0* concept, one of the assumptions set by the team members from the first brainstorming set of the problem-solving process was to envision a drone model that was substantially divergent in the current ways of designing and building their flying body. Our attention was immediately focused on inflatable technology solutions. We became aware of a whole sector of research and development that dates back to the 30/40 s in aviation through archival. Especially in the military field, there are many attempts to develop wings with inflatable aerodynamic profiles. Also, currently, there is a renewed interest in this topic, particularly in the design and production of wings for ultralight aircraft. Exciting was the reading of a paper that analyzes the potential of a *Tensairity*® innovative technology in the UAVs agricultural sector. Unfortunately, we came across this documentation too late for the progress timeline of this chapter. Therefore, the indication relating to the possible engineering constructive redevelopment of our concept, through an upgrade given by the *Tensairity*® technology opportunities, remains limited by way of proclamation of interest. This fact comforts us powerfully in the opportunity of any engineering developments of our prototype, especially in light of the so-called *tensairity technology*. See how our ideas are taking place in high-level research circuits, comforts strongly us in the opportunity of any engineering developments of our prototype in this technologic approach. We believe it is appropriate to cite the full abstract of the research cited, as it explains exactly the appropriateness of the goals we had set for ourselves from the beginning of our design process.

In recent years, the unmanned aerial vehicle (UAV) has undergone rapid development in the field of agricultural plant protection; however, the payload and max-endurance are bottlenecks that limit its further development. Tensairity is a new structure consisting of a bag filled with low-pressure gas, an upper rigid rod and a lower flexible cable. It puts tension pressure on the whole structure through internal gas pressure, provides continuous support to the upper rigid rod, and that is to say, the tensairity improves the stability of the structure in some way. And also, tensairity is a self-supporting and self-balancing system, which means it is a simple, lightweight structure with small storage volume, strong bearing capacity, and low engineering cost. It also has the advantages of gasbag and BSS (beam string structure). The main objective of this research was to introduce the theoretical basis of tensairity and then discussed its development and applications. Moreover, the advantages and disadvantages of

tensairity were summarized, and development prospects of tensairity in agricultural aviation were presented. At last, it can be concluded that tensairity has potentials in agricultural
aviation. If tensairity is applied in agricultural UAVs, it will solve the two major problems
of payload and max-endurance better, and help promoting the development of agricultural
aviation technology in agricultural aviation administration and application. Also, it will help
with meeting the aerial application requirements of high efficiency and maximal environmental protection. This research will provide a reference for improving working efficiency
and economic benefits of UAVs in agricultural plant protection. [43]

The word *Tensairity*®, in some respects similar to the *Tensegrity*® term that was
coined in the late 1940s by the American architect Buckminster Fuller and sculptor
Kenneth Snelson, was proposed first by Dr. Mauro Pedretti, a Swiss civil engineer
of Airlight Ltd., in 2000. *Tensairity*® is a combination of *tension, air,* and *integrity;*
the word illustrates the composition and strength of the structure. Is an evolution
of pneumatic beams. The principle combines an air membrane with compression
elements (steel, wood, or aluminum profiles) and traction elements (cables or profiles
that work in traction). The pneumatic system of *pressure + compression + traction*
integration thus coupled is required, thanks to the inflation of the air chamber, and
at the same time, it is able to transfer also compression or traction forces, making all
materials work efficiently. The result is a very light load-bearing element which, with
the same weight carried, compared to traditional reticular beams, can weigh between
30 and 60% less. Conversely, with the same weight, a *Tensairity*® beam can cover
wider spans than a traditional truss system. The Tensairity system can be applied to
beams, pillars but also arches. The achievable products are therefore flat or curved
roofs, bridges, military pitch tents, and, more generally, large structures. The Italian
company Tensairity®Solutions, is the only depositary of the original patent. This
technology could offer strategic suggestions for the construction of the inflatable
wing in such a way as to create a reinforced, lightweight, and resistant pneumatic
structure [44].

Thanks to all these suggestions, after the geometrization phase that allowed us to
be abstract an aerodynamic shape inspired by the samara design, the sketching and
CAD rendering phase followed. Through various steps of CAD representations, was
defined a characteristic shape and a basic layout that integrates the main components
necessary for the engineering phase of a possible prototype (Figs. 17, 18, 19, 20, 21
and 22).

5.2.2 Block "B"

All the technical aspects related to the sizing and technical wiring of the various
electronic and digital components that make up the FC of our UAV, constitute a
definition of in-depth engineering that would go well beyond the objectives and
competence of this problem-setting definition. Studying a drone organism's basic
functional anatomy was essential to understand the general design and the technical

Fig. 17 First CAD renderings of the *Pherodrone1.0* (Digitals CAD L. Cuccu)

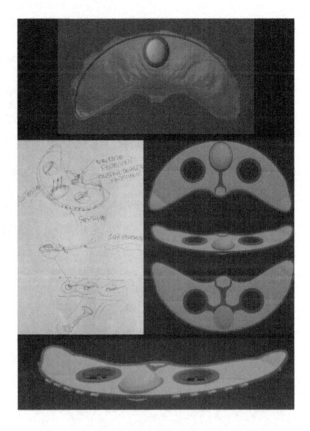

and technological issues involved. The *BionikonLab&FABNAT14's* general educational objectives do not address this technical training area, except through collaborations and assignments to specialized teachers and professionals who can develop specific concept sections. In any case, the *Pherodrone1.0*-UAV must guarantee the operating standards of the state-of-the-art drones in their final engineered design. We just have to define, for the purposes of the present chapter, the different functional blocks and place them in the vehicle as a whole. We are waiting for a subsequent general engineering step, which can only occur if this concept meets the interest of a developer in the sector.

5.2.3 Block "C"

The *Pherodrone1.0* was designed to have a pneumatic, inflatable, shockproof, light, and easily maneuverable body, which combines inflatable parts with rigid and semi-rigid elements, each of which performs specific functions. An essential aspect of this concept is the idea that the aircraft can somehow intercept a chem-plume of pheromones present in the air surrounding a crop, a forest, a garden, and through a

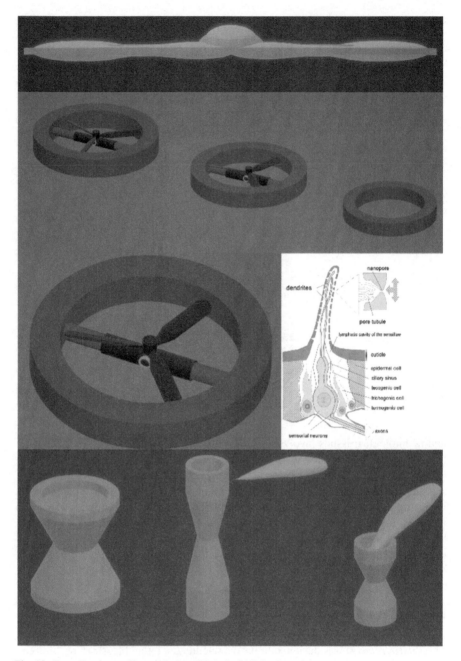

Fig. 18 Propeller & *sensillae* rendering (Digitals CAD L. Cuccu)

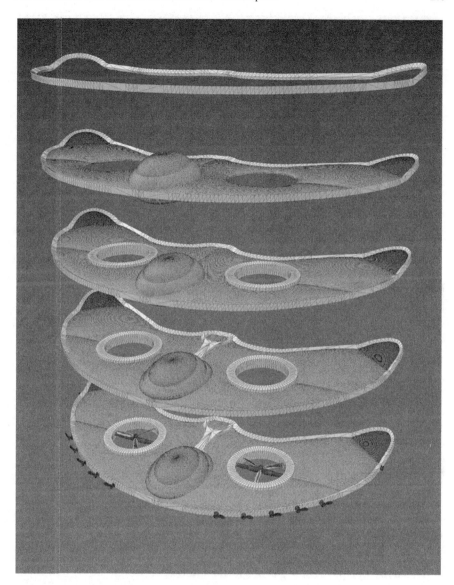

Fig. 19 CAD render (Digitals CAD L. Cuccu)

system of sensitive olfactory sensors that mimic the incredible sophistication of the insect's *sensillae*, entering into competition with them. The incredible development of human nanotechnology towards natural bionanotechnology is continuously updating the state-of-the-art performance of current olfactory sensors. In recent years, technologists have developed sophisticated devices called *electronic noses*. These electronic

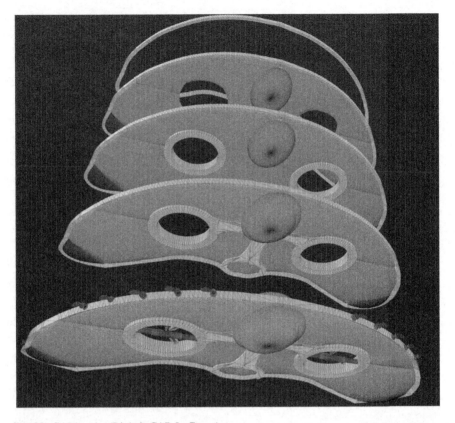

Fig. 20 CAD render (Digitals CAD L. Cuccu)

tools can convert the interaction of airborne and dispersed chemicals into an electrical signal. An artificial nose comprises an electrochemical sensor composed of a sensitive material, which interacts with gas molecules, and of electronics capable of quantifying and processing electrical signals. The essential characteristics of an electrochemical sensor are:

- **high molecular sensitivity;**
- **selectivity to a particular gas of specific interest;**
- **decoding capability;**
- **the ability to return to the initial conditions once exposure to the substances has ended.**

In 1988 Gardner and Bartlett coined the term *electronic nose* and defined it as an "*instrument that includes a set of electrochemical sensors with a partial specificity with the ability to recognize simple and complex odors.*" [45].

Electronic noses, also known as *Artificial Olfaction Devices,* are comprised of a sensor array, signal conditioning circuit, and pattern recognition algorithms.

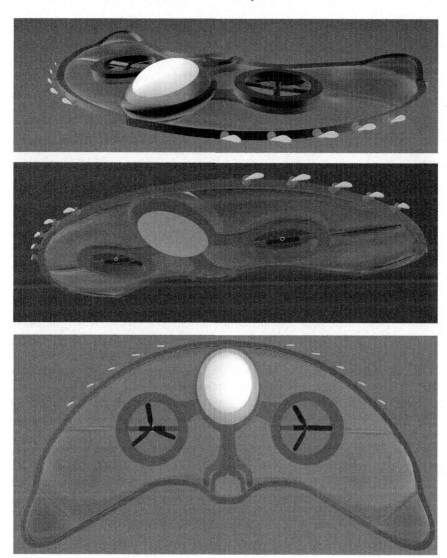

Fig. 21 CAD render (Digitals CAD L. Cuccu)

Compared with traditional gas chromatography–mass spectrometry (GC-MS) techniques, *electronic noses* are noninvasive and can be a rapid, cost-effective option for several applications. By equipping the nose with artificial intelligence, we can imitate what happens in our noses and in insect's *sensillae*; we can make sure that smells are recognized based on their interaction with specific receptors. Upon contact with the sensitive elements of the electronic nose, the odors generate a positive or negative signal or do not interact. By creating a system in which odors can only bind to a

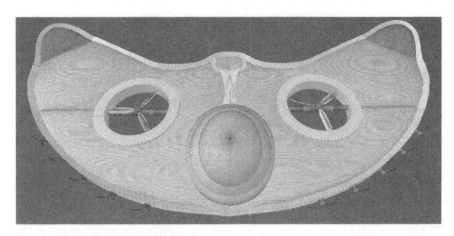

Fig. 22 CAD render (Digitals CAD L. Cuccu)

limited number of receptors and creating a *map* of electrical responses with a specific smell, we can recognize them. The artificial intelligence must then put together the numerous pieces of the puzzle and make the information sniffed appear on our screen. Developments in chemistry, physics, materials engineering, science, and technology will lead in the future to produce an electronic nose capable of *feeling* smells like our nose, or even better. *Electronic noses* will find a place in numerous industrial sectors such as quality control and production, while in everyday life, they can be helpful in the medical and health sector and in general in the nascent sector of the internet of things, the latter closely linked to the use of various sensors. In recent years, due to their simple fabrication and variety, conductive polymer-based sensors have been used in detecting wood decay. The *AromaScan32S®* (Osmetech Inc., Wobum, MA, USA) is a commercial *E-nose* with 32 organic conductive polymer-based sensors that have been used to determine incipient wood decay caused by *fungi* [46].

The chemical sensors miniaturization has been demonstrated to be a highly effective tool for detecting specific molecules in small quantities in the environment, in this way, broadly pursued by scientists. E-nose research is a continuous nano-technology development sector, as it has much strategic applicability to scientific, medical, industrial, agricultural, and environmental sectors. From the heavy, expensive and bulky apparatuses of the 1980s, with exponential speed, technology is developing ever more performing biosensors, tiny and increasingly competing with the extraordinary examples found in nature. The biomimicry that studies these research fields will bring this technology's state of the art in a few years to unprecedented levels. Therefore, it is conceivable without too much forcing that *Pherodrone1.0* could carry biosensors capable of intercepting the very weak (for humans) signals scattered in the air by the females of parasites and harmful insects to crops (Brezolin et al. 2018) [47].

5.2.4 Block "D"

In the samara of Zanonia macrocarpa or Giava's cucumber, the shape's engineering plays a fundamental aerodynamic role, set in its center of gravity position. Its position and weight are strategically barycentric, shifted towards the frontal area of the wing with a relatively flat and lanceolate conformation. In the template we used from the Model Aviation magazine mentioned above, this aerodynamic element is simulated using a paper clip. The sizing that we have given to the cabin that contains the electronic and mechanical elements of the FC and the circuits necessary for the rotors' operation and gyroscopic systems is entirely theoretical and would require a specific prototyping phase in 1:1 scale for all static checks dynamics of UAV flight characteristics. An important feature we have hypothesized is that the case is eventually also produced through 3D printing which uses ecological filaments based on plant fibers such as hemp. In previous projects, we have had the opportunity to test 3dprinting prototypes, using a yarn produced by Kanesis©, an Italian startup (https://www.kanesis.it/10848/) that uses organic material waste from the agri-food industry. These yarns have given excellent characteristics of lightness combined with resistance and low environmental harmfulness.

5.2.5 Block "E"

The aluminum foam is an interesting industrial material, inspired by the conformation of the *trabeculae* in birds' bones, that we have come to know and which, due to its characteristics of great lightness and 100% recycled origin. His innovative technical characteristics are particularly surprising and add value to the already high quality of the starting metallic material, increased by an original production process. The *aluminum foams* are in fact resistant to compression, have a structure capable of uniformly distributing a load and absorbing shocks. They are also able to dissipate heat very quickly, thanks to the increased surface area. The structure of the cells can be closed pores (there is no communication between the air cavities), or open (there is communication between the air cavities). The closed-cell structure allows the foam to be light enough to float. Initially engineered in military and defense technology, metal foams quickly found a place in other sectors of use as heat exchangers in the IoT sector, as ultralight structural panels in the transport sector (rail, aeronautical, and naval). They are also present in the aerospace sector together with other foams such as, for example, ceramic and glass ones and in the water treatment sector as purification and antibacterial filters. They are still used in the biomedical sector for artificial bone implants and in the architecture and interior design sector as cladding panels [48].

The ring-shaped bed that is designed to house the two oscillating propellers could be made by exploiting the characteristics of resistance and extreme lightness of this ultra-technological material. As an alternative to the usual carbon fibers, which are highly polluting, the two propellers could be made of aluminum foam. If this process, in subsequent engineering and economic evaluation tests, shouldn't offer the correct performance or if it resulted too much expensive or complicated to produce them, we have hypothesized the use of hemp or similar filaments for 3D printing.

5.2.6 Block "F"

This component constitutes the underlying body's part of the UAV and it is positioned on the lower edge of the wing. Fasteners provide to the wing-body attack via easy interlockings. It has the function of mechanical connection and of housing the electrical wiring between the parts, offers an element of aerodynamic stability to the wing, houses the set of batteries for the electrical power supply and there is the possibility of attaching any trap device to it when the drone operates in pest capture mode. Its shape is inspired by the ultralight structure of the keeled sternum of birds [49]. Due to its ultralight characteristics it should be made with a hollow aluminum foam mold or alternatively with hemp or similar filaments for 3D printing.

5.2.7 Block "G"

This component has the function of spreading the *pheromones'* plume in flight. It consists of a pod containing the chemicals and a sprayer, all housed in the backside of the wing. From the technological point of view, we have identified a possible solution that could be translated from some working mechanisms like Facial Nano Sprays for cosmetics tracked down in commercial productions in market research. It is a type of device that weights about 50 g and can spread 30 ml of liquid with a nozzle that atomizes 0.3 μm particles powered by the onboard batteries.

5.2.8 Block "H"

In the systems of the fight against parasites with static diffusers, the males' sexual *disorientation* consists in the creation of numerous artificial traces predominant on the appeal of the females. This action requires 20–30 g of pheromone per hectare and about 2,000 diffusers per hectare; monitoring sexual traps within the plot is possible. Instead, the *confusion* is the shielding of the females' call through the saturation of the environment; it requires 170–200 g of pheromone per hectare, 300–800 diffusers per hectare; in this case, monitoring with traps is more complicated. Compared to these scenarios, the use of *Pherodrone1.0* could guarantee greater effectiveness, time saved in placing and removing hundreds of traps and diffusers. IoT technologies guarantee the heart of interaction and digital control of the various onboard sensors. From a functional point of view, the technologies available (*sensillae*) and the programmed management algorithms should allow four action schemes:

- **Sensory imbalance**

Fig. 23 Possible aerial sexual-antagonism scenarios between *Pherodrone1.0* and a couple of pests

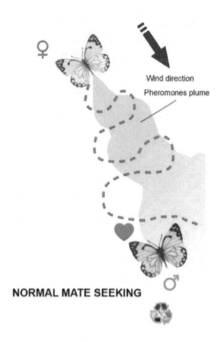

- **Camouflage**
- **False trail following traps capture**
- **Desensitization**.

These techniques are precautions for inhibition in mating (mating disruption). They generally involve the diffusion into the air of quantities of pheromones such as to confuse (sexual confusion) or distract (sexual disorientation) males in the phase of localization of females. In this way, it is possible to inhibit mating by limiting the development of the population. The latter represents today one of the most sophisticated systems in the control of antagonistic insects, capable in particular conditions of eliminating the need for specific insecticide treatments. In particular, sexual disorientation, developed by our concept, involves using low-dose *Pherodrone1.0* to create a series of false traces capable of disorienting males in their search for their partners (Figs. 21, 22, 23, 24, 25, 26 and 27). Compared to other methods, these are already advantageous in small plots or in combination with integrated defense strategies.

5.2.9 Block "I"

Trapping prey is a necessary feature of the food cycle in all terrestrial ecosystems, as a process inserted in the relationships of the food chains between living organisms. In nature, therefore, infinite predation techniques have developed both in the animal and in the plant world. In particular, our attention has shifted to the morphology and functionality of carnivorous plants, and in particular, it was the plants belonging

Fig. 24 Possible aerial
sexual-antagonism scenarios
between *Pherodrone1.0* and
a couple of pests

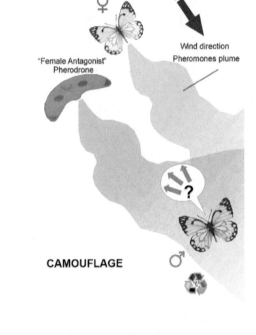

Fig. 25 Possible aerial
sexual-antagonism scenarios
between *Pherodrone1.0* and
a couple of pests

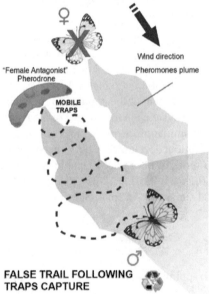

Fig. 26 Possible aerial sexual-antagonism scenarios between *Pherodrone1.0* and a couple of pests

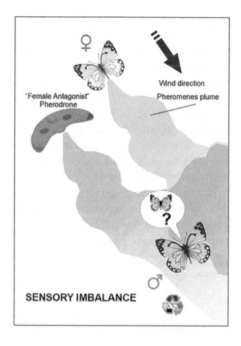

Fig. 27 Possible aerial sexual-antagonism scenarios between *Pherodrone1.0* and a couple of pests

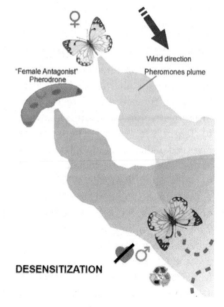

to the *Nepenthes* genus that interested us in possible biomimetic hypotheses. By studying these plants, we have obtained a design hypothesis for constructing a trap that can be attached to *Pherodrone1.0* in the event of an action campaign involving the capture of males. In analogy to the barrel shape of the trapping apparatus, a lightweight canvas bag is impregnated with attractive substances, and thanks to a pot system, once inside it, the insect is no longer able to get out. The trap bag is attached to the ventral body of the UAV. Previously we have analyzed the various scenarios of struggle against the populations of insects harmful to crops based on sexual distraction. One of these alternative possibilities involves the hypothesis of diverting the path of the male individual, attracting him with a trail of pheromones that overlaps the one released by the female. The aim is to capture his sexual attraction to convince him to divert towards the UAV rather than towards the female individual who spread the pheromone plume. The purpose of adopting this technique is to collect many more male individuals than the number of those that can be captured through fixed traps. In case of flight exit of the drone with this purpose of capture, the flying machine is equipped with a trap with keepnet, or glued, obtained by existing re-adapting types.

6 Conclusions

A teamwork of *BionikonLab&FABNAT14* developed the first concept of *Pherodrone1.0* as a result of bionics and biomimetics lectures and design workshop held by Prof. Massimo Lumini during the past two schools' years. Thanks to several brainstorming sets and creative meetings, was generated an original design inspiration for alternative UAV technologies such as those of inflatable toys and high-performance digital technology solutions (*insect sensillae*). We also turned to innovative and still little-known technologies such as those developed in Italy by *Tensairy®Solutions*. The main objective of the students, teachers, and designers involved in this problem-solving process, was to develop a divergent design approach to the UAV sector to evaluate new viable paths in PA. It is necessary to underline that the *BionikonLab&FABNAT14* is not a research center comparable to a traditional University Academic Department or a business or Ministerial Research and Development center. It is bionic and biomimetic research and sustainable design laboratory with a 3D printing fablab added since 2014, located inside a public scientific high school in Sardinia-Italy. The student-researchers are boys and girls aged 14–18, accompanied along the way by some teachers and designers, coordinated by Prof. Massimo Lumini, who has conceived and directed this educational experience since 1996. With this original and innovative research center for *young well-made heads* [50], we have developed numerous didactic experiments relating to the dissemination and use of methodologies in the bionic and biomimetic fields, inspired by the contexts of research and development of the highest level. It is a unique experiment in the Italian and European panorama. The students worked on the project while studying in their typical school curriculum. They did so by following online group conferences, actively participating in creative brainstorming sessions to look for new and divergent ideas, and collaborating to carry out all the analyses and insights needed to learn about the context issues in question. Finally, they made CAD drawings and other material documentaries. Upon delivering the essay, the project reached a theoretical definition of principle. We believe that its strengths can be identified in its inflatable construction technology, its functionality designed to overcome disinfestation through fixed traps, chasing the streams of pheromones scattered in the air, and competing directly in the field with the insects themselves. To carry out these actions of interception and selective recognition of the pheromone flow, the technologies necessary for this seem to have reached a state of the art that can be hypothesized at the mass production level. The sensors, electronic noses, and actuators needed to direct the UAV's flight do not seem particularly expensive and difficult to wire. In practice, they all appear to be active and available technologies on the market at an affordable cost. Many constructive and technological details will have to be tested and prototyped if we want the *Pherodrone1.0* to become an engineered product in all its technical and functional elements, ready to be introduced experimentally in the market of technologists at the service of the PA. Our hypotheses are supported by other academic studies and high-level experiments that we have intercepted on the net. These researches seem to converge in the direction we hypothesized. The future development of this concept can grow through the planning and funding stages of research. Thanks to our contacts, which have extended in recent years internationally, with Research Centers and University departments around the world, we can foresee significant potential developments.

Acknowledgements Prof. Massimo Lumini is the only author of the present chapter but a lot of inspirations, drawings, CAD rendering, and design suggestions, come from the results of several lectures, workshops, and bionics and biomimetics training courses held during the schools-years 2019/2020 e 2020/2021 at BionikonLab&FABNAT14 in Iglesias-SU-Italy. All the infographics, photos, and images accompanying the text are original artwork of the author, except Figs. 9, 10 and 11 (Silvia Musa) and Figs. 18, 19, 20, 21, 22 and 23 (Leonardo Cuccu). The author is thankfull for the BionikonLab&FABNAT14 team, including the teachers Emanuela Manconi and Silvia Musa, and students Riccardo Campesi, Laura Coccolone, Leonardo Cuccu, Michele Di Romano, Paolo Granella, Tommaso Mei, Francesco Matzei, Carlo Saiu, Francesco Soddu.

References

1. Hansson BS, Stensmyr MC (2011) Evolution of insect olfaction. Neuron 72(5):698–711. https://doi.org/10.1016/j.neuron.2011.11.003. PMID: 22153368
2. https://www.nationalgeographic.com/culture/article/neolithic-agricultural-revolution. https://cordis.europa.eu/article/id/430411-neolithic-mothers-and-the-survival-of-the-human-species
3. https://www.cam.ac.uk/research/news/prehistoric-womens-manual-work-was-tougher-than-rowing-in-todays-elite-boat-crews
4. https:www.researchgate.netpublication306096038_Neem_Azadirachta_indica_towards_the_ideal_insecticide

5. https://www.todayville.com/agriculture/the-inception-of-agricultural-chemistry/.   https://ecos.csiro.au/history-of-agricultural-chemicals/. https://www.jstor.org/stable/3493576

6. https://www.airc.it/cancro/informazioni-tumori/corretta-informazione/vero-glifosato-un-erb icida-diffuso-mondo-cancerogeno#:~:text=Forse.-,In%20laboratorio%20il%20glifosato%20p rovoca%20danni%20genetici%20e%20stress%20ossidativo,quella%20dei%20%E2%80%9Cc arcinogeni%20certi%E2%80%9D.   https://www.sciencedirect.com/science/article/abs/pii/S02 73230012000943. https://academic.oup.com/jnci/article/110/5/509/4590280?login=true

7. https://link.springer.com/book/10.1007%2F978-3-319-94232-2.   https://www.aphis.usda.gov/aphis/ourfocus/planthealth/plant-pest-and-disease-programs/biological-control-program. https://www.agric.wa.gov.au/pests-weeds-diseases/control-methods/biological-control

8. https://www.lanuovasardegna.it/regione/2020/12/12/news/allo-scientifico-di-iglesias-il-premio-makeathon-1.39654397. Makeathon is the first #innoismake organized by Abinsula, one of the main Italian players in Embedded, iOT, Web & Mobile and automotive solutions, within the INNOIS-Innovation and ideas for Sardinia project. This making challenge sees makers and FabLab as protagonists and aims to develop solutions and technologies, prototyped with digital manufacturing paradigms, capable of responding to the new needs of communities. A special edition was dedicated to schools. The BionikonLab & FABNAT14 teams of young students aged 14–16 attending the Liceo Scientifico and Liceo Artistico of Iglesias -SU won the first edition and ran for second place in the next two editions. www.innois.it

9. EPRS-European Parliamentary Research Service (2016) Precision agriculture and the future of farming in Europe. Scientific Foresight Unit STOA, PE 581.892

10. Aziz MS, El Sherif AY, Biomimicry as an approach for bio-inspired structure with the aid of computation. https://doi.org/10.1016/j.aej.2015.10.015

11. https://apps.dtic.mil/sti/pdfs/ADA428829.pdf

12. https://cbid.gatech.edu/wp-content/uploads/2018/04/Helms_et_al_2008_Acadia.pdf

13. https://www.mccormick.it/as/precision-farming/

14. https://www.elsevier.com/books/gps-and-gnss-technology-in-geosciences/petropoulos/978-0-12-818617-6

15. del Cerro J, Ulloa CC, Barrientos A, de León Rivas J, Centre for Automation and Robotics (CAR) Universidad Politécnica de Madrid—Consejo Superior de Investigaciones Científicas, 28006 Madrid, Spain. https://www.mdpi.com/2073-4395/11/2/203/htm

16. https://www.routledge.com/Unmanned-Aerial-Vehicle-Systems-in-Crop-Production-A-Com pendium/Krishna/p/book/9781774634370. Maddikunta PK, Hakak S, Alazab M, Bhattacharya S, Gadekallu TR, Khan WZ, Pham QV, Pham5*Unmanned aerial vehicles in smart agriculture: applications, requirements and challenges. School of Information Technology, Vellore Institute of Technology, Vellore, India, https://arxiv.org/pdf/2007.12874. https://it.scribd.com/document/ 499211602/UAV-Applications-in-Agriculture-4-0

17. https://www.faa.gov/uas/.   https://geodetics.com/drone-based-lidar-photogrammetry-systems/? gclid=Cj0KCQiAnaeNBhCUARIsABEee8VhhkDtsHt4tCepyiEmne3amPugGsniLFm2e_Civ6 1UtGWyxE-auekaAorpEALw_wcB. https://www.dronefly.com/the-anatomy-of-a-drone

18. Allegretti M, Unmanned aerial vehicle: tecnologie e prospettive future. Alma Mater Studiorum Università di Bologna-Scuola di Scienze. Corso di Laurea Magistrale in Informatica 2015–2016. https://amslaurea.unibo.it/11979/1/tesi_marcello_allegretti.pdf

19. https://www.sciencedirect.com/science/article/pii/S0960982215009458#:~:text=Modern%20b irds%20achieved%20their%20enormous,in%20the%20Mesozoic%20when%20a. https://evolut ion-outreach.biomedcentral.com/articles/10.1007/s12052-009-0133-4

20. https://www.loc.gov/item/2021668201

21. https://journals.plos.org/plosone/article?id=10.1371/journal.pone.0086506. https://www.nature. com/articles/s42003-018-0029-3. https://www.researchgate.net/publication/227741934_Aerial_ hunting_behaviour_and_predation_success_by_peregrine_falcons_Falco_peregrinus_on_sta rling_flocks_Sturnus_vulgaris

22. https://www.wired.it/scienza/lab/2020/02/18/video-volo-uccelli-bolle/

23. http://www.liceomedi.com/volo/sito/il_volo_negli_uccelli/il_volo_degli_uccelli.htm

24. A clash of wings. Nature 569(7755). Accessed 9 May 2019

25. https://www.ansa.it/scienza/notizie/rubriche/biotech/2011/11/23/visualizza_new.html_1496 2282.html
26. https://cordis.europa.eu/project/id/204513/it
27. https://youtu.be/cJJowVxiaRU
28. https://www.livescience.com/33075-how-bees-fly.html
29. https://youtu.be/e_44G-kE8lE
30. University of Cagliari Department of Experimental Biology Section of General Physiology Ph.D. in Morphological Sciences The activity of motor neurons correlated with the "calling behavior" and the pheromone release mechanism of the night butterfly Lymantria dispar is modulated by octopamine: evidence electrophysiological. Doctoral thesis: Dr. Piera Angioni Tutor: Doctoral coordinator: Prof. Anna Liscia Prof. Alessandro Riva Academic years: 2004–2007
31. https://www.zanonia.de/aurora_evo.php
32. https://www.steelpillow.com/aerospace/tailless.html
33. https://asknature.org/strategy/seeds-with-efficiently-shaped-wings-glide-slowly-to-earth/
34. Model Aviation (1983) rif. http://www.endlesslift.com/foam-zanonia-glider/
35. Lumini M, The invisible factories: nature's technologies and design of artificial innovation. Cuadernos del Centro de Estudios en Diseño y Comunicación N°149-UP Buenos Aires AG. https://fido.palermo.edu/servicios_dyc/publicacionesdc/cuadernos/detalle_publica cion.php?id_libro=932
36. https://www.researchgate.net/scientific-contributions/Jing-Wang-2162768043
37. Denny M, Mc Fadzean A (2011) L'ingegneria dei materiali. Adelphi Edizioni, Milano 2015 trad.pag.169. Engineering animals. How life works. The President and Fellows of Harvard College
38. BugInfo, Pheromons in Insects. Smithsonian. https://www.si.edu/spotlight/buginfo/pheromones
39. https://agronotizie.imagelinenetwork.com/agrimeccanica/2017/09/28/parassiti-castagne-al-sic uro-grazie-ai-droni/55672
40. https://www.researchgate.net/figure/Different-types-of-pheromone-dispensers-used-in-apple-a-Rubber-tube-handmade-dispenser_fig3_305344035
41. https://youtu.be/QK-lxD5pQr0
42. https://www.partylandia.com/palloncini-ecosostenibili-certificati/
43. Zang Y, Xiuyan G, Cheng Y, Review of tensairity and its applications in agricultural aviation. Accessed 31 May 2016. https://doi.org/10.25165/IJABE.V9I3.2510. Corpus ID: 113489439
44. https://www.macotechnology.com/prodotti/sistemi-tensairity/. www.https://www.tensairitysolut ions.com/about/
45. https://www.sciencedirect.com/science/article/abs/pii/0925400594870853
46. Cui S, Ling P, Zhu H, Keener HM, Plant pest detection using an artificial nose system: a review. https://www.mdpi.com/1424-8220/18/2/378/pdf
47. https://www.alice.cnptia.embrapa.br/bitstream/doc/1094178/1/Brezolin2018ArticleToolsFor DetectingInsectSemioch.pdf; Brezolin AN, Martinazzo J, Muenchen DK, de Cezaro AM, Rigo AA, Steffens C, Steffens J, Blassioli-Moraes MC, Borges M, Tools for detecting insect semiochemicals: a review. https://www.alice.cnptia.embrapa.br/bitstream/doc/1094178/1/Bre zolin2018ArticleToolsForDetectingInsectSemioch.pdf
48. http://www.matto.design/it/metallo-schiuma-di-alluminio/. https://cordis.europa.eu/article/id/ 411669-aluminium-recycling-produces-metal-foam-for-industrial-products/it
49. https://shearwater.nl/index.php%3Ffile=kop125.php.html
50. Morin E (2000) A cabeça bem-feita: repensar a reforma, reformar o pensamento [The Well-Made Head: Rethink the Reform, Reform the Thought]. Bertrand, Brasil, Rio de Janeiro

Exploiting the Potential of Nature for Sustainable Building Designs: A Novel Bioinspired Framework Based on a Characterization of Living Envelopes

Tessa Hubert, Antoine Dugué, Tingting Vogt Wu, Denis Bruneau, and Fabienne Aujard

Abstract Living envelopes, such as biological skins and structures built by animals, are functional and sustainable designs resulting from years of evolution, conditioned by biological and physical pressures from the environment. When building a home, animals demonstrate inspiring strategies to protect themselves from predator threats and external climatic conditions. As for human buildings, temperature, humidity, air quality, light, are some of the various factors they have to manage for optimal conditions. Facing the climate emergency, growing efforts to build durable designs have led designers to search for more efficient or alternative solutions by observing Nature. The emerging field of bioinspiration including animal architecture has already brought few but rare exemplary innovations that were integrated into building designs. Data on animal architecture are scattered among various biological domains, from observation of species habitats by zoologists such as entomologists or ornithologists, to bioindicator studies by climatologists. Data collected by scientists is available in eclectic idioms, a challenge to be fully comprehended by building designers. This

T. Hubert (✉) · A. Dugué
NOBATEK/INEF4, National Institute for the Energy Transition in the Construction Sector, 64600 Anglet, France
e-mail: tessa.hubert@u-bordeaux.fr

A. Dugué
e-mail: adugue@nobatek.inef4.com

T. Hubert · T. Vogt Wu
Institute of Mechanical Engineering (I2M), UMR CNRS 5295, Université de Bordeaux, 33400 Talence, France
e-mail: tingting.vogt-wu@u-bordeaux.fr

T. Hubert · D. Bruneau
Ecole Nationale Supérieure d'Architecture et Paysage de Bordeaux, 33405 Talence, France
e-mail: denis.bruneau@bordeaux.archi.fr

F. Aujard
MECADEV UMR CNRS 7179 —National Museum of Natural History, 91800 Brunoy, France
e-mail: fabienne.aujard@mnhn.fr

© The Author(s), under exclusive license to Springer Nature Singapore Pte Ltd. 2022   289
F. L. Palombini and S. S. Muthu (eds.), *Bionics and Sustainable Design*, Environmental Footprints and Eco-design of Products and Processes,
https://doi.org/10.1007/978-981-19-1812-4_10

chapter presents a characterization of living envelopes aiming at facilitating the transposition of some relevant biological features into innovative and sustainable architectural designs. The approach is architecture and engineer oriented, assessing biological functions and strategies, using criteria that are meaningful to building designers: functional and temporal analyses of spaces and materials, physical factors regulated through envelopes, behaviors, and interactions of species. Applied to a sample of species and animal-built structures, the characterized biological role models put forwards multi-functionality and efficiency through relevant construction techniques, the use of local resources, as well as behavioral adaptation. Examples of applications inspired from the characterized species are described, from theoretical proposals to a very practical application of an adaptive envelope skin inspired by the Morpho butterfly.

Keywords Bioinspired design · Adaptive skin · Biological characterization · Animal construction · Living skin · Building envelope · Design framework · Sustainable designs

1   Introduction

Living species have evolved for 3.8 billion years under environmental pressures [1] such as geographic and climatic conditions, competition of material resources, or predation. Observable variations in organisms are called the phenotype and it includes morphology, physiology, and behavior characteristics [2]. Through phenotypic adaptation to maintain viable conditions in regard to the varying external environment, species resulting with the best selective advantages and characteristics have a higher probability to survive.

As the living species found today on Earth are the outcome of a continuous natural selection process while facing varying and extreme conditions, they can prove to be resourceful and very effective, and as such, they have much to teach us in a time where resilience to climate change has become a necessity for our own survival [3]. In Europe, most countries present an ecological deficit, meaning that their population's footprint CO_2 emission with fossil fuel use exceeds the biocapacity of the area available to that population [4]. Concerns about climate change are rising, while new environmental requirements are emerging in the building sector. As it accounts for around 40% of the global energy consumption, sustainable advances in the built environment are essential to reduce greenhouse gas emissions. And more specifically for the building envelope as it has a key role to play in the overall building energy consumption [5].

The envelope is a widely used term to define the interface of a building between its internal and external environment. It includes the roof, façades, floor, openings such as windows or doors, and even transitional spaces such as atria or lobbies. Transposed to Nature, envelopes can be described at different scales. First, on a living organism scale, with biological skin such as the human dermis or the shell of a snail. Second, on a macro scale with constructions built by organisms, such as bird nests or tunnels of ant colonies. Both types of envelopes act as filtering barriers between two fluctuating environments and demonstrate multi-functional properties which could help inspire human designs. The feathers of penguins, for instance, act as a thermal insulation while being superhydrophobic and anti-icing [6], which allows them to daily dive in sub-freezing waters. Combined to a natural oil with water-repelling properties, nano-scaled ridges on the feathers make it impossible for microdroplets to form ice. This envelope could inspire resistant insulated designs while offering water and ice-repellant properties.

Bioinspiration has recently emerged as an interdisciplinary field to lead to innovative, efficient, and durable designs [7]. Research on the area is growing, resulting in several cases of bioinspired building envelopes inspired from species described in the literature by biologists [8–11]. The current description of living organisms would account for 1.7 million species out of 10 million left to describe [12]. Even for quite a small ratio, the amount of potentially inspiring species described in the literature is substantial. However, practice in architecture shows that the selected biological models are often picked among commonly known species, from animal or plant kingdoms [13]. The abstraction of the species also rarely led to multi-functional designs, whereas the feature of the species which provided inspiration usually is involved in several regulatory roles. This pattern could be related to a scarce involvement of biologists during the design process [14], driving the design team to construct inspired designs on biological shallow knowledge.

Yet, even though rarely developed by biologists [15], several tools exist to facilitate the bioinspiration process [16, 17]. More specifically, thesaurus [18], ontologies [19–21], or taxonomies [22] are being developed to help the understanding of the properties and functioning of living species and selecting the relevant case study for the purpose of the application. However, they do not focus on living envelopes hence might not provide a characterization with terminology and semantic oriented toward architects and engineers. Only few methodologies for characterizing multi-regulation capacities of biological envelopes or for proposing multi-functional building envelopes were developed in the literature [10, 23–26]. This chapter presents a characterization of living envelopes derived from the methodologies of [23, 24], further developed using biotic, space, and time-scale criteria. A proposed framework backbone the innovation process, integrating this characterization as a support for the identification and understanding of relevant biological features. They are then explored and transposed into bioinspired concepts. The chapter proposes the description of some general principles for building envelope emerged from this methodology, as well as an actual design.

2   Living Envelopes

2.1   *Building Designs in Needs of Bioinspiration?*

To respond to climatic challenges, extensive research in the building field is focusing on improving the envelope's efficiency. Yet, this is not a recent trend, as the building envelope at various periods of time has been adapted to the occupants' needs, along with construction techniques and technologies. Pioneers of bioclimatic buildings in the 1960s were already targeting reductions in energy consumption of the whole building, considering the overall design of the envelope without the use of separate or additional equipment [27]. Ever since the building envelope was more than a load-bearer structure and a thermal shield, additional layers for specific functions were added, rather than providing a new design based on holistic analysis and resulting in a multi-functional feature.

Consideration of harvesting solar energy using passive systems and influencing the user behavior [28] also goes back and highlights multiple perspectives of the envelope. The approach given above is mainly engineer-oriented, but the building envelope definition changes according to the observer. For the occupants, the envelope elements surrounding them are managing heat and light transfers that result in visual and thermal comfort and delimiting private from public space. For an architect, it can represent a contact surface between the building and the city, while for the energy-focused engineer, the envelope acts as an interior/exterior separation. Therefore, the list of expectations for the building envelope can be broad. Adding to this, the regulations and standards are numerous and complex, and can depend on the countries, but put together they define a regulatory framework that allows characterizing the expected indoor controlled conditions. Those integrate thermal, light, noise, and air comfort for the occupant while being exposed to varying outdoor conditions [29].

In that sense, living envelopes face very similar needs to which they are highly responsive. Despite being inert like buildings, some plants for instance can demonstrate dynamic mechanisms at different scales. They might respond to water through tropism and nastic macro-movement and manage their carbon dioxide needs by opening and closing their stomates [30]. Species have discomfort ranges where their metabolism and physiology are not properly functioning, which will push them to try to reach better or optimal conditions.

The field of examples in biology being very broad, the authors narrowed down their study of biological organisms to samples including structures built by animals and biological skins only, in order to find responsive and multi-functional inspiring properties.

2.2 Biological Skins

Biological skin refers to the envelope separating the internal environment of living organisms from external conditions. As the diversity of biological skins is wide, some research reduced the sampling of their study using various selection criteria [24]. reduced the sampling of their study [13] using various selection criteria. For instance, only the outermost envelopes are taken into account, meaning the study excluded partition envelopes such as blood vessel walls or nervous tissue. Though arbitrary, this choice makes sense since biological skins were selected by analogy with the building envelopes as they are both exposed to similar external conditions (sun radiation, wind, daily temperature variations...) and encompass the whole system. In addition, only pluricellular Eukaryotes are studied; unicellular species, as well as *Bacteria* and *Archaea* domains [2] are excluded because their description in literature is scarce. At last, no living organisms from marine environments were selected as their range of environmental conditions was considered too distinct from the classical terrestrial environment of buildings. Although existing or in their design phase [31], resilient architecture as marine and underwater habitats worthy of Jules Verne is not the norm yet.

2.2.1 Selection Among the Diversity of Skins

The selection criteria described above resulted in a list of 10 types of biological skins [24] defined by their base skin and appendages, i.e., their natural prolongation. The base often includes several layers of tissues, such as the hypodermis, dermis, and epidermis of the human skin. Appendages can be various as shells, scales, feathers, fur, and provide many additional functions to the skin.

For our own characterization, a few species corresponding to those types of envelopes were selected, based on the available data in the literature, on the various climate range of their habitat, but mostly on the diversity of functions their envelopes provide to keep them alive (sample in Table 1).

Chameleons were chosen among the horny scale species for their known color modifications, a property provided by their dermis. This ability, although associated with camouflage and display for social interaction, would help them thermoregulate themselves by changing the configuration of the nano-crystals contained in one of their dermis, reflecting more in the near-infrared (IR) range of the spectrum [33]. Likewise, Silver ants have high reflective properties in the near-IR range, and high emissivity in the mid-IR, thanks to their appendages. Their setae, microscopic hairs triangular-shaped in the case of silver ants, allow them to cool down by a few degrees [34], just enough to find food for the colony by going outside under the desertic Saharan climate. On a nano-scale, the wings of the Morpho butterflies provide self-stabilization in temperature: they emit in the IR when they are too warm, and absorb it again when they cool down [35, 36].

Table 1 Selected species from various types of envelopes

Selected species	Type of envelope [24]	Reign/class/order	Climate (Köppen classification [32])
Chameleon	Skin, scale	Animal/Reptile/Squamata	Temperate/mesothermal: humid subtropical
Garden snail	Skin, shell	Animal/Gastropod/*Stylommatophora*	Temperate/mesothermal
Mouse lemurs	Skin, fur	Animal/Mammal/Primate	Tropical to temperate
Silver ant	Cuticle, setae	Animal/Insect/*Hymenoptera*	Dry climate: hot desert
Morpho butterfly	Cuticle, scale	Animal/Insect/*Lepidoptera*	Temperate to tropical
Pine cone	Plants' bark	Vegetal/*Pinopsida/Pinales*	Various

Mouse lemurs (*Microcebus sp*) are nocturnal species among the smallest lemurs. They have the ability to enter into states of torpor, i.e., they can reduce their metabolism and their body temperature [37]. As appendage, their dense fur of down allows for avoiding water and heat losses. They also have a mechanism of active heating, thermogenesis without shivering, which uses a particular tissue containing a specific protein, commonly called brown fat, and also present in humans [38].

The Garden snails were selected for their high sensitivity to hygrometric variations. When hibernating, they retract inside their shell and seal it using a temporary mucous veil called epiphragm, avoiding desiccation [39]. When active, they can hydrate by absorbing humidity on surfaces through their mucous until they have reached satisfactory hydration [40]. Pine cones are also very hygro-reactive as they open when the ambient air is dry enough to ensure the dissemination of their seeds.

2.2.2 Existing Applications

The species in our sample show functionalities expressed by their base skin, their appendages, or the combinations of both, all relevant in regard to the building envelopes' expectations. But they also come with other regulating functions which will not be all cited here; breathability (Silver ant), hydrophobia (Morpho butterfly), arrangement of matter following the golden angle (Pine cone), etc. When used as inspiring models, it is probable that only one dominant function is transposed into a technology, whereas they demonstrate much more regulation assets.

Several buildings and building envelopes were inspired by biological skins [9–11, 13]. As outlined by [13]—a study on the design process of 30 built bioinspired envelopes including designs inspired by biological skins—few cases address multiregulation apart from coupling light and heat transfers. The latter are interdependent phenomena and quite often simultaneously targeted by the designers.

Likewise, most of the implemented or currently being implemented designs that were inspired by our chosen models were found to result from the abstraction of one function from their biological models. The silver ants inspired a passive radiative cooling under direct solar radiation, using seven layers of HfO_2 and SiO_2 resulting in high reflectivity in the solar spectrum and emissivity in mid-IR [41]. The applied photonic approach is actually very close to the phenomenon at the root of the Chameleon change of color, functioning with periodic nano-structures [42]. In line with photonic crystals, the Morpho butterfly has inspired the design of solar panels for improved efficiency [43, 44] through for instance anti-reflection properties [45]. More generally, the potential on radiative cooling using Morpho-inspired nano-structures has been investigated and could be implemented into building applications [46].

Note that like heat and light mentioned above, functionalities expressed by these designs are sometimes coupled; for instance, the morphological features of the scales of the Morpho wings provide thermal regulation, but also hydrophobic properties. By closely mimicking the structuration, using technologies such as nano-printing, both properties might get achieved even though only one was targeted.

2.3 Animal Constructions

Animal constructions, also called animal architecture in the literature, are structures built by animals. Though the designation *animal architecture* is meaningful for architects, it could be confused with biological fields such as molecular architecture; hence the authors will prefer animal construction over this appellation.

The functions of animal constructions are described in the literature as protection from a hostile environment such as predators or climatic conditions, trapping preys, and intraspecific communication [47]. These structures usually are destined for themselves, for their offspring or family, but they can be taken over by predators, co-occupied with other species, or even re-adapted later on by other species for new purposes. Constructions abound in nature under diverse forms—burrows, nests, webs, tubes, caves—and embed sophisticated functions.

By analogy, these envelopes built by living organisms resemble in many ways to human buildings: they serve a similar function of shelter while being exposed to uncontrolled varying factors from the outside environment. And while humans build structures with very energy-consuming techniques and most of the time imported materials, animals only use local resources that are processed with soft chemistry and are biodegradable (Fig. 1).

2.3.1 Primary Functions

Endothermic animals, such as mammals, can maintain their body at a certain temperature, mostly through metabolic mechanisms. By opposition, ectothermic animals

Fig. 1 Examples of animal constructions: from left to right, wood ducks, owls, and wasps. *Credit* Public domain, images from Davie [48] and Abbott et al. [49]

such as reptiles, also known as *cold blood* species, have none or very few internal heating sources, hence they rely on environmental sources to be at favorable temperatures. Regardless of their physiological resources, they all have optimum temperatures. Many of them were reported to create shelters to filter, store or dissipate heat. These shields come with managing other vital factors—air renewal, humidity level, water flows such as floods—while undergoing environmental variations.

Animal constructions are also exposed to predation, and threats can be managed in different ways. Camouflage helps animals stay hidden; for instance, some species are suspected to use branches with lichen to blend in the background when constructing their nests [50]. Another technique would be preventing the invasion either by using signals or defensive elements; the nest Chimpanzees build every day as a platform to sleep would include paralytic compound and thorny leaf stems as protection and dissuasive elements from potential predators [51].

Storage and cultivation in constructions help animals endure seasons or periods of time where food is scarce. Kangaroo rats would store seeds in their mounds, as it would have a more favorable environment to make them moldy [52], a seed characteristic they would prefer [53]. Animals sometimes even build special storage rooms but at a higher cost, mostly because of maintenance and a higher threat from predators and thieves [47].

For this study, communication (such as display by the Male Satin bowerbirds build to impress females) and transportation constructions were excluded. Although resourceful and instructive, they appeared out of our scope during our research.

2.3.2 Perimeter of Study

Specific references in the literature give overviews on multiple animal constructions [47, 54–60]. Most of them contain descriptions, drawings, photographs of construction processes, used materials, and operating principles of animal constructions. The descriptive and narrative discussions are all specific to their authors and are mostly qualitative in their characterizations. To find specific data, e.g., the hygrothermal properties of some construction envelopes, it was necessary to search specific literature about the builder species. Despite this process, very little quantitative data was found and they usually mostly cover the thermal aspects of the nests.

The authors selected 50 constructions from species with diverse needs and constructions techniques, to have a diversified sample. For this, they chose various animal classes, including marine species. The resulting sample contains constructions made through all four processes mentioned above, with various shapes, materials, and features (synthesis in 3. Take-aways from animal constructions).

2.4 Taxonomic Bias

The distribution of our samples does not reflect the diversity of described nor estimated species on Earth. To approximate research activities according to 24 taxonomic classes, a study calculated the occurrence of data from the Global Biodiversity Information Facility (GBIF) database, i.e., the biggest biodiversity data repository available proportionally sampled to its number of known species [61]. They found that land plant classes and animal classes including birds, mammals, amphibians, and reptiles, are over-represented, while arachnids, fungi, and insects are under-represented, the latter having the worst occurrence. They underline the fact that classes such as birds were known as over-represented in many disciplines for a long time [62] and the gap with the rest of the species is barely decreasing since the 1950s.

The authors' sampling of biological skins is limited as it excluded many species, hence it cannot be properly confronted with the expected taxonomic biases investigated in the cited references. On the other hand, the sampling of animal constructions confirms some arguments: birds and fungi are respectively over and under-represented, which is consistent with the literature. As for insects, arachnids, and mammals, the trend seems to be reversed and matches the authors' occurrences of animal-built structures in the literature: indeed, construction for birds and insects were presented as the most occurent by Hansell [47].

The difference of occurrence in research and publications—hence of taxonomic biases—between species and built structures of species—is not surprising. First, all species do not build homes; arthropods apart from arachnids and insects are not much reported as builders, neither are reptiles, which explains their absence in the sample. Secondly, research has established a strong link between societal preference and scientific productions [63]. It is understandable that most of the built structures found in the literature are either built from birds (already popular) or burrows from arthropods as they might attract curiosity. As for mammals, the examples of constructions might be fewer since they are endothermic and can manage their internal conditions with sole metabolisms and behavior strategies. The difference with birds, endothermic as well, would be their size: following the square-cube law [64], the heat losses of birds are higher than larger animals, since they have a higher surface-to-volume ratio; hence more surface area to lose heat compared to their volume.

3 An Engineer/Architecture-Oriented Characterization

The use of bioinspiration in the building sector has led some researchers to focus on design processes, frameworks, and tools to facilitate the design of technological solutions based on biological models [8, 10, 65]. In support, an ISO standard was proposed to homogenize existing design processes [66], and methods and tools such as databases, thesauri, or taxonomies were recently developed [16]. However, practices of bioinspiration in architecture have shown that most of the bioinspired designs do not rely upon these tools and design processes proposed in the literature, and that bioinspired designs are most of the time monofunctional [10]. Many concepts are never implemented in buildings and stay stuck at the stage of concept or patent [67]. It underlines the challenge of the transposition from the biological models to technical solutions, which can then be applied to a building. More advanced interdisciplinarity during the design process would certainly help the transferability of a biological principle into technologies, by providing a better understanding of the phenomena at stake.

One of the challenges is that although a broad range of papers proposes descriptions of living envelope features (behavioral biology, anatomy...), the information is not structured to clearly identify the causes or triggers of the involved phenomena, and what it can accomplish in physical terms. Moreover, existing tools to describe these phenomena lack the involvement of biologists as they are usually developed for and by engineers [15]. Thus, structuration of the available data on living envelopes is necessary, as this is the biological starting point that will then be exploited during the design process.

3.1 Existing Tools for Abstractions

Structuration of biological data to support bioinspiration is found in abstraction tools. Abstraction is crucial in the early steps of the bioinspiration design process as it consists in analyzing the biological systems and creating parallels between biology and their own domain. The way models are abstracted plays a key role in the conceptual phase; hence it requires relevant tools to ensure reliable understanding and connection to our own design domain.

Biology covers a very broad field of life sciences and can be divided into multiple subcategories such as mycology, botany, zoology, themselves including other fields of study (entomology, ornithology, etc.). Even in the biological domain, heterogeneous information and vocabulary and semantic issues do exist [68]. As highlighted in [69], the structuration of knowledge using tools such as ontologies is becoming valuable in bridging the gap between different domains and constructing integrative biology across its subcategories. Likewise, it could help structure knowledge across multiple domains such as engineering or architecture.

Whether they are biological or technological, representing systems through functions is already commonly used to abstract problems in engineering. It provides a systemic way of specifying the functionalities of one model while contributing to its continuous improvement. Several tools tackle functional modeling in the literature, proposing ontologies, thesauri, and methods in the frame of bioinspiration processes. Table 3 presents tools mostly using structural and functional representations or descriptions. It puts forwards various approaches, such as behavior, strategy, and causality which could make sense for the characterization of complex structures interacting with their occupants, such as animal-built structure.

Only one of the listed tools is dedicated to an ontology for bioinspired architecture, but the project is currently focusing on the energy component although they plan to extend their study to more parameters such as water management or structural properties of buildings [70].

3.2 Systemic Approach

No tool was found to help build systemic descriptions and representations of animal-built structures in the literature. To structure their biological characterization, the authors first set a list of commonly accepted requirements for the building envelope and for the indoor conditions to provide comfort.

Human perception of comfort is related to three main factors: physiology, behavior, and psychology. The physiological is related to the metabolism, i.e., the body's mechanisms to keep acceptable conditions such as perspiration or metabolic heat. Behavioral mechanism represents actions taken by the person to reach a more comfortable state, such as getting dressed or doing physical activity. The last factor has been shown to be very variable according to the people, their culture, and their geographical origins; for instance, the habit of a certain type of climate increases the body's tolerance to this same climate in terms of temperature. As such, it is hard to set standards on indoor environmental quality (IEQ). However, optimal conditions can be approximated using the following factors:

- Stable temperature usually appreciated in a range from 17 to 22 °C [81];
- Relative humidity in the 40%–65% range [81];
- Illuminance during the day between 300 lx [81] and 500 lx, and limited illuminance during the night;
- Carbon dioxide concentration under 1000 ppm and VOCs (Volatile Organic Compounds) under their respective TLVs (Threshold Limit Values);
- Airspeed higher than 0.1 m/s for renewal and below 0.3 m/s to avoid thermal discomfort;
- Background noise below 35 dB [82].

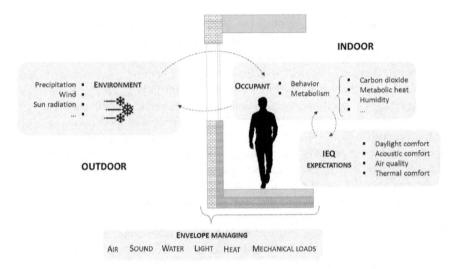

Fig. 2 Connections between outside and inside inputs on the IEQ expectations

The variations of these internal conditions are either due to environmental inputs, such as the temperature or relative humidity from the external ambient air or to the occupants and users. Indeed, the occupants unintentionally unbalance conditions through exchanges between their own body with their close media (heat exchanges, perspiration, breathing). Also, they usually can action commands that operate the building (HVAC systems, openings, etc.) and can modify conditions by their behavior. Figure 2 illustrates environmental factors and occupants as influencing indoor conditions. They highlight the regulation functions the building envelopes should manage: air, water, light, and heat transfers, as proposed by [23], completed by [24] with sound transfer and resistance to mechanical loads.

This representation of the interaction between building comfort and functionalities of the envelope foresees the complexity of characterizing living envelopes, whether they are organisms or built structures. They are part of a system whose inputs and outputs are interconnected. This angle directly refers to a systemic approach, i.e., an analysis method to handle systems from a global point of view. Commonly used in biology or engineering without being cited as such, it aims at understanding complex systems through three main concepts [83]. First, a system is governed by loops balancing or unbalancing each other. In the building, these loops were identified as indoor variations due to the occupants and environmental factors, in time scales of hours, days, or seasons. Second, all parts of a system are connected to each other. This can relate to the interconnection of the system elements: for instance, the temperature felt in a room depends on many factors, such as the air velocity, depending on itself on the shape of the room, on the user movements, on the airtightness of the envelope, and so on. Third, the different parts of a system working together result in synergy. The combined effect is greater than the sum of their separate effects. This concept also implies the dynamic and evolutionary properties of the system.

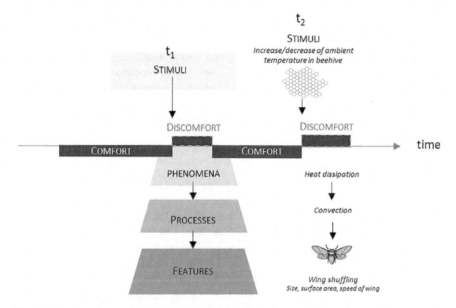

Fig. 3 Structure of proposed characterization for living envelopes. The "Comfort" zone in dark blue stands for the optimal range for one specific biological need. Disruptive elements at time t_2 are an illustration of the concept illustrated for time t_1

All three concepts are complementary and helped the authors define characterization criteria for understanding and describing biological skins and animal constructions. The chosen structure for the characterization is presented in the scheme of Fig. 3. Stimuli, such as environmental factors or interaction with other species, unbalance the system. The response of the system can be either the direct effect of stimuli on the system (e.g., the sun radiation leads to a rise of body temperature of a species) or readjustment of the system to return to comfortable conditions (e.g., discomfort in the rise of temperature induces the species to reach a shaded area).

Responses of various stimuli occur along different temporalities and locations of the system. For a bee, stimuli can have an impact either on their wings, their scales as a sub-part of the wings, or a completely different area as the thorax. Responses to these disturbances can be described through physical phenomena which are the very same heat, light, air, water, sound transfers, and structural properties expected to be managed in the building envelope.

3.2.1 Physical Phenomena and Features

Physical phenomena can be expressed through functions, physical processes, and features (see Table 4). For instance, the wool of some animals prevents them from losing heat with low conduction. This is due to the composition of matter, keratin, and its arrangement, as it traps air which has low conductivity compared to most solids. This phenomenon can then be described under the structure: HEAT (retain) > Conduction > Composition and arrangement of matter.

The list of features given in Table 4 is not comprehensive, as living organisms have developed extensive behavioral, physiological, or morphological [84] adaptation strategies. They are the essence of the multiple functionalities found in nature. As they are categorized similarly as an architect or engineer would describe the envelope properties, this depiction of physical phenomena should assist the design of bioinspired systems.

3.2.2   Stimuli

The description of biological models is not complete without determining the elements that influence and interact with them. These elements are part of their direct surrounding environment and can be dissociated into two main categories: biotic and abiotic factors, listed in Table 5 for biological skins and animal constructions.

The abiotic factors [85] are non-living elements of the environment that affect living organisms. As the ambient temperature can affect the indoor conditions of a building, it also affects the comfort conditions of living species. Rain, wind, solar radiation are all factors that living organisms need to comply with.

The biotic factors [86], complementary to abiotic factors, are the living elements affecting living organisms. For instance, behavior mechanisms or metabolism are biotic factors affecting the species' internal conditions. A mechanical behavior from the organism such as the pilomotor reflex, commonly called goosebumps, makes hair raise and traps air for a thin layer of insulation when some mammals are cold. This mechanism has a direct impact on the species, as well as on the indoor condition of a volume, as the occupant provides heat, air, water gains into the air (see Fig. 2). Inter and intraspecies relationships, e.g., mating or predation, are also considered biotic factors as they unbalance living systems.

3.2.3   Synthesis

The resulting characterization is aimed at being elaborated into a functional structure that fits the architect and engineer ideation language, structuration, and process. The characterization is schematized in Fig. 4 and consists of a systemic approach, determining for each case the following elements:

– **A system** (biological skin or animal construction) **and its sub-systems**: these are either layers of the biological skins (such as the "skin" or base coupled with an appendage as described in Sect. 2.2.1), parts of the organism envelopes (wing vs. thorax for *Lepidoptera*), or partitions of the animal constructions. For the latter, in the same way that humans compartmentalize buildings, some animal constructions are dissociated into partitions providing food production, waste

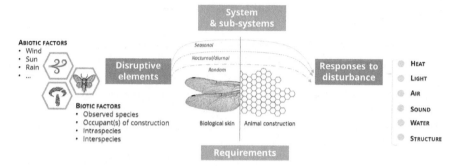

Fig. 4 Schematic representation of a system of animal architecture: disruptive elements and functioning predictions

disposal, or storage. Note that chambers ("incubation area", "inner construction", "room") were spaces described predominantly in the literature, mainly for parturition, brood, and juvenile growth implying very specific temperature ranges requirements.

- **Requirements for the system**: they correspond to the biological needs of the living organisms (needs in water, internal temperature, etc.), to the indoor comfort conditions for animal constructions with ranges that meet the biological needs of the construction users, or to the sociological needs of the species (interactions, e.g., communication or reproduction).

- **Stimuli** as abiotic and biotic factors impacting the internal conditions of the system. Biotic factors are more numerous for animal constructions, as it adds up with the biotic factors of the structure occupants.

- **Responses to the stimuli** to stay in or return to optimal conditions for survival: these responses can be described using processes, and most importantly features (Sect. 3.2.1), which are key to understanding the underlying functioning of the system. Note that it is common that multiple phenomena take place at the same time or co-operate at different levels: hence, their features and corresponding spatial scales must be thoroughly described. Likewise, disturbances and responses operate at different time scales: random stimuli might not trigger the same mechanisms than expected disturbances, such as daily or seasonal events.

The description of time and spatial scales for the various physical phenomena and their linked features provides a multidimensional vision of the analyzed system. Although not easy to grasp, this complexity is what brings emergence to the whole. Inspiration from biological models can happen at different levels or scales, and it is up to the designers to choose what is relevant for their own purposes.

4   Take-Aways from Animal Constructions

For a better understanding of the proposed characterization, the next section illustrates an example of the characterization process through the prairie dog burrows.
Applied to the entire sample (list in Table 2), several main take-aways have been
identified regarding multi-functionality of building envelopes as well as construction modes. Those are worth addressing as guidelines or as research areas to further
explore toward more efficient, responsible, and durable buildings and will be briefly
described in the following.

4.1   Example of Characterization: The Prairie Dog Burrows

Prairie dogs, from the Genus *Cynomys* of the squirrel Family (*Sciuridae*), are
mammals from the *Rodentia* Order. They are so-called *prairie dogs* because of their
habitat—open vegetation with herbs and grasses mainly—and their alarm calls when
they see danger, which are similar to the barking of a dog. This endothermic species
mainly lives in North America under two subgroups recognized by mammologists,
black-tailed & white-tailed, and they co-habit in groups in burrows (photograph of
a burrow entrance in Fig. 5). While some of them are believed to hibernate continuously during winter, others would facultatively enter torpor during winter, spring,
and summer [87]. The Black-tailed prairie dogs, for instance, rather choose winter
rest over hibernation, i.e., they will have a rest interrupted by numerous awakenings,
with moderate but not drastic hypothermia.

   Although some areas they live in can get as warm as 38 °C in the summer and as
cold as—37 °C in the winter [88], they do not rely on hibernation to survive winter,
but depend on their burrows for protection from the weather and predators. Hence,
the characterization of the prairie dog burrows might bring relevant insights on how
to ensure thermal comfort and sufficient air renewal using elaborate tunnels.

   Following our established characterization process, various elements were investigated in literature documents (note that the information provided here is not
comprehensive):

   Systems and **sub-systems**. The burrows are dug straight down in soil by the prairie
   dogs. They are made of openings, circulation areas, rooms, and an envelope, i.e.,
   the ground in which there are dug. The openings are several entrances, "back
   doors", and plugged emergency exits in case of intrusion. The entrances are mound
   shaped, to be a lookout and to redirect water during heavy rain. The rooms serve
   diverse functions: chambers for the offspring and adults, rooms for feces, or for
   storing food.
   Comfort requirements. The prairie dogs require an internal temperature of
   around 36 °C. Their body temperature can decrease down to 10 °C during deep
   torpor episodes [87]. They mostly eat vegetation from the prairies they live in and
   hydrate through the moist of the eaten food, i.e., they do not specifically drink

Table 2 Selection of building species ($n = 50$)

Animal class	Animal order	Family, genus, or species (Scientific or common names)
Actinopterygii (Ray-finned fishes)	*Gobiiformes*	*Periophthalmodon schlosseri* (Mudskipper)
Arachnida (Arachnids)	*Araneae*	*Argyroneta aquatica* (Water spider), *Micrathena gracilis* (Orb-weaver)
Insecta (Insects)	*Blattodea*	*Allomerus decemarticulatus, Amiterme meridionalis* (Compass termite), *Cubitermes, Macrotermes natalensis, Macrotermes subhyalinus, Noditermes lamanianus, Odontotermes transraalensis, Procubitermes, Thoracotermes macrothorax*
	Hymenoptera	*Vespidae, Formica rufa* (Red wood ant), *Harpegnathos saltator* (Indian jumping ant), *Iridomyrmex* (Rainbow ant), *Brachymyrmex* (Rover ant), *Leguminivora glycinivorella, Macrotermitinae, Vespa orientalis* (Oriental hornet), *Zeta abdominale*
	Lepidoptera	*Eriogaster lanestris* (Small Eggar), *Gynaephora groenlandica* (Arctic woolly bear moth), *Tinea pellionella* (Case-bearing clothes moth)
Aves (Birds)	N/a	*[1]Early breeding bird species in Europe
	Apodiformes	*Swiftlet birds, *Calypte anna* (Anna hummingbird), *Glaucis aeneus* (Bronzy hermit)
	Passeriformes	*Ploceus philippinus* (Baya weaver), *Spermestes bicolor* (Black-and-white manakin), *Campylorhynchus brunneicapillus* (Cactus wren), *Petrochelidon pyrrhonota* (Cliff swallow), *Delichon urbicum* (House martin), *Picathartes gymnocephalus* (White-necked rockfowl), *Turdus merula* (Common blackbird), *Furnarius rufus* (Red ovenbird), *Aegithalos caudatus* (Long-tailed tit), *Philetairus socius* (Sociable weaver), *Ploceus cucullatus* (Village weaver)
	Procellariiformes	*Thalassarche melanophrys* (Black-browed Albatross), *Thalassarche chrysostoma* (Gray-headed Albatross)
	Charadriiformes	*Himantopus mexicanus* (Black-necked Stilt)
	Bucerotiformes	*Ocyceros birostris* (Gray hornbill)
	Galliformes	*Leipoa ocellata* (Mallee fowl)
	Columbiformes	*Treron vernans* (Pink-necked green-pigeon)
Mammalia (Mammals)	*Primates*	*Pan* (Chimpanzee)
	Rodentia	*Cynomys* (Prairie dog), *Burrowing mammals

(continued)

Table 2 (continued)

Animal class	Animal order	Family, genus, or species (Scientific or common names)
Polychaeta (Bristle worms)	*Terebellida*	*Pectinaria koreni*

[1] The asterisk * stands for unspecified families in the literature

Table 3 List of bioinspired tools for helping abstraction

References	Bioinspired tools	Principle
Chakrabarti and Sarkar [71]	SAPPhIRE (State change, action, part, phenomenon, input, oRgan and effect model)	Structural and functional description of biological and technical systems Based on FSB (Function, Behavior, Structure) [72]
Vattam et al. [73]	DANE (Design analogy to nature engine) interactive tool	Functional representation: box diagrams linked together by behavioral causal explanations Based on FBS [72]
Rosa et al. [21]	Uno-BID	Hybridization of SaPPhIRE [71] and DANE [73]
Goel et al. [74]	Ontology of biomimetics	Functions Referring to the TRIZ method [75]
Yim and Wilson [76]	Ontology for bioinspired design	Use of bioinspired designs Physical architectures, behavior, function, strategy
Chayaamor-Heil et al. [70], BiomimArchD—UMR MAP 3495 [77]	Ontology of BiomimArchD	Ontology focusing on building energy
Nagel et al. [78]	BioM (Biological modeling)	Functional representation of biological systems Models with different levels of granularity Using engineering-to-biology thesaurus [79]
Helms and Goel [80]	Four-box method	Description of a model of biological and technical systems: operational environment, functions, specifications, and performance criteria

Table 4 Regulation factors, principles, and features

Phenomena	Processes	Features
Heat (Gain, retain, transmit, prevent)	Conduction, convection, radiation, metabolic rate, electromagnetism, phase change	Macro-geometry Surface states Composition of matter Arrangement of matter Movement Surface-to-volume ratio Fluid properties Property of mechanical wave […]
Light (Absorb, reflect, transmit, diffuse)	Transmission, reflection, absorption, refraction	
Sound (Absorb, transmit, dissipate)	Transmission, absorption, reflection	
Water (Conserve, filter, gain, loose)	Diffusion, gravitational action, capillarity, sorption, absorption, condensation	
Air (Move, retain)	Pressure difference, induced flow, air renewal, convection	
Structure (Balance, integrity, comfort)	Absorption, transmission	

Adapted from Badarnah and Kadri [23], Cruz [24]

Table 5 Abiotic and biotic factors, identified as disruptive elements

Abiotic factors (environmental conditions)	Biotic factors		
	Interspecies	Intraspecies	Individual
	Behavior	Behavior	Behavior
Wind Sun radiation Environment radiation Precipitation Relative humidity Medium tectonic Mechanical load Flooding Electromagnetic radiation Material	Competition Predation Symbiosis Mutualism Commensalism	Competition Predation Reproduction Resources Space Cooperation Cohabitation Mating Brood/juvenile protection	Feeding Resting Mating Hibernation Thermoregulation Metabolism Metabolic rate Hibernation Thermoregulation

Fig. 5 Prairie dogs above and in front of their burrow entrance. *Credit* Pixabay, Licence CC0

water. They are diurnal species but are more active in the morning and evenings during hot summer days. Their burrows are meant for rest, protection from cold or hot weather, and are located next to vegetation for food supply.

Social requirements. They are social animals. They live in families or colonies with a range of up to 68 individuals per hectare for some species [89]. They breed once a year, and the litter of 2 to 8 pups requires a lot of attention as they are born blind and without fur. They are kept hidden in low chambers until they are strong enough to venture out. In addition to abiotic protection to meet their comfort requirements, the burrows require protection from predation, either by behavioral mechanisms from the prairie dogs, or using structural adaptation in the burrows, such as partition walls or narrowing the tunnels.

Disruptive elements are listed in Table 6 and are categorized into abiotic and biotic factors.

Table 7 shows examples of some relevant **physical responses** from the burrows and their occupants. To ensure thermal comfort for themselves and their offspring, prairie dogs demonstrate several behavioral adaptations: snuggling to keep warm, using vegetal fibers in chambers as insulation material, storingage feces for heat production through organic decay, but also choosing their resting rooms for optimum temperature according to the altitude. Those are strategies that aim at benefiting from the environment when physiological responses alone might not have been enough.

The distribution of the strategies between all parts of the systems also helps understand what features are to be described: the shape of the tunnels, the porosity of the envelope, the added materials into the chambers, etc. To mimic the general thermoregulation strategies of the prairie dogs, one would have to consider a combination of heterogeneous heat sources and their spatial distribution. For the inspiration of one feature only, the use of inputs, spatial and time scales should help grasp the elements at stake: for instance, adapting the altitude for optimum temperature involves multiple phenomena: air convection from bottom to top of burrow varying

Table 6 List of potential disruptive elements and inputs

Abiotic factors	Biotic factors		
	Interspecies	Intraspecies	Individual
	Behavior	Behavior	Behavior
• Sun radiation, wind, heat energy from soil, precipitation, relative humidity, air • Mechanical loads, medium tectonic	• Predation (hawks, eagles, snakes) • Matter (vegetal: seeds, fibers, straws, rotten roots) • Other species (soil) • Cohabitation (squirrels)	• Cooperation (resources, nursing, snuggling, detecting threats) • Cohabitation (other families) • Brood/juvenile • Mating	• Lifespan (5–8 years) • Feeding (vegetarian) • Detecting predators • Maintaining burrow • Mating (once a year)
			• Metabolism
			• Metabolic rate • Hibernation, winter rest, torpor

Table 7 Example of strategies characterized following the approach described in Sect. 3.2

Disruptive elements (inputs)	Response	Phenomena and Processes	Features		Temporalities
Heat (thermal regulation)					
Dormitories					
Air temperature rairie dogs (intraspecies)	Specific number of individuals in one room Snuggling [89]	Heat (gain)	Convection	Macro-arrangement Ratio of number of users per volume	Diurnal
Air temperature Prairie dogs Thermal energy from the soil	Adaptation of altitude for optimum temperature [90]	Heat (gain)	Conduction Convection	Spatial variation in the thermoregulatory benefits	Diurnal and seasonal
Dormitories and other rooms					
Vegetal matter Feces	Decay of matter for heat increase (up to 5 °C) [91]	Heat (gain)	Metabolic rate conduction	Organic decay (micro scale) fibrous material	Diurnal
Air (renewal)					
Tunnels and entrances					
Wind Indoor and outdoor air	Ventilation in the tunnels with the pressure difference between entrances [89, 92]	Air (move)	Pressure difference	Tunnels (15 cm of diameters, Rim crater (1 m-1, 5 m)	Diurnal
Water (protection from flood)					
Bottom					
Precipitations Prairie dogs	Protection from flooding with draining tunnels at bottom of burrow [93]	Water (transport)	Gravitational action	2–5 m width, 30–100 m long	Random

(continued)

Table 7 (continued)

Disruptive elements (inputs)	Response	Phenomena and Processes	Features		Temporalities
Structure (physical integrity)					
All parts					
Prairie dogs	Continuous repair of cracks [94]	Physical integrity	Physical equilibrium	Macro sealing	Diurnal

with daily temperatures, conduction from the soil toward the burrow internal surface, and input geothermal energy, with seasonal variations.

Morphological features can also be mentioned on another scale. For maintenance, prairie dogs regularly repair cracks they find on the inner surface of their burrows, along with creating new tunnels and closing others. Structurally, their know-how on the integrity and hold of soil would benefit to research and experimentation on materials little used today, such as mud bricks, but which were once very common.

The air renewal of prairie dog burrows depends on ingenuity in the construction of burrows. Relying on chimney effect, the burrow entrances are built at different altitudes. With a slight air movement, the air is engulfed in the lower entrance then pulled out through the higher opening which has lower pressure. A concept of façade was found in the literature based on this principle to minimize cooling needs and urban heat island effect [95]. Also inspired by the Voronoi-like growth of Barnacles (from Group *Crustacea*), it combines a ventilated air gap with Voronoi pattern-shaped opening vents. To implement the principle found in prairie dog burrows, they combined it with a feature based on another biological model.

4.2 Toward Multi-functional Envelopes

Multiple properties for heat, air, water, and structure were identified in our analysis of animal constructions. Table 8 lists a couple of them and proposes their potential technical analogies for the building envelopes. These transpositions are commonly already applied in the building field but could be improved by exploring how biological species implement them with little resources and in a sustainable way. The authors do not propose practical solutions here but point out some features of interest for designers, architects, and engineers which emerged from our research.

Note that few features related to light management (daylight comfort management) were found in the literature for our sample—and when it was, it was mostly as optical features resulting in thermal properties—none was reported for sound regulation. However, it would seem feasible that species pay attention to acoustics in regards to predation and the danger of being loud. We can only assume that constructions are not only insulated for thermal properties but also to prevent noisy litters from

Table 8 Envelopes features, classified by animal classes and regulation factors

	Features found in animal constructions and their occupants	Example of transposition to human building envelopes
HEAT (Thermoregulation)		
Mammals *Birds* *Insects*	• Insulation with collected vegetal fibers [51, 89, 96] • Use of smaller leaves for higher insulation [51] • Brewing the soil for nest heat regulation [97] • Feathers added as a layer on envelope [50, 60] • Use of turf at bottom of habitat [56] • Multi-layering with silk to retain heat [98] • Use of electric charge in silk cap to heat up [98]	• Use of on-site vegetal resources • High insulation properties with dense material • Warmer envelope by enhancing conduction with an external source • Use of air-trapping material • Localized insulation reinforcement with on-site mineral material • Reduce heat loss with reflective material and convection • Integrate thermoelectric properties in the envelope
Air (Air renewal)		
Birds	• Air renewal from the environment by brewing the soil [97] • Air exchange ensured through clay envelope • Entrance enlarged when the breed is old enough	• Air renewal by extraction of air contained in the ground • Use of porous mineral material for breathing envelope • Adaptation to air renewal needs with modular openings
Water (Watertightness, relative humidity)		
Mammals *Birds* *Insects*	• Dome for rain protection at the entrance of burrow [96] • Interlocked forest-edge plant on envelope [50] • Rain protection with feathers or cuticula (Chitin /protein-based materials) [56] • Redirect drops of rain using thin needles [56] • Constant vapor flow through the envelope by homeostasis	• Water redirection for entrances on top (troglodyte or semi-buried habitats) • Use of dense fine-branching material to ease dry out • Use of hydrophobic material to redirect water (redirection, collection, or watertightness) • Water redirection with needle-shaped and water drop scaled material • Use of pressure gradient to control vapor flows
Structure (Physical integrity, mechanical resistance)		
Mammals *Birds & insects*	• Counterweight to keep the habitat leveled [60] • Saliva used as a material cement and joint [99] • Vegetal fibbers added to mud and or on-site chitin growth using fungi [100]	• Counterbalance compensators in seismic regions • Study of secretion composition as natural cement • Structure rigidity from the combination between mud (mineral) and vegetal or commensalism with fungi

being spotted by potential threats. The use of soft materials, such as vegetal fibers, or collected secreted materials as feathers or hairs, might bring acoustic absorption that was not interpreted as such in our readings.

Among the strategies we identified in living species, many are related to thermoregulation, the most common being the collection and use of local materials, i.e., vegetal, mineral, or secreted organic materials. As they are air-trapping materials, feathers appear to be the best insulators, compared to plant materials [101]. Localized insulation can also be used by species such as the *Formica rufa* ants forming a crown of poles reinforced with peat and turf in their half-buried nest made of twigs [56]. The Malle-fowl (*Leipoa ocellata*) bird has found a more active way to gain heat; when incubating their eggs in a large nest mound, they scratch the soil on or off their nest to keep optimal temperature by retrieving heat from plants they have buried [97]. This method relies on the same principle described for prairie dogs (see Sect. 4.1) using metabolic heat from plant decay and does not require an ingenuous structure to operate.

Bird nests envelopes can also manage functional air renewal need using a relevant choice of porous materials such as clay or by modifying the construction to adapt it to the new air renewal needs. The red ovenbird, (*Furnarius rufus*) for instance, is able to remodel the entry of its mud-clay nest without it collapsing when the offspring are of the age of leaving the nest. For watertightness, species are able to combine materials into watertight linings; for instance, the *Harpegnathos saltator* ants [56] can cover the internal envelopes of their galleries with a mixture made of plants, insect cuticles, and cocoons. The *Formica rufa* ants will cover their nest with needles to have the water run away from the structure [56].

Several strategies were found regarding the management of structural loads on the animal construction: secretion of saliva as only construction material [99], on-site growth material using fungi [100], counterbalance of a hanging structure with mud pellets [60], and specific selection of material for stiffness [51]. The latter is rather complex as the builder, the chimpanzee, builds basketweaves from twigs and branches [102] platforms on very specific trees; they would have small distances between branches, allowing a higher number of interlocking during the weaving, providing better integrity [51]. Chimpanzees would also select trees with repelling properties against mosquitoes, implying a pathogen avoidance strategy. The *Allomerus decemarticulatus* ants co-operate with spores living on the plant *Hirtella physophora*; the fungus possesses the enzymes necessary for the digestion of chitin, and rapidly grows on the plant, feeding on the remains of insects. As such, it produces filaments penetrating and binding together the tube, solid enough to be a trap device for the ants [100].

Other commensalism strategies in animal constructions such as the *Macrotermitinae* termites growing fungus for nutrients [56] were found in the literature. It highlights potential additional characterization elements to be integrated into our database which is rather focused on efficiency: food culture, pathogen avoidance, commensalism such as symbiosis, are approaches that might be in the future in the front line of the requirements for closed spaces with the recent sanitary crisis of the Covid19.

4.3 The Construction of Animal-Built Structures

The proposed characterization has revealed some morphological features provided by the used materials or macro-geometry of the constructions. They indicate specific site selection, orientation, choice of materials, and constructive modes.

4.3.1 Site Selection and Orientation

Successful nests, in terms of reproductivity, were found to be related to the choice of a safe nesting regarding predators [103], proper exposition to environmental factors [47], and available resources such as food or construction materials.

To minimize predation threats, animals might build their nest at places difficult to reach, such as high locations from the ground [51, 99], crevices on cliffs [104], on spiny cacti [50], or underground [105]. Others demonstrate more aggressive strategies such as positioning thorny leaf stems and jagged trunks at entrances of construction to discourage intruders and predators [51] or making associations with other species, more threatening to predators [106]. In opposition, constructions are also camouflaged, using surrounding materials such as lichen on trees [50], or just adaptively located in "safer" locations where predators or parasites even if it sometimes implies lower thermoregulatory benefits [103].

Despite this trade-off between optimal microclimates site for their offspring and predation abundance, animals chose their construction site orientation for best exposition to wind, precipitations, and solar radiation. For instance, "magnetic" *Amitermes meridionalis* termite mounds obtain a uniform temperature with a south orientation [107]. Likewise, the south orientation of webs benefits best to the *Micrathena gracilis* spider thermoregulation in shaded sites, while west orientation is better for opened sites [47]. An inappropriate choice of orientation can be very detrimental in regards to wind, as it could for instance end up with the fall of all eggs contained in the weaverbird nest because of a slight draught [108].

4.3.2 Material and Processes

Animals have to take advantage of the materials they can find in their surroundings. They have to be maneuverable by the builder during the design process for construction purposes; appendages of arthropods allow them to manipulate vegetal or mineral and claws of mammals help them dig burrows. However, it is not always possible to deduce the nest shapes with anatomical specializations. An example used in [54] shows that birds with very similar beaks might build distinct nests and conversely, differentiation is due to feeding habits.

The choice of material depends on several factors. Despite a careful selection of a site for nesting, trade-offs can be made because of predation (see Sect. 4.3.1) and the natural resources available are not always precisely optimal for the intended construction. The animals might trade materials because of competitions, or choose materials by opportunism, randomly finding particular materials. As an example, with the increase in human activities, incorporation of debris in bird nests has been observed. Songbirds have for example used fibers from cigarette butts as lining; if it reduces the ectoparasite load, it is also genotoxic for the species [109]. Rather than collecting local resources, some species even develop their own building materials that are required to be specific for the intended structure: with silk for precise shape, with saliva for cement, joint or assembling materials, with integuments (such as hairs and feathers for thermal insulation, or cuticle for watertightness), etc.

The four main construction methods observed in the literature are sewing, secreting, excavating, assembling, or a combination of them. Assembling materials, whether they were collected or secreted by animals requires fastening methods; without using adhesive, species were found to use the Velcro method, stitches, entanglement [60], and weaving [47], the latter one described as the most complex fasten behavior, though common with some birds and arthropods.

Table 9 lists some processes used by a selection of species and a transposition of the resulting properties toward generic technological topics applicable to the building: drying out of materials and mechanical resistance, reinforcement with load addition of materials or weaving techniques, etc.

It also includes species relying on commensalism from living organisms other than plants, i.e., fungi for structural properties. This use is already explored in building applications such as self-repair concrete using bacteria [110] but could benefit from the inspiration of the *Allomerus decemarticulatus* ants.

As for the synthesis proposed in Sect. 4.1, this overview points out criteria that could be added to the characterization. First, the time needed for the construction could be an indicator of the complexity of the building process: for instance, swiftlet birds and *Vespidae* wasps can build dry structures made of secretion and secretion mashed with vegetal fibers with well-managed drying times. Also, temporality on the construction phases throughout the entire existence of the constructions might bring insights into our own renovation and densification strategies. The nests of sociable weavers end up forming a massing nest which actually provides better insulation than individual nests [111]. It gradually extends and is rebuilt with new material after any severe damage that could be caused by the weather or attacks of predators [112]. However, this gradual extension of the nest eventually reaches the limits of the tree to which it is attached.

4.4 Discussion

The information provided in this take-aways section is all qualitative. Our readings in the literature have shown that studies on more than one parameter among heat, air, daylight, sound, water, and structures are scarce, and not always adapted to our characterization. Using rodents as an example [113], explains that the microclimate of burrows lacks long-term temperature measurements and that they should

Table 9 Examples of construction processes and morphologies

Illustration	Species Used materials	Process Morphology features	Transposition to technology
1	**Black-and-white manakin** *Plant:* *Forest-edge plants*	*Weaved* Open cup Dense interlocked panicles	Quick dry out and protective dense structure
2	**Chimpanzee** *Plant:* *Ironwood sticks, thick foliage*	*Weaved* Open cup-shaped Crisscross pattern, small internodes Smaller/denser leaf distribution	Force stress decrease on pressure points
–	**Bronzy hermit** *Plant:* *Rootlets, moss, liverwort*	*Weaved* Cone-shaped Attached on a branch beneath living leaves	Quick dry out after precipitation by loose open walls
1	**Sociable weaver** *Plant:* *Dried grass straw, twigs*	*Assembled* Non-woven, thatch-like Huge communal nest made of several nests and materials Attached on tree	Variation of specific materials in superstructure
1	**Red wood ant** *Plant:* *Twigs, spruce needles* *Mineral:* *Soil, mud*	*Assembled & excavated* Mud and twigs used instead of needles on the lower part of the nest	Load resistance reinforcement
3	**Swiftlet birds** *Secretion:* *Saliva, feathers*	*Secreted* Half-bowl shape Adhesion to rock	Cement for material assembly or joint to rock-alike structures
3	**Water spider** *Secretion:* *Silk*	*Secreted and weaved* Bowl web. Pattern Hierarchical arrangement of fibbers	Mechanical stress resistance
–	*Allomerus decemarticulatus* *Plant:* *Hirtella host shrub* *Fungi:* *Spores* *Other:* *Nectar, preys*	*Weaved* Tube-trap with a multitude of ant-sized holes Surfaces coated with fungi which grow with the remains of insects	Structural rigidity from openings using living fungi
3	**Cactus wren** *Plant:* *Twigs, grass* *Secretion:* *Feathers*	*Assembled* Build-in cacti Prolate spheroidal, inside nest lining with feathers	Heat retainer with feathers and dense structure

Credits [1] CC BY 2.0 license, Rui Ornelas, Gailhampshire, Festive Coquette, [2] CC BY-ND 2.0 Richard Toller, [3] Public domain be performed with occupied nests. Indeed, many studies are either done on the body temperature or the burrow climatic conditions when they are empty, hence their data do not reflect real conditions.

The characterization of biological skins shows that finding information on the species itself is not an easy task either [24]; the amount of data is increasing every year with biology and biomimetic research and it is scattered in many research fields. We can conclude that there is a need for an access to structured, detailed information on both biological skins and animal constructions, that can be exploited for its integration in such a characterization.

5 Integration into a Design Process for the Building Envelope

The use of the characterization in the frame of a design process for building envelope designs can be considered following two approaches described in the literature. The first one, "technology-pull", starts from a technological problem and tries to solve it by learning from biological organisms. Quite logically, it is commonly used in the industrial area as it is more suited to provide solutions to existing problems. The second one, "biology push", consists in relying on discovery or established knowledge of a biological model for innovation. Both approaches are explored through multiple design processes, many of them being destined to the world of engineering and industry [74, 114–116].

In building design practice, the limit between technology-pull and biology push approaches is not so defined as the designers make iterations during the design process, between finding inspiring biology strategies and defining new functional requirements for the expected design [13]. When initiating the project, our approach was more technology-pull oriented, as living envelopes were selected by analogies with existing requirements and expectations regarding both the building envelope and the indoor conditions. However, the resulting characterization can be handled in both approaches, depending on whether designers are targeting one or several specific managing functions for the building envelope, or simply want to explore biological models for fresh new ideas. The framework described in the next section is suitable for the two approaches.

5.1 Technology-Pull Framework

The characterization was integrated into a full design process, with the final aim of the proposition and assessment of a multi-functional building envelope system. The idea was to confront the characterization of bioinspired practices and generate feedback

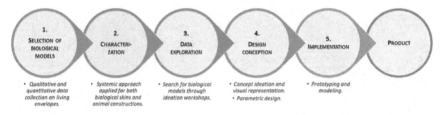

Fig. 6 Proposition of a design process using living envelopes as models

from professionals and actors in designing buildings, and from the assessment of the resulting designs. The design process is illustrated in Fig. 6 and includes five main steps.

Steps 1 and 2 were described in the previous sections. They consist in choosing biological models that could inspire technical solutions for the building envelope as a preliminary work of data exploration. As explained in 1. Living envelopes, the biological models were limited to living envelopes and chosen for their diversity regarding their taxa, their functionalities, and also the availability of data in the scientific literature. They were then characterized using building-oriented criteria as described in 2. An engineer/architecture-oriented characterization. The abstraction models that emerged from this characterization were stored in a database available to designers for Step 3.

The data exploration was performed during Step 3 by engineers, architects, and designers. They were asked during workshops to rely on the provided database to propose concepts of envelopes managing at least several functions for the envelope or indoor comforts of their choice. Ideas that emerged from this step were then further investigated through technical and feasibility considerations, and parametric design. The final step consisted in implementing at least one design, by the assessment of a prototype using measurements and modeling.

An adaptive bioinspired envelope, managing heat, air, and light transfers toward the building and inspired by the wings of the Morpho butterfly, emerged from this framework. The concept was prototyped and is currently being assessed (Step 4 of the process). The following section presents the process of Steps 3–5 that led to this design.

5.2 A First Case Study: The Morpho Butterfly

The Morpho butterfly is an insect from the *Lepidoptera* Order (Fig. 7a). The *Morpho* term stands for the Genus which includes a variety of subspecies [117]. Found in tropical forests of Central America and South America, they are known for their intense blue color, yet they contain no blue pigment. The reason is structural; at a nano-scale, light beams hit overlapping rows of scales, themselves forming ridges, and creating

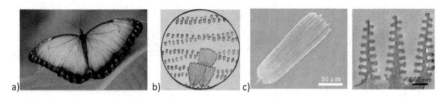

Fig. 7 **a** Morpho species photography, Credit: Pixabay Licence. **b** Drawing of the overlapping scales [118]. **c** Optical images of a butterfly wing scale and a transverse section of the scale showing ridges with lamella structures. *Credit* Adapted from He et al. [119], Licence CC BY-NC 3.0

constructive interferences (Figs. 7b and 7c). Specific wavelengths corresponding to the blue color are reflected, while others are canceled out.

As mentioned in Sect. 2.2.1, they were chosen to be characterized as they are part of one of the ten envelopes categorized by [24], i.e., a combination of cuticles and scales. They also live in temperate to tropical climates, meaning they probably have developed strategies and functions to survive warm external conditions. Finally, although insects are under-represented in the scientific publications (see Sect. 2.4 Taxonomic bias), morpho has recently attracted the spotlight of media [120–122] and research, hence data was more easily available in the literature than other insects from the same type of envelope.

5.2.1 Our Characterization

The anatomy of the morpho follows the anatomy of *Lepidoptera* from the phylum arthropods: it includes the head, with feeding organs and many sense organs, the thorax on which are attached the legs and the wings, and the abdomen. Our readings mostly focused on the wings as it is probably their colors that sparked people's curiosity about this species in the first place.

As performed for the prairie dog, a list of requirements, stimuli, and responses was first established. The resulting characterization is shown in Table 10.

5.2.2 Data Exploration During Workshops

This characterization was then introduced in a workshop, gathering different profiles: architects, engineers, and architect-engineers. The participants received a 2 h training on bioinspiration and were shown existing technological designs in architecture so they would be more familiarized with bioinspired processes. They were then given the database resulting from the characterization and asked to propose envelopes concepts inspired from one to several species skins, or animal constructions.

The database was provided as datasheets including general information on the species and the phenomena processes generated by various stimuli. Figure 8 illustrates the datasheet provided for the morpho species. Decision-support tools, e.g.,

Table 10 Characterization parameters applied to the wings of the morpho butterfly

Stimuli	Response	Phenomena and processes	Features	Temporalities	
Heat (thermal regulation)					
Wings (up and bottom)					
Daylight beam Sky Environment Morpho (behavior)	Orientation of wings for long-wave radiation toward the sky or near environment [123]	Heat (gain)	Radiation	Different matter arrangement and composition (pigments) between up and bottom sides	Diurnal
Air temperature Sun radiation Environment	Higher emission in near-infrared when overheating [35]	Heat (dissipate)	Radiation	Matter arrangement & matter composition (multi-scale organization)	Diurnal
Wings and thorax					
Air Metabolic heat from flight Morpho (behavior)	Wings shuffling for forced convection on thorax [123]	Heat (gain)	Convection	Movement of wings, macro scale	Random
Water (water repellent)					
Wings (up and bottom)					
Precipitation Water from environment	Self-cleaning surface [124]	Water (loose)	Gravitational action	Surface texture, matter arrangement and composition, hydrophobia from nano-structuration [125]	Random
Light (visual effect)					
Wings (up)					
Daylight beam	Structural blue color for iridescence with air/chitin [126]	Light (reflect)	Reflection	Surface structure	Surface structure
Structure (physical integrity)					
Wings (up and bottom)					
Morpho envelope	Flexible and ductal material, multi-structuration [127]	Physical integrity	–	Chitin-made [128]	Diurnal

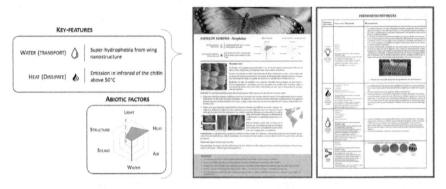

Fig. 8 Datasheet given to participants including graphs, texts, and tables. The left part is a zoom-in of the datasheet

illustrations, key features were included to help designers explore the characterized models and sort them according to their preferences. Among them, radar charts suggested by [24] indicated the phenomena involved in the skin functioning: for a scale from 0 to 3, it specifies low to the high contribution of the system or subsystem in managing heat, air, light, sound, water, and structural properties. This way, designers willing to consider specific factors for the building (e.g., water and air) can easily identify biological models to explore in priority.

5.2.3 Design Conception

After one week apart, we gathered with the workshop participants and brainstormed the emerged ideas of building envelope concepts.

- The emerged concepts had various operating spatial scales. For instance, proposing the rotation of human hands-sized or morpho wings-sized elements is considered a macro-operating scale. On the other hand, proposing the integration of an envelope element with emissive properties from its nano-structuration implies a nano-operating scale. The differentiation of these scales permitted the designers to understand what technical means would be required for the implementation of their ideas, and as such to link them with existing technical solutions (rotating systems, coatings, nano engraving, etc.).
- Then, these operating scales were confronted with the scales of the initial features they were inspired from. By doing so, it helped investigate other degrees of contextualization of the biological models, and so to think of new bioinspired ideas. As an example, a possible concept based on the Morpho is a building envelope with similar adaptability of its emissive properties. Obtaining these properties on a man-made element could imply reproducing the exact nano-structuration found on the wing scales, and therefore using advanced technologies such as nano 3D printing. This "low" abstraction level, although based on a thorough understanding

of the morphological features of the wings, could be transposed into alternative technological solutions less gray energy-consuming in their design and more low-tech if explored at different scales; for instance, manually alternating envelope elements whose emissive properties are different.

In parallel to the morpho, the same investigation occurred on concepts from other biological models. One of them caught our attention as it is inspired by the chameleon and proposes a deformable skin to manage daylight and heat transfers toward the building. As mentioned in Sect. 2.2.1, camouflage, display, and probable thermoregulation mechanisms in the chameleon are to change color: to do so, it turns into an excited state which makes the crystalline network contained in one dermis looser (see Fig. 9). The result is a change of interaction with light beams, reflecting the red color instead of the blue. Mixed with yellow pigment, the chameleon either appears in green or orange tones.

This feature of crystalline-network expansion affects the light flow hence the thermal properties were added to the final concept. As hybridization of two models, the resulting concept from the workshop combines the following principles: rotating elements to adapt to transient thermal conditions (macro), a smart-coating on these elements for similar adaptation (micro or nano), deformable and openable mesh (meso to macro), partial control on these adaptabilities given to the occupants of the building (macro). The smart-coating and rotating elements were both chosen to experience diverse scales of operations and means of implementation. The principle is represented in Fig. 10.

Merged into one principle, the concept is a deformable mesh of small opaque elements bonded by an elastic mesh. By stretching the mesh toward one side or the other of the surface envelope, the elements are pulled apart. It lets light and air go through these newly created apertures modifying the overall behavior of the envelope to external weather conditions (sun, air, rain, etc.).

The unitary elements are made of two layers; a base and a rotating flap (respectively gray and blue hexagons in Fig. 10). When the elements are pulled away from each other by the deformation, each of them initiates individual rotation of the flaps.

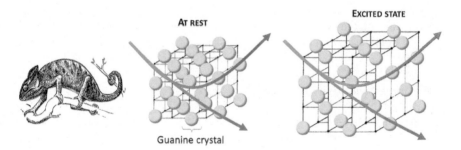

Fig. 9 Representation of the Chameleon crystalline network at rest and in an excited state. On the left network, all colors are absorbed except for the blue which is reflected due to interactions. On the right network, crystals are pulled apart, absorbing all colors except for the red

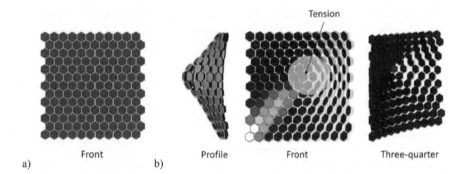

Fig. 10 Building envelope concept bioinspired on the morpho wing features and the chameleon: Flat (**a**) and deformed (**b**)

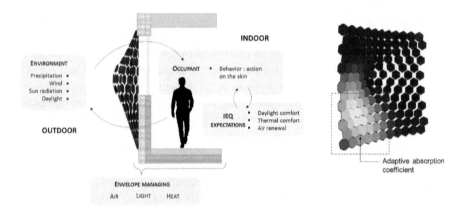

Fig. 11 (Left) Integration of the bioinspired concept into a building and (right) adaptive absorption coefficient represented by color change

It generates shadows on the surface and new orientations for the flaps. Additional functionality is achieved through a coating on the external surface of the unitary elements, i.e., on top of the flaps, that provides an auto-reactive behavior. Expected managed functions and impacts on the occupant comfort are schematized in Fig. 11.

5.2.4 Implementation

After parametric studies to determine shapes, sizes, and configurations of various elements of the design, we prototyped a flat but deployable version of the concept called Stegos. Manufactured to be modular, it was designed as an assembly of hexagonal aluminum pieces and 3D-printed material for the rotation systems.

The auto-reactivity of the Stegos, a feature inspired by the adaptive emissive properties of the butterfly, was technologically translated using a thermochromic paint, i.e., whose color changes at a temperature threshold of 45 °C. From color blue to white, the coating absorptivity naturally decreases in the visible spectrum (40% average decrease between 400 and 700 nm). Since this wavelength range accounts for almost 40% of the total solar spectrum energy [129], such a change on a thermochromic-coated material should have a significant impact on its solar thermal absorption. Therefore, when exposed to constant solar radiation, the expected behavior of a coated sample is a slowdown in the increase of temperature around the threshold temperature (Fig. 12).

Measurements of heat and light transfers through the test-box envelope were performed on the prototype and confronted to simulations. A gray-box approach is currently ongoing, meaning that partial theoretical models are coupled with data from performed measurements to obtain a fully descriptive model of the phenomena taking place in the Stegos. It should help determine parameters such as the solar intakes.

A deformable version of the Stegos concept is planned to be prototyped and tested as well in a near future in terms of heat, light, and air transfers. A version with a rain-collector system added on the flaps combined with an evaporative-cooling system, maybe using porous material for the hexagonal bases, is also considered. A follow-up of these designs will be a generic methodology for the characterization of dynamic envelope elements such as the Stegos, based on calibration of a model against measurements on similar protocols. It will be applied to the new version of the Stegos allowing to validate an overall model, characterize the performance, and as such propose more efficient alternative designs.

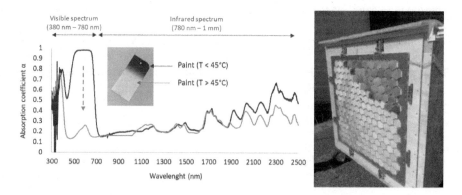

Fig. 12 (Left) Absorption coefficient of paint, measured at temperatures below and above 45 °C. (Right) Final prototype integrated into a test box. The difference of colors on the hexagons is due to differences in flaps orientations hence temperatures

5.3   Feedbacks and Analysis for This Framework

As mentioned in this chapter, it is uncommon to observe several factors managed by bioinspired designs apart from heat and daylight combination. Therefore, addressing several regulation factors was a key approach for the authors during the design process. The principle of the Stegos emerged from a workshop that had no specifications for the building envelope, and which provided illustrated datasheets as a support tool. Other workshops with new formats were organized afterward in order to evaluate the biases and benefits caused by the way the characterization was handed over to participants. The new participants were provided the following:

1. a characterization of our sample structured in a table, with a keyword search and a taxonomy based on the 6 managed functions by the envelope as support;
2. both the datasheets and the table, gathering all information.

If the first method seems more functional and is consistent with the current research conducted on ontologies and tools to help abstraction (see Sect. 3.1 Existing tools for abstractions), it appeared less ergonomic and illustrative for participating architects, who stated they preferred to rely on datasheets. However, engineer profiles reported finding the database useful though not illustrated enough as some biology terms were too technical. Hence, the main challenge in the use of the database relies on the sorting of the biological species, not on their understanding and abstraction.

Emerged concepts from these workshops, in particular those based on the characterization of animal constructions, will not be discussed in this chapter. However, they might be transposed in the near future into technological solutions for the building envelope, as done for the Stegos, which will permit us to bring new insights and improvement in our characterization.

From the emerged concept and the ongoing modeling and prototyping of the Stegos, we can also conclude that the characterization performed on the selected species and built structures by animals helped build a bridge between biology and architecture/engineer fields, and facilitated the proposals of simple to complex multiscale bioinspired concepts.

6   Conclusion and Future Work

This research aims at facilitating the design of efficient and durable building envelopes, inspired by features found in Nature. Because we identified the challenges of bioinspired design processes mostly in the abstraction and transposition phases, we chose to focus on the description and structuration of biological features in a way that would help designers ideate multi-functional and transposable concepts.

Our field of biological study is based on an existing work on biological skins and was enriched with animal-built structures. We intended to propose a characterization framework based on a systemic approach. For this, it includes several criteria that the authors consider necessary to comprehend multi-functional features in living organisms: systems and sub-systems, their respective requirements for optimal conditions, stimuli, and responses. This characterization was illustrated in detail through the prairie dog burrows and put forward key guidelines and strategies from animal constructions that could be relevant for sustainable buildings.

In parallel, we developed a framework in order to experiment bioinspired design process and assess all its steps. This framework integrates the characterization as a functional database to explore for ideation. Workshops with mixed profiles (architects, engineers) led to several multi-functional principles, including one inspired by the wings of the Morpho butterfly. To assess the concept and improve the first design corresponding to the current version, measurements on a prototype and modeling are ongoing.

The proposed biological characterization helped combine multiple features into one main concept. The chosen criteria eased the transposition of the described features into one solution. Indeed, they suggested designers frugal to complex implementation solutions for their concepts. For instance, behavioral mechanisms described in the characterization inspired designers to assign to the building users some control over the envelope adaptability. A combination between the wings 'orientation and their nano-structuration also pointed to the use of both deployable low-tech systems and thermochromic paint. It resulted in a new principle of auto-shading elements coupled with adaptive properties of absorption. Though not yet entirely implemented or assessed, the prototyped concept is an innovative solution for the building envelope, which goes against the paradigm of a static and solid envelope.

Future work includes a new iteration of the workshops, and emerged concepts based on animal constructions will be driven to implementation, as it was done for the Stegos. Feedbacks from participants will help us work on the format of the database to provide a more user-friendly and versatile medium. Finally, new criteria for the characterization will be integrated (pathogen avoidance, construction time), as they were put forth throughout our analysis and appear relevant in the frame of sustainability and resilient building designs.

References

1. Capra F (2016) The systems view of life, Reprint edn. Cambridge University Press, Cambridge
2. Mason KA, Losos JB, Singer SR, Raven PH (2017) Biology, 11th edn. McGraw-Hill Education, New York, NY
3. Masson-Delmotte V, Zhai P, Pirani A, Connors SL, Péan C, Berger S, Caud N, Chen Y, Goldfarb L, Gomis MI, Huang M, Leitzell K, Lonnoy E, Matthews JBR, Maycock TK, Waterfield T, Yelekçi O, Yu R, Zhou B (2021) Climate change 2021: the physical science basis. In: Contribution of working group I to the sixth assessment report of the intergovernmental panel on climate change. IPCC
4. Open Data Platform. https://data.footprintnetwork.org/#/. Accessed 29 Nov 2021
5. Herzog T, Krippner R, Lang W (2017) Facade construction manual. DETAIL

6. Wang S, Yang Z, Gong G et al (2016) Icephobicity of Penguins *Spheniscus Humboldti* and an artificial Replica of Penguin feather with air-infused Hierarchical rough structures. J Phys Chem C 120:15923–15929. https://doi.org/10.1021/acs.jpcc.5b12298
7. Ricard P, Le biomimétisme, s'inspirer de la nature pour innover durablement, p 138
8. Knippers J, Nickel KG, Speck T (2016) Biomimetic research for architecture and building construction. Springer International Publishing, Cham
9. Gruber P, Gosztonyi S (2010) Skin in architecture: towards bioinspired facades. Pisa, Italy, pp 503–513
10. López M, Rubio R, Martín S, Croxford B (2017) How plants inspire façades. From plants to architecture: biomimetic principles for the development of adaptive architectural envelopes. Renew Sustain Energy Rev 67:692–703. https://doi.org/10.1016/j.rser.2016.09.018
11. Kuru A, Oldfield P, Bonser S, Fiorito F (2019) Biomimetic adaptive building skins: energy and environmental regulation in buildings. Energy Build 205:109544. https://doi.org/10.1016/j.enbuild.2019.109544
12. Strain D (2011) 8.7 Million: a new estimate for all the complex species on earth. Science 333:1083–1083. https://doi.org/10.1126/science.333.6046.1083
13. Cruz E, Hubert T, Chancoco G, et al (2021) Design processes and multi-regulation of biomimetic building skins: a comparative analysis. Energy Build 246:111034. https://doi.org/10.1016/j.enbuild.2021.111034
14. Graeff E, Maranzana N, Aoussat A (2018) Role of biologists in biomimetic design processes: preliminary results, pp 1149–1160
15. Graeff E, Maranzana N, Aoussat A (2019) Biomimetics, where are the biologists? J Eng Des 30:289–310. https://doi.org/10.1080/09544828.2019.1642462
16. Wanieck K, Fayemi P-E, Maranzana N, et al (2017) Biomimetics and its tools. Bioinspired Biomimetic Nanobiomater 6:53–66. https://doi.org/10.1680/jbibn.16.00010
17. Fayemi P-E, Innovation par la conception bio-inspirée: proposition d'un modèle structurant les méthodes biomimétiques et formalisation d'un outil de transfert de connaissances, p 247
18. Hirtz J, Stone RB, McAdams DA et al (2002) A functional basis for engineering design: reconciling and evolving previous efforts. Res Eng Des 13:65–82. https://doi.org/10.1007/s00163-001-0008-3
19. Kozaki K, Mizoguchi R, An ontology explorer for biomimetics database, p 4
20. McInerney S, Khakipoor B, Garner A, et al (2018) E2BMO: facilitating user interaction with a biomimetic ontology via semantic translation and interface design. Designs 2:53. https://doi.org/10.3390/designs2040053
21. Rosa F, Cascini G, Baldussu A (2015) UNO-BID: unified ontology for causal-function modeling in biologically inspired design. Int J Des Creat Innov 3:177–210. https://doi.org/10.1080/21650349.2014.941941
22. Deldin J-M, Schuknecht M (2014) The asknature database: enabling solutions in biomimetic design. In: Goel AK, McAdams DA, Stone RB (eds) Biologically inspired design. Springer, London, pp 17–27
23. Badarnah L, Kadri U (2015) A methodology for the generation of biomimetic design concepts. Archit Sci Rev 58:120–133. https://doi.org/10.1080/00038628.2014.922458
24. Cruz E (2020) Multi-criteria characterization of biological interfaces: towards the development of biomimetic building envelopes. MNHN-CEEBIOS
25. Kuru A, Oldfield P, Bonser S, Fiorito F (2020) a framework to achieve multifunctionality in biomimetic adaptive building skins, p 28
26. López M, Rubio R, Martín S, et al, Adaptive architectural envelopes for temperature, humidity, carbon dioxide and light control, p 12
27. Nicolas F, Vaye M (1977) Recherches sur les enveloppes bioclimatiques
28. ADEME (2010) Histoire de la recherche sur l'enveloppe du bâtiment
29. Papadopoulos AM (2016) Forty years of regulations on the thermal performance of the building envelope in Europe: achievements, perspectives and challenges. Energy Build 127:942–952. https://doi.org/10.1016/j.enbuild.2016.06.051
30. Esau K (1977) Anatomy of seed plants, 2nd edn. Wiley, New York

31. The Jacques Rougerie Fondation. https://www.fondation-jacques-rougerie.com/homepage. Accessed 30 Nov 2021
32. Hufty A (2001) Introduction à la climatologie: le rayonnement et la température, l'atmosphère, l'eau, le climat et l'activité humaine. Presses de l'Université Laval; De Boeck université, Saint-Nicolas, Québec: [Paris]
33. Teyssier J, Saenko SV, van der Marel D, Milinkovitch MC (2015) Photonic crystals cause active colour change in chameleons. Nat Commun 6:6368. https://doi.org/10.1038/ncomms7368
34. Shi NN, Tsai C-C, Camino F, et al, Keeping cool: enhanced optical reflection and radiative heat dissipation in Saharan silver ants, p 5
35. Berthier S (2005) Thermoregulation and spectral selectivity of the tropical butterfly Prepona meander: a remarkable example of temperature auto-regulation. Appl Phys A 80:1397–1400. https://doi.org/10.1007/s00339-004-3185-x
36. Tsai C-C, Shi N, Pelaez J, et al (2017) Butterflies regulate wing temperatures using radiative cooling. In: Conference on lasers and electro-optics. Optical Society of America, p FTh3H.6
37. Vuarin P, Dammhahn M, Kappeler PM, Henry P-Y (2015) When to initiate torpor use? Food availability times the transition to winter phenotype in a tropical heterotherm. Oecologia 179:43–53. https://doi.org/10.1007/s00442-015-3328-0
38. Génin F, Nibbelink M, Galand M et al (2003) Brown fat and nonshivering thermogenesis in the gray mouse lemur (*Microcebus murinus*). Am J Physiol-Regul Integr Compar Physiol 284:R811–R818. https://doi.org/10.1152/ajpregu.00525.2002
39. Machin J (1968) The permeability of the epiphragm of terrestrial snails to water vapor. Biol Bull 134:87–95. https://doi.org/10.2307/1539969
40. Prior DJ (1985) Water-regulatory behaviour in terrestrial gastropods. Biol Rev 60:403–424. https://doi.org/10.1111/j.1469-185X.1985.tb00423.x
41. Raman AP, Anoma MA, Zhu L et al (2014) Passive radiative cooling below ambient air temperature under direct sunlight. Nature 515:540–544. https://doi.org/10.1038/nature13883
42. Luo C, Narayanaswamy A, Chen G, Joannopoulos JD (2004) Thermal radiation from photonic crystals: a direct calculation. Phys Rev Lett 93:213905. https://doi.org/10.1103/PhysRevLett.93.213905
43. Un papillon solaire | CNRS Images. https://images.cnrs.fr/video/4869. Accessed 1 Dec 2021
44. Heilman BD, Miaoulis Ioannis N (1994) Insect thin films as solar collectors. Appl Opt 33:6642–6647. https://doi.org/10.1364/AO.33.006642
45. Lou S, Guo X, Fan T, Zhang D (2012) Butterflies: inspiration for solar cells and sunlight water-splitting catalysts. Energy Environ Sci 5:9195. https://doi.org/10.1039/c2ee03595b
46. Didari A, Mengüç MP (2018) A biomimicry design for nanoscale radiative cooling applications inspired by Morpho didius butterfly. Sci Rep 8:16891. https://doi.org/10.1038/s41598-018-35082-3
47. Hansell MH (2005) Animal architecture. Oxford University Press, Oxford, New York
48. Davie O (1889) Nests and eggs of North American birds. Introd. by J. Parker Norris. Illus. by Theodore Jasper and W. Otto Emerson. Hann & Adair, Columbus [O.]
49. Abbott CC, Blanchard E, Conger PS, Duncan PM (1870) The transformations (or metamorphoses) of insects (Insecta, Myriapoda, Arachnida, and Crustacea): being an adaptation, for English readers, of M. Émile Blanchard's "Metamorphoses, moeurs et instincts des insects;" and a compilation from the works of Newport, Charles Darwin, Spence Bate, Fritz Müll by P. Martin Duncan. Claxton, Remsen, and Haffelfinger, Philadelphia
50. Collias NE, Collias EC (2016) Nest building and bird behavior. Princeton University Press Two Rivers Distribution [Distributor, Princeton; Jackson]
51. Samson DR, Hunt KD (2014) Chimpanzees preferentially select sleeping platform construction tree species with biomechanical properties that yield stable, firm, but compliant nests. PLoS ONE 9:e95361. https://doi.org/10.1371/journal.pone.0095361
52. Moroka N, Beck RF, Pieper RD (1982) Impact of burrowing activity of the Bannertail Kangaroo Rat on Southern New Mexico Desert Rangelands. J Range Manag 35:707. https://doi.org/10.2307/3898244

53. Frank CL (1988) The effects of moldiness level on seed selection by dipodomys spectabilis. J Mammal 69:358–362. https://doi.org/10.2307/1381386
54. Hansell M (2009) Built by animals the natural history of animal architecture. Oxford University Press, Nueva York (Estados Unidos)
55. Gould JL, Gould CG (2007) Animal architects: building and the evolution of intelligence. Basic Books, New York
56. Corbara B (2005) Constructions animales. Delachaux et Niestlé, Paris
57. Rennie J, Bliss E, Wood JG, et al (1830) Insect architecture. Lilly & Wait
58. von Frisch K, von Frisch O (1974) Animal architecture, 1st edn. Harcourt Brace Jovanovich, New York
59. Caras RA (1971) Animal architecture, 1st edn. Westover Pub. Co, Richmond, Va
60. Hansell MH (2000) Bird nests and construction behaviour. Cambridge University Press, Cambridge, New York
61. Troudet J, Grandcolas P, Blin A, et al (2017) Taxonomic bias in biodiversity data and societal preferences. Sci Rep 7:9132. https://doi.org/10.1038/s41598-017-09084-6
62. Bonnet X, Shine R, Lourdais O (2002) Taxonomic chauvinism. Trends Ecol Evol 17:1–3. https://doi.org/10.1016/S0169-5347(01)02381-3
63. Martín-López B, Montes C, Ramírez L, Benayas J (2009) What drives policy decision-making related to species conservation? Biol Conserv 142:1370–1380. https://doi.org/10.1016/j.bio con.2009.01.030
64. Allen D (2014) How mechanics shaped the modern world, 1st ed. Springer International Publishing, Imprint, Springer, Cham
65. Mazzoleni I, Maya A, Bang A, et al (2011) Biomimetic envelopes: investigating nature to design buildings. In: Proceedings of the first annual biomimicry in higher education webinar, The Biomimicry Institute Webinar Document, pp 27–32
66. ISO/TC 266 Biomimétique (2015) ISO 18458 2015—Terminologie, concepts et méthodologie
67. Jacobs SR, Nichol EC, Helms ME (2014) "Where Are We Now and Where Are We Going?" The BioM innovation database. J Mech Des 136:111101. https://doi.org/10.1115/1.4028171
68. Donovan MP (1997) SCST: the vocabulary of biology and the problem of semantics: "Dominant," "Recessive," and the puzzling role of alleles. College Science Teaching 026
69. Courtot M, Juty N, Knüpfer C, et al (2011) Controlled vocabularies and semantics in systems biology. Mol Syst Biol 7:543. https://doi.org/10.1038/msb.2011.77
70. Chayaamor-Heil N, Guéna F, Hannachi-Belkadi N (2018) Biomimétisme en architecture. État, méthodes et outils. craup. https://doi.org/10.4000/craup.309
71. Chakrabarti A, Sarkar P, Leelavathamma B, Nataraju BS (2005) A functional representation for aiding biomimetic and artificial inspiration of new ideas. AIEDAM 19. https://doi.org/10. 1017/S0890060405050109
72. Gero JS, Design prototypes: a knowledge representation schema for design, p 11
73. Vattam S, Helms M, Goel AK, Biologically-inspired innovation in engineering design: a cognitive study, p 41
74. Goel AK, McAdams DA, Stone RB (2014) Biologically inspired design: computational methods and tools. Springer, London, New York
75. Vincent JFV, Mann DL (2002) Systematic technology transfer from biology to engineering. Philos Trans R Soc London Ser A 360:159–173. https://doi.org/10.1098/rsta.2001.0923
76. Yim S, Wilson JO, Development of an ontology for bio-inspired design using description logics, p 11
77. BiomimArchD—UMR MAP 3495. http://www.map.cnrs.fr/?portfolio_page=biomimarc hd-4. Accessed 2 Dec 2021
78. Nagel JKS, Nagel RL, Stone RB (2011) Abstracting biology for engineering design. IJDE 4:23. https://doi.org/10.1504/IJDE.2011.041407
79. Nagel JKS (2014) A thesaurus for bioinspired engineering design. In: Goel AK, McAdams DA, Stone RB (eds) Biologically inspired design. Springer, London, pp 63–94
80. Helms M, Goel AK (2014) The four-box method: problem formulation and analogy evaluation in biologically inspired design. J Mech Des 136:111106. https://doi.org/10.1115/1.4028172

81. Molina F, Yaguana D (2018) Indoor environmental quality of urban residential buildings in Cuenca—Ecuador: comfort standard. Buildings 8:90. https://doi.org/10.3390/buildings807 0090

82. Organisation mondiale de la Santé (1980) Critères d'hygiène de l'environnement 12: LE BRUIT. Genève

83. von Bertalanffy L (2009) General system theory: foundations, development, applications, Rev. ed. paperback print. Braziller, New York, NY, , p 17

84. Badarnah Kadri L (2012) Towards the LIVING envelope: biomimetics for building envelope adaptation. Delft University of Technology

85. Hogan CB (2010) Abiotic factor. In: Encyclopedia of earth

86. Society NG, Society NG Biotic Factors. http://www.nationalgeographic.org/topics/resource-library-biotic-factors/. Accessed 21 Jul 2021

87. Lehmer EM, Savage LT, Antolin MF, Biggins DE (2006) Extreme plasticity in thermoregulatory behaviors of free-ranging black-tailed prairie dogs. Physiol Biochem Zool 79:454–467. https://doi.org/10.1086/502816

88. Chace GE (1976) Wonders of prairie dogs. Dodd, Mead, New York

89. Hoogland JL (1995) The black-tailed prairie dog

90. Lovegrove BG, Knight-Eloff A (1988) Soil and burrow temperatures, and the resource characteristics of the social mole-rat *Cryptomys damarensis* (Bathyergidae) in the Kalahari Desert. J Zool 216:403–416. https://doi.org/10.1111/j.1469-7998.1988.tb02438.x

91. Begall S, Berendes M, Schielke CKM et al (2015) Temperature preferences of African molerats (family Bathyergidae). J Therm Biol 53:15–22. https://doi.org/10.1016/j.jtherbio.2015. 08.003

92. Vogel S, Ellington CP, Kilgore DL (1973) Wind-induced ventilation of the burrow of the prairie-dog, Cynomys ludovicianus. J Comp Physiol 85:1–14. https://doi.org/10.1007/BF0 0694136

93. Elliott L (1978) Social behavior and foraging ecology of the eastern chipmunk (Tamias striatus) in the Adirondack Mountains

94. Oliver F Exerpts from: "our comic friend the Prairie Dog and the story of Prairie Dog Town, Texas !" https://ci.lubbock.tx.us/storage/images/53jgCSj7PR8BIjAkRutTlTgkJyUNC6znO WQuo5AE.pdf. Accessed 5 Dec 2021

95. Paar MJ, Petutschnigg A (2017) Biomimetic inspired, natural ventilated façade—A conceptual study. FDE 4:131–142. https://doi.org/10.3233/FDE-171645

96. Cooke LA, Swiecki SR, Structure of a white-tailed prairie dog burrow, p 3

97. Weathers WW, Seymour RS, Baudinette RV (1993) Energetics of mound-tending behaviour in the malleefowl, Leipoa ocellata (Megapodiidae). Anim Behav 45:333–341. https://doi.org/ 10.1006/anbe.1993.1038

98. Ishay J (1973) Thermoregulation by social wasps: behavior and pheromones. Trans N Y Acad Sci 35:447–462. https://doi.org/10.1111/j.2164-0947.1973.tb01518.x

99. Viruhpintu S, Thirakhupt K (2002) Nest-site characteristics of the Edible-nest swiftlet Aerodramus fuciphagus (Thunberg, 1812) at Si-Ha Islands, Phattalung Province, Thailand. Natural

100. Dejean A, Solano PJ, Ayroles J et al (2005) Insect behaviour: arboreal ants build traps to capture prey. Nature 434:973–973. https://doi.org/10.1038/434973a

101. Mainwaring MC, Deeming DC, Jones CI, Hartley IR (2014) Adaptive latitudinal variation in common blackbird *Turdus merula* nest characteristics. Ecol Evol 4:851–861. https://doi.org/ 10.1002/ece3.952

102. Fruth B, Hohmann G (2010) Ecological and behavioral aspects of nest building in Wild Bonobos (Pan paniscus). Ethology 94:113–126. https://doi.org/10.1111/j.1439-0310.1993. tb00552.x

103. Mainwaring MC, Hartley IR, Lambrechts MM, Deeming DC (2014) The design and function of birds' nests. Ecol Evol 4:3909–3928. https://doi.org/10.1002/ece3.1054

104. Velando A, Márquez JC (2002) Predation risk and nest-site selection in the Inca tern. Can J Zool 80:1117–1123. https://doi.org/10.1139/z02-091

105. Karels TJ, Boonstra R (1999) The impact of predation on burrow use by arctic ground squirrels in the boreal forest. Proc R Soc Lond B 266:2117–2123. https://doi.org/10.1098/rspb.1999. 0896

106. Quinn JL, Ueta M (2008) Protective nesting associations in birds: protective nesting associations in birds. Ibis 150:146–167. https://doi.org/10.1111/j.1474-919X.2008.00823.x

107. Grigg GC (1973) Some consequences of the shape and orientation of 'magnetic' termite mounds. Aust J Zool 21:231–237

108. Collias NE, Collias EC (1962) An experimental study of the mechanisms of nest building in a weaverbird. Auk 79:568–595. https://doi.org/10.2307/4082640

109. Suárez-Rodríguez M, Montero-Montoya RD, Macías Garcia C (2017) Anthropogenic nest materials may increase breeding costs for urban birds. Front Ecol Evol 5. https://doi.org/10. 3389/fevo.2017.00004

110. Vijay K, Murmu M, Deo SV (2017) Bacteria based self healing concrete—A review. Constr Build Mater 152:1008–1014. https://doi.org/10.1016/j.conbuildmat.2017.07.040

111. White FN, Bartholomew GA, Howell TR (2008) The thermal significance of the nest of the sociable weaver philetairus socius: winter observations. Ibis 117:171–179. https://doi.org/10. 1111/j.1474-919X.1975.tb04205.x

112. Maclean GL (1973) The sociable weaver, part 2: nest architecture and social organization. Ostrich 44:191–218. https://doi.org/10.1080/00306525.1973.9639159

113. Burda H, Šumbera R, Begall S (2007) Microclimate in burrows of subterranean rodents— Revisited. In: Begall S, Burda H, Schleich CE (eds) Subterranean rodents. Springer, Berlin, Heidelberg, pp 21–33

114. Yen J, Helms M, Goel A et al (2014) Adaptive evolution of teaching practices in biologically inspired design. In: Goel AK, McAdams DA, Stone RB (eds) Biologically inspired design. Springer, London, pp 153–199

115. Fu K, Moreno D, Yang M, Wood KL (2014) Bio-inspired design: an overview investigating open questions from the broader field of design-by-analogy. J Mech Des 136:111102. https:// doi.org/10.1115/1.4028289

116. Lepora NF, Verschure P, Prescott TJ (2013) The state of the art in biomimetics. Bioinspir Biomim 8:013001. https://doi.org/10.1088/1748-3182/8/1/013001

117. Callaghan CJ (2004) Atlas of neotropical Lepidoptera. Volume 5A: checklist: part 4A: Hesperioidea-Papilionoidea/edited by Gerardo Lamas ; by Curtis J. Callaghan [and others]. Scientific Publishers [for the] Association for Tropical Lepidoptera, Gainesville, Florida

118. Kellogg VL (1904) American insects. Holt and Company

119. He J, Villa N, Luo Z et al (2018) Integrating plasmonic nanostructures with natural photonic architectures in Pd-modified Morpho butterfly wings for sensitive hydrogen gas sensing. RSC Adv 8:32395–32400. https://doi.org/10.1039/C8RA05046E

120. (2015) Butterfly wings & the rise of color. In: Rainforest expeditions. https://www.rainfores texpeditions.com/butterfly-wings-the-rise-of-color/. Accessed 7 Dec 2021

121. What Gives the Morpho Butterfly Its Magnificent Blue? In: KQED. https://www.kqed.org/ science/24552/what-gives-the-morpho-butterfly-its-magnificent-blue. Accessed 7 Dec 2021

122. (2014) Morpho butterfly; color without pigments | bionicinspiration.org. http://bionicinspir ation.org/282/. Accessed 7 Dec 2021

123. Van Hooijdonk E, Berthier S, Vigneron J-P (2012) Contribution of both the upperside and the underside of the wing on the iridescence in the male butterfly *Troïdes magellanus* (Papilionidae). J Appl Phys 112:074702. https://doi.org/10.1063/1.4755796

124. Bixler GD, Bhushan B (2013) Rice- and butterfly-wing effect inspired self-cleaning and low drag micro/nanopatterned surfaces in water, oil, and air flow. Nanoscale 6:76–96. https://doi. org/10.1039/C3NR04755E

125. Mejdoubi A, Andraud C, Berthier S, et al (2013) Finite element modeling of the radiative properties of *Morpho* butterfly wing scales. Phys Rev E 87:022705. https://doi.org/10.1103/ PhysRevE.87.022705

126. Chapman RF, Simpson SJ, Douglas AE (2012) The insects: structure and function, 5th edn. Cambridge University Press, New York

127. Urry L, Cain M, Wasserman S, et al (2016) Campbell biology, 11th edn. Pearson, New York, NY
128. Niu S, Li B, Mu Z et al (2015) Excellent structure-based multifunction of Morpho butterfly wings: a review. J Bionic Eng 12:170–189. https://doi.org/10.1016/S1672-6529(14)60111-6
129. Bhatia SC (2014) Solar radiations. In: Advanced renewable energy systems. Elsevier, pp 32–67

Bio-inspired Approaches for Sustainable Cities Design in Tropical Climate

Miguel Chen Austin, Thasnee Solano, Nathalia Tejedor-Flores, Vanessa Quintero, Carlos Boya, and Dafni Mora

Abstract The remarkable growth of urban areas is a scenario faced by many cities due to the high rate of population that migrates to these zones, increasing the

M. Chen Austin · T. Solano · N. Tejedor-Flores · V. Quintero · C. Boya · D. Mora (✉)
Research Group in Energy and Comfort in Bioclimatic Buildings (ECEB), Faculty of Mechanical Engineering, Universidad Tecnológica de Panamá, Vía Centenario, Panama City 07098, Panama
e-mail: dafni.mora@utp.ac.pa

M. Chen Austin
e-mail: miguel.chen@utp.ac.pa

N. Tejedor-Flores
e-mail: nathalia.tejedor@utp.ac.pa

V. Quintero
e-mail: vanessa.quintero1@utp.ac.pa

C. Boya
e-mail: carlos.boya@uip.pa

N. Tejedor-Flores
Centro de Investigaciones Hidráulicas e Hidrotécnicas (CIHH), Vía Centenario, Panama City 07098, Panama

V. Quintero
Faculty of Electrical Engineering, Universidad Tecnológica de Panamá, Centro Regional de Panamá Oeste, La Chorrera, Guadalupe 07071, Panama

C. Boya
Dirección de Investigación, Universidad Interamericana de Panamá, Vía Ricardo J. Alfaro, Panama City 0819, Panama

M. Chen Austin · N. Tejedor-Flores · V. Quintero · D. Mora
Centro de Estudios Multidisciplinarios en Ciencias, Ingeniería y Tecnología (CEMCIT-AIP), Vía Centenario, Panama City 07098, Panama

M. Chen Austin · N. Tejedor-Flores · V. Quintero · C. Boya · D. Mora
Sistema Nacional de Investigación (SNI), Clayton Ciudad del Saber Edif. 205, Panama City 07144, Panama

© The Author(s), under exclusive license to Springer Nature Singapore Pte Ltd. 2022
F. L. Palombini and S. S. Muthu (eds.), *Bionics and Sustainable Design*, Environmental Footprints and Eco-design of Products and Processes,
https://doi.org/10.1007/978-981-19-1812-4_11

heat stored in the built environment creating insurmountable microclimatic conditions within the metropolitan area for pedestrians. Such microclimatic conditions might cause the unfeasibility of using natural ventilation for indoor passive cooling, increasing the air conditioners usage, and by overlapping to the previous heat stored the risk of overheating rises. Tropical regions have presented increased floods, extreme winds, earthquakes, and tropical-heat waves. To address such climate related challenges, a review on bio-inspired designs strategies at city scale, although not widely implemented in situ, is presented. On the other hand, developing countries in tropical regions recently started to develop energy regulations for the built environment, making it difficult to visualize a short-term implementation of any bio-inspired design at the city scale. As a result, most studies remain in a preliminary research project status. The evaluation and comparison of the sustainability of various tropical region cities through the Green City Index is presented. This evaluation led to assess in detail a Case study in Panama City considering the three critical aspects in the built environment: the conditioning of indoor spaces for cooling, transport, and lighting. Based on ecosystem services, a set of indicators are proposed and evaluated to measure regeneration at the city scale. Finally, to evaluate the proposed solutions, a SWOT analysis is presented. The use of a regenerative methodology in cities would mean a greater consideration of nature in planning goals and an improvement in urban ecosystem relations.

Keywords Biomimicry-based strategies · Green City Index · Ecosystem services · Panama City · Regenerative planning · Sustainability evaluation · SWOT · Tropical climate

1 Introduction

During the last decades, the idea of creating sustainable cities has been one of the main objectives to which many regions aspire, for this reason, it is important to do a general review through the different approaches and strategies of a biological nature, which have as their primary objective to improve the design of cities. This review identifies three main aspects and is structured to follow a general approach to improving cities' design based on one of the biomimetic approaches. It starts with the urban metabolism that allows studying the different systems leading to the identification of problems, followed by the biomimetic approach applied to case studies, and ends with considerations toward regenerative cities.

Within the different points that evidence that the causes of climate change are anthropogenic, we can mention the accelerated increase in the consumption of natural resources [61], with cities and urbanizations causing most of it, considering that they consume 40% of final energy and are related to 70% of global greenhouse gas (GHG) emissions [58]. In tropical areas, this consumption increases mainly due to cooling requirements for comfort and refrigeration, demanding more capacity and time of use of these systems [74].

Within the concept of analysis through resource flow, new theories emerge, such as urban metabolism (UM), one of the key processes, first coined by Wolman in 1965, who determined it fundamental for developing sustainable cities in the future. Others, however, date back to 1883, when Marx described it as the exchange of materials and energy between society and the ecosystem [60]. In fact, the material flow analysis (MFA) methodology covered 32% of a total of 165 case studies that have been evaluated by The Sustainability Institute and the United Nations—Environment. Likewise, the most prominent locations that conducted studies of this type were counted, with the northern hemisphere responsible for 56%, followed by China with 23%, the southern hemisphere, 12%, and in different cities at the same time or global scope 9% [60].

Current research shows a great deal of overlap with various disciplines that can help provide answers in the study of city metabolism [60]. The model proposed in Fig. 1 [60] considers the different interactions that occur continuously with society and its governments at local, regional, and global levels. In the MU analysis, inputs and outputs are considered passive and active, while biogeochemical and socioeconomic context cycles are fulfilled, being key to the environmental dispersion of materials, heat, water, and air in the process.

Among the methods for their evaluation, one can find (i) counting strategies [31, 33], (ii) input–output analysis [80], (iii) life cycle analysis [60], (iv) ecological footprint analysis [81], (v) simulation methods (dynamic and agent-based) [60], and (vi) hybrid methods [19, 31].

The analysis of MU in tropical zones differs from its analysis for other zones in the individual internal person-building interaction and the joint external person-structures interaction. In the latter, it is recommended to favor natural ventilation to

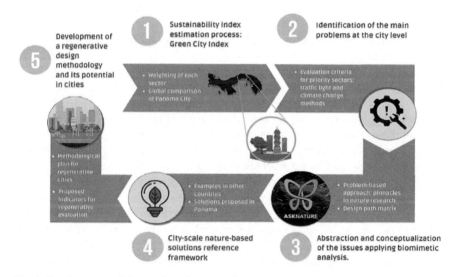

Fig. 1 Roadmap toward the creation of regenerative cities [64]

counteract the heat-humidity set, whereas, in the former, mechanical ventilation is resorted to without the need for heating. There are exceptions to the latter, such as some districts found at significant altitudes, e.g., Boquete and Volcán, in Panama (classified as Awi, according to Köppen).

The constant search for strategies leads us to identify the importance of biological knowledge and the solutions this knowledge allows us to solve. Two approaches are introduced in biomimetic design: the problem-based approach and the solution-based approach [5]. The first seeks to relate a similar problem presented by an organism and how it has managed to solve it. In the second, the design is based on biological and scientific knowledge, which represents a complex process involving multiple disciplines, including biology, architecture, and engineering, to avoid remaining only in the idea for its future application.

Biomimetic design is widely used in other rather distant branches of science; however, its methodology in the problem-based approach is unique, following four phases [5].

In biomimetics, we find three primary levels adapting to the design problem to be solved. These levels include form, process, and ecosystem, by analyzing a species or ecosystem, to solve through nature a human problem [75]. All these thanks to the study of biology, making it possible to innovate with technologies in various fields of application through biomimetics.

A concrete example at the building level is the Eastgate Center in Harare, Zimbabwe, whose challenge was to create a ventilation system that would keep the building at comfortable temperatures for the occupants. The inspiration came from the temperature regulating models of termite mounds that allowed the internal climate to remain stable due to their physical structure and passive internal airflow, thus avoiding the use of air conditioning [28]. In [27], it is shown innovative concepts arise for the development of building materials thanks to bio-inspired ideas thanks to the study of biology.

On the other hand, studies in material properties and how they influence the building envelope result in an applicable example in appearance. Such is the case of the retro-reflective properties present in flower petals and how they manage to decrease the heat reflected by radiation in buildings that are close by reducing the urban heat island effect (ICU) [38]. Other bio-inspired alternatives can be found, such as passive cooling, an example of this case we have in [17] whose multi-author SWOT analysis of how passive cooling through a series of indicators; such as practice, health, energy, among others, can have its advantages and disadvantages for urban planning at a macro, meso, and micro level. In [71], an investigation is carried out on the design of a biomimetic façade capable of adapting, providing a solution to improve the energy efficiency of buildings with high glazed surfaces for the hot and humid climate of Lahore, Pakistan. The numerical results indicate that after retrofitting the designed facade, the building's existing energy load decreases by 32%. In this way, it is shown that with a minimal reduction in visual comfort, favorable results are achieved in energy consumption thanks to the biomimetic strategies implemented in highly glazed facades.

There have been notable examples of how biomimicry has solved specific problems at the building level, but taking it to a larger scale at the district or city level involves integrating a whole system, so it is convenient to handle this concept. The system level creates an integrated system that efficiently manages energy and available materials in a continuous cycle, just as natural ecosystems do [51]. We can relate it in our environment to creating communities or buildings with zero energy consumption, where the energy supply comes from renewable energy and the community has no dependence on the power grid system from other types of energy production. This type of level is one of the most complexes to perform since it considers not only energy consumption but also the life cycle of materials, the way to reuse them, and other factors, providing a solution to major current problems at the level of urbanization or cities.

Effectively, biomimicry at the ecosystem level moves us toward regenerative urban design, where an understanding of ecosystem patterns and their functioning is used as models for urban space design. In 1960, McHarg worked with these concepts and formally proposed that human-designed landscapes should replicate the performance and logic of natural systems [7]. These regenerative systems examine several contemporary examples of relevant character, where strategies are put into practice that involve the use of biomimetic technologies and architecture, which facilitate the process of adaptation to climate change in the urban built environment and be a favorable factor for the ecological health of the ecosystem [16, 58].

By definition, a regenerative design seeks to address the ongoing degradation of the different factors that ecosystems provide to restore the capacity that ecosystems have to function optimally. This design challenges the orthodoxy of current green building practice and the design tools that support it, considering the building as an important factor in giving back to the environment more than it receives while reducing the aspects causing climate change by increasing biomass, thus increasing carbon capture and storage, while contributing to better social and natural capital over time [18, 58].

It is necessary to mention that one of the most relevant criteria for regenerative design, and which seeks to approach its actual application, is the use of ecosystem services [58]. The benefits that people obtain, directly or indirectly, from ecosystems, greatly promote comforts to the physical, psychological and economic level of the inhabitants and are usually divided into [58]: provisioning services (food and medicine), regulating services (pollination and climate regulation), supporting services (soil formation and solar energy fixation), cultural services (artistic inspiration and entertainment).

2   Biomimicry-Based Approaches to Improve Districts and Cities Designs in Tropical Climates

The concept of developing a global community within healthy ecosystems has been investigated and described in [79], which explores abundant case studies and new practices that go beyond outdated notions of sustainability and green design.

We know that biomimicry could be important in cities' regenerative and sustainable design to achieve harmony between human and non-human life. However, we must be clear that approaches vary for each city, just as it varies in nature, but we can find adaptive aspirations for cities with similar climates.

The effect of the human footprint on the natural and built environment has demonstrated the need for a change in the way cities are planned and built, setting as a goal the design of sustainable cities where biomimicry has become the key concept [30]. Therefore, biomimicry could be important in the regenerative and sustainable design of cities to achieve harmony between human and non-human life, but we must be clear that approaches vary for each city, just as it varies in nature. Still, we can find adaptive aspirations for cities with similar climates; however, although great efforts are devoted to the study and research in the field of biomimicry for the creation of sustainable architecture and cities, the specific knowledge of tropical climate is still behind that of temperate climate [68]. Nevertheless, recent research has studied some cases where biomimetic strategies have been implemented to develop cities in tropical climates.

Case 1. Indonesia, classified as a humid tropical country according to Köppen-Geiger, experiences high levels of UHI [68], due to its increasing urbanization and reduction of green space. These studies determined that, by implementing urban greening through planned open green spaces, evaporation and humidity at street level increases, as well as the reduction of heat exposure through shading, in addition to a series of elements focused on passive design.

Case 2. Singapore, one of the cities with a tropical climate, offers a landscape with approaches worthy of replication, where nature is integrated into the landscape and does not clash with the processes of urban life. Within these approaches is connectivity, which is based on a landscape that allows access to multiple points of the city, improving human movement through pedestrian bridges and, in turn, allowing wildlife connectivity with the construction of ecological bridges. Another approach used was shade and shelter, which was achieved with tree planting in pedestrian areas, drastically impacting urban heat islands. On the other hand, this city's orientation and public transportation, with the use of various layers of public transportation and signage or cartographic maps that allow access and experience nature. Finally, shared benefits are mentioned, including creating an extensive park system reaching many parts of the city [10].

Case 3. In India, the private city of Lavasa Hill is referred to as a case at the level of biomimetic cities, planned by Biomimicry 3.8 specialists. The land-use planning in this area was based on concepts such as "Walk to Work," "Walk to School" and "Walk to Park," where the strategy is based on locating the places of interest of its residents in such a way that the distance they need to walk to reach them is short, which prompted in 2009, to enhance these strategies with the construction of pedestrian walkways, resulting in safe walkways with good structure, which offer an improvement in the efficiency of pedestrian movement through the city [2].

Caso 4. Qatar, which experiences a harsh hot climate with high temperatures most of the year, can be considered a key example of tropical cities that experience dry seasons. The studies conducted focused on the sustainable development patterns of urban settlements, making a comparison between the grid patterns of today's modern cities against the organic pattern, characteristic of traditional settlements, which, consequently, are consistent with microclimatic realities and cultural and spatial identities [30]. The research similarly encompassed other studies of biomimicry and urban development toward eco-district strategies in the Gulf region, Greater Doha, analyzing the relationships in architectural design and biomimicry of Qatari cities to validate a model to guide current and future cities, thereby mitigating urban heat islands. This research highlights the biomimetic field at two levels: system integrator and detailing applications. It additionally indicated four main advantages for urban development in Qatar, which are obtaining lessons from Qatar Flora for eco-integrated systems, knowing the similarities between native plants and traditional Qatari settlements, the need for a biomimicry database for Qatari flora with adaptive characteristics, carrying out three models (a set of design principles and strategies) related to self-shading, wind channeling, and eco-urban density. The fully sustainable city includes three perspectives: environment, economy, and society.

It can be concluded that all cities under a tropical climate scheme must deal with high temperatures and similar problems such as UHI, which occur more intensely in cities of tropical environments [21]. Such as the case of Indonesia and Singapore, which use strategies in arborization, and creation of planned green spaces. This leads to thinking in ways to improve cities' design is a natural model centered on the forest, as originally proposed by Braungart and McDonough when they called to imagine a building as a tree and a city as a forest [24]. There is still much to investigate in this field of biomimetics applied to the design of cities.

3 Sustainability Evaluation: Case Studies

Every day we manage to be clearer about everything that comprises the concept of sustainability, and it gives us a clearer idea to define it. The Brundtland Report captures its essence in broad terms: meeting the needs of people now without devastating the life-supporting ecosystems for future generations [11]. Indeed, sustainability indicators are essential in advancing the science and practice of sustaining cities systems. In the literature on sustainability indicators, a distinction is often made among the terms of data, indicators, and indices, which together form a conceptual hierarchy or an indicator pyramid. Data are the basic components of an indicator, and multiple indicators comprise an indicator set or a composite index [41]. These days, many cities worldwide now routinely generate suites of indicators, using them to track and trace performance, guide policy formulation, and inform how cities are governed and regulated [45]. One of these indices, the Green City Index was used to understand the sustainability problem for Panama's metropolitan area [64]. In this study, the authors used ecosystem services and developed a set of indicators to measure regeneration over the years at the city scale. The data used to calculate the Green City Index in Panama were obtained from scientific publications, statistics, regulations, and plans or studies generated by the Panamanian Government. The authors found that the Green City Index score for the Pacific Metropolitan Area of Panama (1.5 million inhabitants) was adequate, however, with other variables such as transport and air quality, the scores were not as good compared to other sister cities, such as Quito (2.1 million), Curitiba (1.8 million), Montevideo (2 million), and Porto Alegre (1.4 million). In Table 1, we add the results from another study of a tropical country conducted by [37]. This study indicates that the environmental performance of the City of Depok in Indonesia (2.6 million inhabitants) has an average percentage of all Green City Index categories of 50.2%. Table 1 shows that the overall results of the performance of the Cities of Quito, Panama, and Depok in heading toward the green city, are average in the total results, compared with the other tropical cities.

A study in Brazil evaluates sustainability from a health indicator perspective to identify conditions and trends in environmental sustainability and the well-being of the societies [22]. In this context, sustainability represents harmony in the system that seeks to maintain equity between the present and future generations, where aspects such as economic sustainability (employment, income, inequalities) and environmental sustainability (ecosystem services) come to light, together with human well-being [8].

Table 1 Comparison of Panama City and other tropical cities from the Green City Index

Category	Far below average (0–20%)	Below average (20–40%)	Average (40–60%)	Above average (60–80%)	Well above average (80–100%)
Energy and CO_2		(d), (e)	(a), (b), (f)	(c)	
Land use and buildings	(d)	(b)	(a), (c), (e), (f)		
Transportation		(a), (e), (f)	(d)	(b), (c)	
Waste		(d), (f)			(a), (b), (c), (e)
Water		(d)	(b), (e)	(a), (c), (f)	
Sanitation		(b), (d)	(e),(f)	(a), (c)	
Air quality	(a)	(d)	(e), (f)	(b)	(c)
Environmental governance		(e)	(a), (b), (f)	(a), (c), (d)	
Total results		(d)	(a), (b), (e), (f)		(c)

Cities as: (a) Panama, (b) Quito, (c) Curitiba, (d) Montevideo, (e) Porto Alegre, and (f) Depok

Adapted from Quintero et al. [64], Hakim and Endangsih [37]

This type of sustainability assessment uses the Millennium Ecosystem Assessment (MEA), which emphasizes the linkages between ecosystems and human well-being. It recognizes that the actions people take that influence ecosystems result from concern about human well-being and considerations of the intrinsic value of species and ecosystems [56]. Regarding human well-being, the conceptual framework of the MEA includes:

1. Basic materials: where access to resources is promoted to sustain a dignified life.
2. Health: It implies an adequate diet, without diseases, and living in a healthy environment.
3. Security: refers to living in a safe environment, where they are not exposed to situations of ecological impacts and stress.
4. Good social relations: the ability to observe, study and learn about ecosystems and thus achieve express values that highlight mutual respect and community integration.
5. Freedom of choice and action: an opportunity for individuals to achieve what is valuable to them.

According to make this study possible, the databases and information sources were obtained from the Rio de Janeiro Information and Data Center Foundation. The authors of this study found that by comparing a series of data, indicators, and indices, it was possible to demonstrate how direct and indirect driving forces have degraded the ecosystem services, even while (paradoxically) the economy and population have grown, life expectancy has increased, and child mortality has decreased. It is necessary to emphasize that these processes are not always beneficial in terms of sustainability despite certain traditional indicators of health and well-being, since it has detected degradation of the environment and ecosystem services in Rio de Janeiro [22].

Like Sustainable Urban Development (SUD), sustainability assessment is increasingly being viewed as an important tool to monitor the human–environment interaction at different temporal and spatial scales [77]. It provides valuable information to assess the performance of the existing economic, social and environmental policies, plans, and programs. In this way, indexing can be an appropriate systematic tool for sustainability assessment in sophisticated decision-making situations involving simultaneous consideration of multiple qualitative and quantitative factors. In [52], a systems approach was adopted to develop a composite index called Sustainable Urban Quality Composite Index (SUQCI) to assess a city's potential to become sustainable. In this study, a multicriteria model was applied to select and prioritize sustainable urban environment quality indicators. Each indicator is a pointer that makes the elements of causation as well as the consequences of policies more understandable. The SUQCI is composed of 10 components containing a total number of 16 indicators and was evaluated in Tehran (Iran). To make this index, the authors used the major sustainability indices/indicators in terms of their applicability to city sustainability (United Nations Organization for Economic Co-operation and Development, World Bank, World Health Organization, European Commission, and Asian Development

Bank and an inventory of urban environment quality indicators). The obtained results indicated that the city of Tehran has a least three districts (22 districts in total) that are in ideal sustainability conditions. These areas are among the prosperous places in the city. About 59.09% of all districts were below the average SUQCI.

In [45], the authors also developed a composite index and presented it in a dashboard project. These types of projects often seek to make urban processes and performance more transparent and improve decision-making. The power of indicators, benchmarking, and dashboards reveal in detail and very clearly the state of play of cities. We added this kind of study to show the importance of presenting a sustainable urban index as a source of useful, contextual information, facilitating coordination, integration, and interaction across departments and stakeholders by providing a common, trusted, and authoritative data set for a city. The utility and value of composite index and dashboard initiatives provide detailed spatial and time-series data about various aspects of cities enabling longitudinal studies of socio-spatial, economic and environmental processes.

Environmental Integrity Index (EII) for decision-making is a SUD index that supports planning for sustainable cities by promoting environmental integrity and balancing biological and ecological components in a highly intricate urban system. The EII is suitable for measuring environmental health in rapidly growing and highly complex urban systems. The EII approach addresses natural stresses and anthropogenic ones through the selected set of indicators [70]. Three main indicators have been selected for this index: landscape fragmentation, urban climate, and vulnerability to an environmental hazard. These main indicators have several sub-indicators which have been quantified and integrated through standardization. This approach aims to set a multi-scaled, flexible, simple, adjustable, and policy-relevant index. The proposed indicator landscape fragmentation indicates the nature of alteration of urban land due to anthropogenic activities. The second indicator, urban climate, represents how to land alteration affects water bodies and vegetation, increasing urban heat. The third indicator, vulnerability to environmental hazard, calculates the state of urban air and water quality. The majority of the data for the proposed indicators are usually based on remotely sensed satellite data and other digitized data. It will be interesting to apply this index in Latin American cities, like Panama.

In the "Hedonic Price Indices to Understand Ecosystem Service Provision from Urban Green Space", the authors focus on using real estate prices to provide an estimate of the monetary value of the ecosystem services provided by urban green space across five Latin American megacities: Bogota (Colombia), Buenos Aires (Argentina), Lima (Peru), Mexico City (Mexico), and Santiago de Chile (Chile) [23]. Hedonic price indices are based on correlations between prices in existing markets (i.e., the real estate market) and specific ecosystem services (i.e., air quality) or bundles of ecosystem services, as, for example, provided by urban green spaces [76]. This study used Google Earth images to quantify urban green space and multiple regression analysis. They evaluated the impact of urban green space, crime rates, business density, and population density on real estate prices across the five mentioned Latin American cities. In addition, for a subset of the data (Lima and Buenos Aires), the authors analyzed the effects of landscape ecology variables (green space patch size, connectivity, among others) on real estate prices give us a clearer idea of how the diversity of ecological attributes that are generated in the creation of urban green spaces impact the diversity of benefits that we obtain from nature in different urban contexts in Latin America. It can be said that there is a direct connection where the existence of green spaces influence 52% of real estate prices in the five megacities studied, indicating that people who live in Latin American megacities are more attracted to living near green spaces, expressing their preferences through the real estate market. However, there is still a significant margin that gives rise to new research that allows us to answer the questions of how, when, and why the ecological attributes of urban green spaces influence the different ecosystem services [72].

4   Application of Biomimicry-Based Strategies Toward Regenerative Planning: A Case Study in Panama

Hereafter, a case study is presented for Panama City through the city sustainability evaluation via the Green City Index, based on the urban metabolism, which lets to identify several categories not reaching adequate ranges.

First of all, it is relevant to indicate that issues such as sustainable mismanagement and problems related to climate change can be addressed through biomimetic solutions, in conjunction with what is known as ecosystem services analysis (ESA). This approach is promising because it addresses many underlying issues in urban environments that need to be re-evaluated. A proposed trajectory toward regenerative cities is presented in Fig. 1 and applied to the case of Panama City, based on the results of the Green City Index.

When implementing the concept of regenerative urbanizations, it seeks to create a built environment that is harmoniously related to ecosystems, avoiding negative impacts that tend to degenerate it, and for this, it is necessary to implement regulations and public policies focused on promoting urban developments that contribute more than they consume the ecosystems and at the same time, remedy past and current actions in terms of environmental damage. This would allow progress toward truly regenerative actions.

Since it is impossible to replace all buildings and infrastructure for regenerative development, an alternative would be to provide ecosystem services on one's own to reduce the existing pressure on local ecosystems. This would mean improving their current state, supporting their capacity to become healthier, and maintaining their biodiversity.

In this way, with a healthier ecosystem, they will be able to continue providing humans with those advantages or services that the urban environment cannot simulate.

Because of this, the application of the concepts involved in architectural design will focus on a more dynamic design over the years, involving additive design and disassembly techniques. By incorporating some level of overlap on the complexity added to system design, it will evolve, increasing the ability of the built environment to respond on its own to new existing conditions [78].

By using the GCI, a list of more specific intervention issues that are considered a priority in the city is presented in Table 2.

Once the main problems at the city scale have been identified, the problem-based biomimetic approach was applied for abstraction and conceptualization. Such an approach led to selecting various pinnacles based on their life-nature strategies (see Table 3 in [64] for more details).

In the case of Panama City, strategies inspired by nature can be adopted, we consider the most predominant variables of biomimetic abstraction, we can rely on models, such as.

- Shading: by trees, roofs, cantilevered elements, and blinds.
- Pigments: trees, microalgae, plants on roofs, and vertical gardens.
- CO_2 reduction: filters, vegetation, green hydrogen.
- Solar use: photovoltaic and solar panels.
- Routes or branches: sidewalks and green corridors.
- Morphology: focused on buildings, lattices, and sidewalks.
- Passive Behavior—Found on buildings, rooftops, bus stops, and sidewalks.
- Dynamic behavior: found on green roofs, green corridors, microalgae filters, sequestration of emissions, non-motorized mobility, and electric mobility.

These characteristics that are identified as the most dominant from the selected pinnacles will form part of the roadmap toward a regenerative solution for Panama City. Table 3 describes these proposals and their successful applications in other parts of the world.

Some of these proposed solutions have already been evaluated for Panama City via dynamic simulation and modeling, others experimentally. For instance, the influence of arborization [4] and green roofs [59] on buildings energy performance was evaluated for two building typologies (residential and office) via dynamic simulations employing typical meteorological data. The latter, following two parametric studies focused on the vegetation layer parameters and the roof construction comparing green (with four endogenous plants) and no green roofs, encountered that such green roofs performed better than non-green roofs (even with different insulation degree and thermal mass content), where the vegetation height, leaf area index, reflectivity, and stomatal resistance are key parameters. The former highlighted that the inclusion of trees around a building improves its cooling performance reducing heat gains; this inclusion should focus on trees distribution and dimensions instead of quantity [4]. The effect of arborization on outdoor comfort was also studied at an urban scale [1], by including trees in regions with higher temperatures within the urban area. The comfort indicator, physiological equivalent temperature (PET), employed showed no significant changes, but the air temperature levels were reduced significantly. This led to conclude that buildings distribution and arrangement within this urban settlement in this tropical climate had a greater impact than the strategic inclusion of trees in view of the PET indicator. Similar results were obtained using the universal thermal climate index (UTCI).

Table 2 Priority intervention topics in Panama City

Energy, CO_2, and transportation (*Mobile sources, residential energy, and services*)	High per capita fuel consumption in the transportation sector is linked to the monocentric city—constant use of diesel, gasoline, and kerosene in industry, mobile, and air transport
	Increased emissions from mobile sources
	Electricity consumption in the service and residential sectors has increased, and emissions are above the average for ICES cities. There is a high electricity emission factor
	The decarbonization of the energy matrix is required, seeking other energy sources such as renewables and the adaptation of autonomous networks to reduce dependence on fossil fuels for energy generation
Partners—Territorial (*Land use*)	Growth of high purchasing power over natural areas (occupation of mangroves) and areas of high ecological value, complex coast-city relationship
	Discontinuity in the urban footprint; there is a high percentage of vacant land
Infrastructure (*Land use and transportation*)	Deficient, complex, and segregated road network
	The Metro Bus system shares infrastructure with private vehicles, generating very high travel times and making it unattractive for a possible modal shift
	Pedestrian infrastructure is in poor condition, discontinuous, and in many cases nonexistent. Road projects and urban developments do not adequately consider the space and facilities that should be provided to pedestrians and cyclists
	Some drainage systems are often poorly maintained and clogged with debris, increasing pollution so that falling water is not properly captured, treated, and discharged. As a result, localized flooding occurs, water collection and reuse are not optimized, and water resources' quality decreases
	There is no real rainwater management in buildings or its use in urban areas
Waste	Inadequate waste management, lack of CO_2 capture in landfills, and lack of consideration of associated CH_4 emissions
	Optimization and improvement of current waste-to-energy systems in landfills

(continued)

Table 2 (continued)

	Improvements are needed in the collection service and collection centers for recyclable materials to avoid the amount of garbage disposed of in improvised dumps, rivers, streams, and others
Air quality	Lack of adequate implementation of national regulations for mobile and stationary sources
	High emissions levels, including high levels of atmospheric particulate matter, NO_2, SO_2, and particulate matter
	There is no implementation involving air quality monitoring in the region
Global effects	Deficient network of recreational and leisure areas in the city (parks, green areas, sports complexes, bicycle paths). Deficit of public open spaces and unequal distribution of green areas and spaces for socializing and coexistence
	Heat islands effect produced by the lack of urban vegetation and constructions with heat-absorbing and retentive materials
	Center-periphery operation

Moreover, the walkability and bikeability within other urban areas in Panama City was assessed by many studies for both walk/bike to work and walk to park solutions, among them are [6, 12, 35]. Other passive and low-consumption solutions, such as bioclimatic-based strategies and design [14, 49], demonstrate strong potential for net zero energy buildings [13, 39]. Others are radiative sky cooling [53] and occupant behavior [15].

Finally, Panama City is also concentrating efforts at government scaled toward the energy transition and carbon neutralization with a large implementation of automobile electric-charge stations and electric transport [55, 66], as well as the use of air purification filters in transport and people high-density areas such as the bus terminal station [54].

4.1 Methodological Plan for the Design and Evaluation Proposal of Regenerative Cities

In the previous section, an evaluation of the different sustainable alternatives based on nature was made, allowing the regeneration capacity of Panama City.

Based on studies conducted by [78], incorporating ecosystem services is considered a method to move toward more regenerative cities. In Table 4, each of the ecosystem services is described. In addition, the abstracted biomimetic solutions are part of these services and will be classified into their corresponding function. Finally,

Table 3 Summary of proposed solutions for the implementation of cities based on the biomimetic analysis [64]

Solution	Description	Application around the world
Solar roofs	It lies mainly in the use of photovoltaic solar energy as a primary or secondary source for buildings. It seeks to adapt the roofs of residential, commercial, and industrial buildings to place solar panels or collectors using 20–30% of the available space. Consideration should be given to using larger and more efficient energy storage	In many countries, rooftop solar power is booming, Vietnam is a clear example of this, which saw an increase in rooftop solar installations from 378 MW in 2019 to 9,583 GW, that is, an increase of 2.43% since 2019. Currently, the numbers exceed 100,000 systems in total [36]. Energy storage strategies based on biomimetic approaches have been implemented that show us the way toward the production and sustainable use of distributed renewable energy [26]
Panels at bus stops and pedestrian walkways	The implementation of solar panels at bus stops to contribute to the reduction of energy from the public network. This also aspires to take advantage of energy capture, in pedestrian walkways, educational, government, commercial buildings, and public spaces in general, where this type of strategy can be implemented	The implementation of this type of initiative has been implemented in countries such as China, Brazil, the USA, India, and more, with successful results, as is the case of the Indian Institute of Technology in Kharagpur with the construction of a section of 70 m of a pedestrian walkway with solar panels [57]
Solar sheets	Innovation in the way we capture solar energy has led to the creation of solar sheets, which consist of triple laminated amorphous silicon glass. The advantage it offers is the little space it requires on the roofs for its installation	This type of solar sheet has been used at the Instituto Canario Superior de Estudios in Las Palmas de Gran Canaria, Spain. The window façade of this building has been used for the placement of these solar sheets [3]
Flectofin blinds	Adaptation of blinds for buildings, inspired by nature (bird of paradise flower). Its operation does not require the use of hinges and has 90° displacements. Its adaptive shading efficiently covers buildings from solar radiation	At the Expo 2012 pavilion in Yeosu, Korea, a kinematic façade designed by Soma architects was built, where individual kinematics were applied to control the incidence of daylight [40]

(continued)

Table 3 (continued)

Solution	Description	Application around the world
Trees	Its implementation is based on the arborization of urban areas to improve the microclimate thanks to the specific characteristics of trees with dense foliage, light green, thick and rough leaves, which at the same time help to maintain optimal levels of comfort by blocking solar radiation	Studies carried out in Taipei, Taiwan, demonstrate the effects of twelve tree species in subtropical urban areas. The results revealed the most effective species, which are *Ulmus parvifolia, Pterocarpus indicus* and *Ficus microcarpa*. Furthermore, the importance of leaf color, foliage density, leaf thickness, and leaf surface roughness were concluded [50]
Sierpinski ceiling	It is a biomimetic model where the leaves of the trees are emulated on the surface of the roofs, allowing the temperature to be distributed and thus reducing it. Its use is suggested in spaces such as terraces, gazebos, social areas, among others	The National Museum of Emerging Science and Innovation (Miraikan) in Tokyo, Japan, built a fractal roof (Sierpinski forest) and compared the fractal prototype with a part of the roof made of PVC panels, concluding that the fractal surface had a much lower temperature [67]
Green hydrogen	The use of hydrogen as a fuel replacement reduces emission levels. This alternative can be taken advantage of to a great extent by the geographical position, which allows the investment of this type of projects and the opportunities offered by the Panama Canal	Hydrogen refueling stations currently exist in countries, such as Japan, the United States, and Germany. H_2 Energy Applications (in) Valley Environments (for) Northern Netherlands abbreviated HEAVENN [44]
Green walls and roofs	Is based on the implementation of green roofs and walls using native plants in buildings to cool the surrounding air through evapotranspiration from the plant. Allows you to transform rooftops that have more than 1,300 square feet or at least 20% of the available space (for green or solar roofs) into green roofs. This would mean using natural barriers for CO_2 reduction, divided into three types of roofs: intensive, semi-intensive, and extensive	In Cordoba, Argentina, a law was enacted in 2016, making it mandatory to convert any rooftop more significant than 1300 square feet, including new or existing, into green roofs [25]

(continued)

Table 3 (continued)

Solution	Description	Application around the world
Purifying plants	By creating green spaces in cities where there are large areas with trees and plants, it is possible to trap volatile organic compounds by opening and closing the pores of the leaves. This process contributes to the reduction of toxic pollutants in the atmosphere of urban areas and improves air quality	In many places, mention is made of different species that could work as the most effective for sequestering emissions, such as VOCs, formaldehyde, benzene, carbon monoxide, and trichloroethylene; these species are *Spathiphyllum*, areca palm, tiger tongue, and *Chlorophytum comosum* [47]
Bio-filters	The implementation of algae-based biofilters is useful for areas where there are no large areas to plant trees, so it is possible to use them on high traffic streets. Additional can be considered in port areas where there are large logistical movements both at sea and on land, industrial areas, etc	There is the Biourban filter from Mexico. Its technicians assure that the models: BioUrban 2.0 and Bio Urban Industries, can supply the same amount of oxygen as 368 mature pine trees in a year, equivalent to the daily breathing of 2890 people. This filter has been incorporated in the Bus Terminal in Albrook Mall, in Panama City [73]
Photocatalytic cement	The use of this titanium dioxide compound has generally been implemented as cladding on avenues or sidewalks of public spaces and buildings with large surfaces exposed to sunlight, which allows its use in different types of buildings, especially in places with high contamination/odors	In Milan, a 7000 m^2 road surface was built with photocatalytic cement, obtaining a 60% reduction in the concentration of nitrogen oxide (NO_x) at street level [42]
Walk to school, walk to park, walk to work	The implementation of this strategy promotes mobilization without the use of vehicles, creating facilities for the population to walk to different destinations, within the city center	Lavasa Hill, India, has a biomimetic design, which has land-use planning based on these concepts. Mobilization of residents through walking to their workplaces, education, leisure, or socio-cultural activities is implemented [2]
Sidewalks	It is about building roads or sidewalks whose main characteristic is their capacity for permeability. It is an easy measure to implement and includes the stripes for the signs, lights, and other elements typical of the urban space	Studies carried out on sidewalks in Barcelona, Cuenca, and Prague, consider the importance of using materials to construct sidewalks that can produce friction (cobblestones or concrete) [34]

(continued)

Table 3 (continued)

Solution	Description	Application around the world
Green corridors	Its main function in urban areas is to serve as a connection to different points using vegetation along its extension	In cities, such as Madrid, Mexico City, New York, and Seoul, these corridors have been implemented along with trees, flowers, shrubs, walking paths, and bicycle lanes, and thus, improving the area's average temperature [43]
Electric transport	It covers the use of electric vehicles by creating facilities that include charging points in key areas of the country, encouraging the use of this type of transport	In Europe, for instance, the e-mobility advancement was foreseen with ABB (the main technology partner and supplier of IONITY, together with Audi, BMW, Ford, Porsche, and Volkswagen), focusing on operating a 400 fast charging points network in 24 European countries by 2020 [29]
Bicycle lanes	It is about creating an infrastructure that includes or improves bicycle lane spaces along the main avenues of the city with the purpose of creating connections in such a way as to increase the efficiency of urban area mobility free of emissions	The countries with the most extended distances covered by cycling (600–900 km) are Belgium, Denmark, and The Netherlands. Moreover, in Germany, between 1994 and 2017, distance cycled per capita increased by over 150 km, consistent with an increase of over 50% [69]
Routing algorithm	Algorithms based on nature minimized resources usage, helping create future networks of the Panama Metro. Its use is currently considered in those logistic services existing in Panama City	In La Paz and El Alto cities in Bolivia, the ant colony algorithm and Dijkstra's algorithm (minimum paths algorithm) were applied for the combinatorial optimization of transportation flow patterns [65]
Wireless sensor networks (WSN)	Timely monitoring of the health of structures contributes to extending the life of the structure, detecting damage, reducing routine inspections, and safeguarding lives	In, WSNs have been employed to monitor environmental, structural vibrations and look for deviations from a baseline response to evaluate the structure's condition under observation. On the other hand, in [63], the use of WSNs to evaluate the stiffness characteristics of a bridge and hence its capacity has been established. In [48] proposed improving the network's energy efficiency by selecting the optimal route by employing an optimization algorithm based on the behavior of ant colonies

Table 4 Ecosystem services for the urban environment and applicability to the case study

Ecosystem service	Applicability in the urban environment	Description [78]	Applicability to the case study [64]
Habitat Provision	Medium	Allows shelter and protection of organisms, as for nutritional needs. It has the function of protecting young organisms. They are relevant for permanent as well as transient populations of organisms and are extremely important for maintaining biodiversity and thus most other ecosystem services	Priority was given to ways of creating habitats in a way that is adapted to cities; this is achieved by incorporating native trees or species that have been fully studied to grow without problems in urban environments. This indicator is based on the function of providing habitats just as nature does
Nutrient Cycling	Medium	These can be added to cities through imports of food and materials and lost through exports, and the inability to recover and reuse materials through processes such as dumping and wastewater being discharged into the oceans or other regions	For this ecosystem service, one of the actions that help regenerate the urban environment and reduce its impact was chosen: the reuse and recycling of garbage. Currently, this collection is carried out by non-profit organizations, small SMEs, actions of the Zero Garbage program, and informal collectors

(continued)

Table 4 (continued)

Ecosystem service	Applicability in the urban environment	Description [78]	Applicability to the case study [64]
Climate regulation	High	It regulates processes related to the atmosphere's chemical composition, the greenhouse effect, the ozone layer, precipitation, air quality, and the moderation of temperature and weather patterns. Globally, it encompasses the ecosystems capacity to emit and absorb carbon as other compounds. In contrast, locally, it considers vegetation to reduce temperatures in urban environments and remove pollutants from the air	For this ecosystem service, this analysis focused on one of the great benefits of reforestation, which is the absorption of compounds as a method to minimize emissions from the different sectors of the metropolitan area of Panama
Air purification	High	They encompass systems that keep the air, water, and soil clean. Urban vegetation is an effective in removing pollutants from the air, but some building materials and filtration systems, can do a similar job and may be better suited for integration into some types of construction, particularly in medium- or high-density areas. Examples are porous metal–organic framework materials, mesosilica materials, titanium dioxide materials, air ionizers, particulate absorption filters, and other materials	For the atmospheric purification service, the indicator focused on natural solutions, such as implementing green areas in the urban environment. Green areas and corridors are considered the main measure to regenerate air and soil in cities

(continued)

Table 4 (continued)

Ecosystem service	Applicability in the urban environment	Description [78]	Applicability to the case study [64]
Water supply	High	It includes the regulation of hydrological flows, as well as the storage, purification, and retention of water. Water is used for the consumption of human and animal needs; therefore, large quantities are used for irrigation or other agricultural purposes and some industrial processes. Some aspects that have a direct impact on the water supply service in local ecosystems are water retention, volume management, the timing of eventual runoff, aquifer recharge, flood control, and drinking water quality	In the case of water supply services, the annual precipitation for Panama City was analyzed, considering the existing potential for rainwater harvesting in the different types of buildings in the area. This analysis focuses on saving water resources and conserving them in the city
Energy provision	High	Biomass and renewable energy usage is essential as ecosystem service. Knowing the use of energy will serve as feedback for analysis of human behavior and the degradation caused to ecosystems. However, attempting to artificially replace lost ecosystem services will increase energy, and thus, leading to further ecosystems degradation. Because of this, the implementation of energy efficiency is vital over the overproduction of renewable energy, as a considerable increase in energy production would have a detrimental effect on biodiversity or other ecosystem services in the long term	For the service of the nature of affordable energy production for all living beings, the equivalent in cities was selected: electric energy, focusing on the fact that this is produced in a renewable way to recognize the level of regeneration present in the energy sector

other design methods that improve the urbanized environment will be presented, in their regenerative capacity and the environmental benefits provided by each of them (Table 4): Nutrient cycling, climate regulation, air purification, water provision, and energy provision.

As shown [32], the Urban Regeneration Model focuses on a strategy that considers a model that offers actions in sectors such as energy, mobility, and information and communication technologies (ICT).

As a technical proposal to reduce energy consumption to minimum values, the installation of cleaner generation will be essential to obtain greater efficiency. It is essential to replace many conventional automobiles with cleaner cars, such as electric vehicles. The implementation of these must go hand in hand with adequate charging infrastructure while providing solutions using ICT in the city, implementing traffic management systems, or smart grids to improve users' movement. Additionally, to replicate this model later, these applications will involve citizen participation activities, whose objective is to achieve a high impact.

Ecosystem services analysis provides a starting point for creating a measurable regenerative design. This is crucial to establish the credibility of regenerative designs [78]. As a strategy to measure the sustainability of urban regeneration and ensure the principles of sustainable development, a set of proposed indicators focused on the ecosystem services described above were evaluated.

This system of indicators will aim at a positive environmental gain, evaluating the necessary actions on these aspects. They are related to international goals such as SDG 11: Sustainable Cities and Communities, SDG 13: Climate Action, and SDG 15: Life of Terrestrial Ecosystems. It also follows the targets established in the last Nationally Determined Contribution (NDC) in the forest and biodiversity sectors. An important point to cover during the evaluation process is the assumption of a static and isolated condition in urban regeneration. The most relevant ecosystem services that can be measured in cities were selected.

The proposed indicators took the same name as the ecosystem service and resulted as follows for evaluation in Panama City [64]:

(a) Habitat provision indicator (HPI)

Is an important indicator for calculating the number of existing trees in the land area of Betania, Bella Vista, Calidonia, San Felipe, San Francisco, and Santa Ana. Here, the input variables are the number of trees and the total territorial surface area (data taken from local institutions), resulting in an HPI value of 1815.77 trees/km^2. For this, the data obtained in the inventories belonging to the Arborization Plan of the Municipality of Panama were considered, excluding the rest of the district of Panama due to its lack of data, for the years 2016–2018. It is related to the goals of the Arborization Plan of the Municipality of Panama.

(b) Nutrient cycling (recycling indicator)

Determine the ratio of tons of recycled garbage to waste disposed of in landfills for the capital city. This resulted in a 2% only. For this, the nature of the garbage or materials found is not considered, nor the recycling category for each material (plastic, paper, cardboard, glass). Recycling by private industry is also not taken into account. For this evaluation, data from 2016 was used. It is related to the goals of the Solid Waste Management Plan 2017–2027 and the Zero Garbage Plan. In addition to the initiatives established in the last Nationally Determined Contribution (NDC) in the circular economy.

(c)   Climate regulation (Emission absorption indicator)

Measurement of the number of emissions per year absorbed by hectares of the forest without change of use of the metropolitan area under study. This resulted in 5.31 tCO$_2$e/ha. It was developed in the metropolitan area, covering the districts of Panama, San Miguelito, La Chorrera, and Arraiján, for the year 2015. Hectares are not distinguished by soil type; areas covered by mature, secondary, and mangrove forests are considered. Agricultural land use is not considered, such as crops and pasture. REDD + Plan 2017, National Forestry Strategy, Forestry Incentives Law of 2017, Alliance for the Million Hectares.

(d)   Air purification

Measuring the number of green areas in Panama City over the total urban area excludes forests, mangroves, water bodies, and agricultural land. This resulted in a 0.97%, supporting the air quality score of the Green City Index assessment (Table 1) falling into a notably below average ranking. For this assessment, 2015 was considered based on available data. Different land uses are not considered for the "Urban Surface" variable. Water surfaces are excluded. It can be related to the National Land Use Planning Policy and the Arborization Plan of the Mayor's Office. It would follow the goals established in the last Nationally Determined Contribution (NDC) in resilient human settlements and sustainable infrastructure.

(e)   Water supply

Refers to the rainwater volume used in buildings in residential areas, shopping centers, economic centers, and other urbanizations. This resulted in 517,965 million liters of water that could significantly reduce the city's all-purpose drinking water usage dependency. Only 2015 was used for this assessment. It is not considering vacant lots, open areas, green areas such as parks and squares because there is greater feasibility in installing rainwater collection structures in buildings. It is related to the goals of the Water Security Plan and environmental goals within the National Climate Change Strategy 2020–2050 and the Natural Disaster and Climate Change Vulnerability Reduction Program II. It can be used in the targets established in the last Nationally Determined Contribution (NDC) in the circular economy, resilient settlements, and sustainable infrastructure sectors.

(f)   Energy provision (Renewable Energy Indicator)

It shows the calculation of the amount of renewable energy consumed compared to the total energy consumed in the province of Panama. This resulted in 11.98%, leaving a great opportunity to expand the renewable energy market.

For this evaluation, data was only available for the year 2020. Data from the energy distribution companies in the province of Panama (EDEMET and ENSA) was considered, excluding the rest of the provinces assigned to these companies. It is related to the goals of the National Energy Plan 2015–2050. It can be considered in the goals established in the last Nationally Determined Contribution (NDC) in the energy sector.

It was necessary to apply indicators with a greater focus on ecosystems to measure the capacity of these solutions quantitatively, compared to the GCI indicators, which were based on sustainability. The similarities and differences between them are analyzed hereafter.

In the aspect of land-use evaluation in the city, the Green City Index only considers the indicators of "Green spaces per capita (m^2/inhabitant)" and "Population density (hab/km^2)". Meanwhile, in the ecosystem services analysis (ESA), the equivalent indicators for this sector were "Habitat provision or HPI (trees/km^2)" and "Air purification or IP (% of green areas)." Among these indicators, they have in common the use of vegetation in general. However, they differ in that the GCI takes data based on an individual scale, taking the city's population as a reference in per capita values; in contrast, the regenerative indicators consider data within a study area in km^2, and these seek to emphasize more specific issues such as biodiversity, the number of trees, and their capacity to regenerate the urban environment. The IP indicator (% of green areas) shows how much the city contributes to the use of urban spaces dedicated to green areas, contemplating more aspects in comparison to the GCI.

In the case of indicators that measure emissions, the GCI has three indicators for air quality: concentration levels of nitrogen dioxide, sulfur dioxide, and suspended particles ($\mu g/m^3$ per day); on the other hand, the counterpart indicator in ecosystem services would be "Climate regulation (Iabs)" expressed in tCO_2e/hectare-year. For the first indicator, air quality (GCI), they focused on describing the emission levels of a city, while the Iabs is based on quantifying the tons of emissions that are absorbed and returned to the soil, thanks to the hectares of forested areas and forests without change of use within Panama City.

Considering garbage disposal, both indicators consider the weight of garbage, where the GCI uses data on how much garbage is disposed of on a kg/inhabitant/year ratio. On the other hand, the SWA indicator focuses on "Nutrient cycling (IR)" using the amount of garbage recycled per resident in relation to the total garbage disposed of, in tons per year of both data, thus obtaining a percentage indicator of how much garbage is recycled in Panama City. The GCI also describes the proportion of waste collected and properly disposed of in a percentage analysis. In addition, although the GCI does consider waste recycling and reuse policies, it does not have an indicator that quantitatively measures performance in this aspect, as compared to the SWA indicator. It is important to mention that both indicators fail to analyze the nature of the waste material collected, considering whether it is organic or inorganic.

The GCI performs a percentage measure in both indicators on issues such as access to potable water in the city and the number of leaks generated in its facilities. In addition, it also estimates the water consumption of the average citizen, measured in L/person/day. Accordingly, the data available for Panama shows that drinking water supply is a highly studied and developed sector in Panama City; however, there is not much support or progress in the area of rainwater management and use, resulting in major floods and natural disasters. For this reason, the amount of rainwater that is usable in residential and non-residential cities could be taken into account, considering its potential use. This is measured by the indicator "Water supply (VT)" based on the SWA of water cultivation as a way to regenerate. This indicator relies on Panama City's annual rainfall to estimate the liters of rainwater per year obtained within the study area in square meters (m^2).

Estimating the energy consumed, both the GCI and the ESA have indicators that use values in kWh or MWh to evaluate electricity consumption. In the case of the GCI, data was used for the whole country in kWh/person since there is no information distributed in cities or provinces. For this reason, both the GCI and ESA indicators were calculated as estimates.

For the calculation of the "Energy Provision (IER)" indicator belonging to the SWA, it was carried out only in the province of Panama, being this data the closest to the capital city. Since the energy information was limited, we proceeded to estimate the necessary data according to the available data by type of electricity distribution company in the province of Panama (EDEMET and ENSA). This was done by calculating the electricity consumed of renewable nature among the total electricity consumed to obtain a percentage value (%).

4.2 Proposal Evaluation for Regenerative Cities via SWOT Analysis

With the result of the biomimetic analysis, the most relevant biological concepts were synthesized and adapted to the proposed solutions. Accordingly, this section will be devoted to evaluating their potential for adaptation in Panama City, where the strengths, weaknesses, opportunities, and threats will be explored through a SWOT analysis for each approach studied. Knowing the aspects covered in a regenerative design in the urban context is convenient. The aspects analyzed through SWOT are based on the direct objectives covered in strategic plans to comply with the agenda toward 2030 and 2050 found in the National Energy Plan 2015–2050 and the Nationally Determined Contribution.

Based on the antecedents and projections of Panama City, it is important to emphasize that the current position of the city was the result of variables, such as the construction of the Panama Canal, which limited the way it expanded. This occurred thanks to the constant activity that indirectly pushed the metropolitan area eastward. The urban growth expanded to the northeast, increasing considerably in population. Another influential factor was the physical obstacles such as the hills: Ancon, Sosa, and Cabra, resulting in a diffuse urban space. Because of this, today, the metropolitan area has different types of morphology, which can be seen in the design of its avenues and blocks. Considering other factors part of the urban growth, some studies point to the real estate explosion as an important cause, beginning in the city's central banking and financial area (Bella Vista and San Francisco). However, it has been advancing to the north and east of the capital [20].

It could be considered that the cause of the disturbance in the urban space of Panama is the large amount of energy needed to execute its functions within the framework of urban metabolism, in aspects such as mobility, buildings, and services. This results in a possible connection between areas where there is only a complicated network of roads, deteriorated highways, dense streets, and unfinished train lines. For this reason, there is plenty of motorized vehicles with constant congestion within the hours of entry and exit of working days. There is considerable travel time at the various entrances to the city center.

In addition, mass transportation in the metropolitan area is not very efficient without considering the poor design of the city since its inception. Due to insecurity and the poor quality of infrastructure for pedestrians and bicycles, many users resort to the automobile as their first option for getting around the city.

When considering other conflicts within the city, it is important to note that rural areas have low levels of connection to sanitary sewerage and sewage treatment, where maintenance is still poor. In addition, some areas are not connected to the potable water system and have cistern tanks and aqueducts. In general, the sanitary sewer system and the sewage pumping stations are very deteriorated.

During the analysis, the city's urban metabolism, considering its emissions, is mostly caused by mobile sources (motorized transport), which again shows a lack of planning to reduce these consequences, affecting the quality of life of the environment's inhabitants.

The urban area resources that allow achieving the objectives considered in the area's social structure and physical conditions are considered as a strength. On the contrary, a weakness focuses on the limitation preventing the project from achieving the results or objectives in the urban environment. Aspects such as economic points, land use, citizen participation, among others, are analyzed.

If the external points are considered, then it can be mentioned that an opportunity focuses on the undesired result that may occur with the project, directly affecting the physical environment and the citizenship when applying urban regeneration. At the same time, the threats represent those obstacles to developing the objectives set by the project, which could diminish its profitability.

On the other hand, a significant challenge that is still open is how biomimicry techniques applied to improve energy efficiency in buildings can impact Smart Grids issues and their subtopics, such as Distributed Generation, Demand Side, and Electricity Consumption Forecast. Given the climatic variability due to the seasons or the producer–consumer (prosumer) approach a building may face, it is crucial to study this relationship [9]. The climatic dynamism affects passive techniques, as reported in [46], in which energy savings are generated in summer, but consumption increases on colder days. Also, the implementation of materials for the generation of photovoltaic electric energy and its relationship with facades [62]. As a future study, it is important to integrate Smart Grids in biomimicry-based or inspired implementations to face the generation-consumption dynamism in buildings and smart cities.

5 Conclusions

As for Panama City's current potential in terms of sustainability, the Green City Index provided a starting point to describe the needs and key issues using criteria of importance to society and the environment. The economic issue was not evaluated.

Some indicators used in the GCI had a score of 0.00, due to a lack of reliable data sources, as in the case of the sulfur dioxide level indicator. On occasion, values that exceeded internationally established standards were also obtained. Other indicators also exceeded the norm, such as the length of the public transport network, per capita electricity consumption, and per capita water consumption. The latter is very noticeable in the city, due to the large water losses and waste in industrial activities, as well as by the citizens themselves.

By observing the framework of biomimicry-based solutions, it was possible to understand nature's opportunities for urbanizations. It offers an efficient way for humans to advance sustainability by emulating biological designs. For example, designs that require minimal resources and at the same time adapt to human needs. Utilizing biomimicry's principles, especially ecosystem-based biomimicry, would be a step on the path to regenerative cities.

It can be concluded that major obstacles arise for architects, engineers, and urban designers due to the need to control the negative flows resulting from our activities, such as GHG emissions, particulate matter, water losses, exploitation of water resources, deforestation, and increase of the urban footprint, poorly treated solid waste and wastewater disposal in waterways and seas, among other consequences of a linear and inefficient urban metabolism.

Through evaluation strategies such as the Green City Index, it was possible to obtain a general overview of the problems currently found in Panama City. In this regard, three priority approaches can be mentioned, which were revealed by the index's weighting. Regarding the quantitative and qualitative analysis of the sectors in their respective policies, the problems were classified into energy, atmosphere, and mobility. These were selected because they present alarming situations without disregarding the waste, water, and sanitation sectors regarding their importance in society. However, they were not considered in the focus of this study.

The identified conflicts were abstracted to find their corresponding solution by keeping biomimicry in mind. It can be affirmed that, although it is not possible to reach perfection the level of adaptation achieved by organisms in nature, it is still necessary to act toward designs of this type. Difficulties may be encountered in applying biomimicry, in the case of representing biophysical knowledge correctly and the challenging abstraction of principles. However, they would contribute to the revision, re-evaluation, and planning of urban functions through green infrastructures that consider their environment.

Ecosystem biomimicry relies on a design criterion based on sustainable and regenerative principles; therefore, the support of professionals with this knowledge is important. By applying this philosophy, and on a large scale, it will be possible to implement buildings whose capacities can be comparable to what nature offers, becoming a fundamental factor for improving ecosystems.

In the framework presented in Sect. 4 as proposed solutions for Panama City, the following points can be concluded:

- The indicators of "air quality" and "transportation" show the importance of encouraging purification strategies. This can be achieved using and planning adaptations of nature such as tree planting, green roofs, green facades, green corridors, biofilters, and others.
- Due to the heat island, a product of cities, it is necessary to optimize the inhabitant's comfort, considering the importance of tree leaves for temperature regulation of the soil and air. Simultaneously, the opportunities are recognized for heat regulation and energy savings presented by the application of prototypes and/or biomimetic designs in buildings and communities.
- Solar energy usage through photovoltaic systems is also considered a vital pillar in the energy transition and distributed generation, which would increase the percentage of renewable energy in the energy matrix.
- Panama can establish a hydrogen distribution hub, thanks to its geographical position and the opportunity generated by the Panama Canal and existing port centers.
- The population must be conscious of the impact and emissions caused by motorized transport for a modal shift that considers sustainable transport (walking, bicycles, and electric alternatives).
- Incorporating greener infrastructure and solutions in Panama, government initiatives will be needed to create laws that encourage the application of local photovoltaic systems, green roofs and facades, and other solutions in buildings, communities, and cities. Here, including a monitoring system for such infrastructures, such as sensor networks, would be vital to understand their health both structurally and in their indoor environment.

In more detail, the transition to a regenerative city can be addressed, starting with implementing LEED certification in buildings to conserve resources. Then, it is possible to move forward with the Living Building Challenge certification principles as a first step to start regenerating at the building level. As the last step for applying regenerative designs, biophilia and bioethics are taken into account, where biomimicry is used as a strategy.

The use of a regenerative methodology in cities would mean a greater consideration of nature in planning goals and an improvement in urban ecosystem relations and the quality of life of living beings. Using these concepts, certain indicators were determined to evaluate the potential in the context of Panama City to adapt more regenerative practices. The benefits of these ecosystem services will be reflected in the local climate, the environment and will improve the comfort of the inhabitants and the urban environment.

References

1. Araque K, Palacios P, Mora D, Chen Austin M (2021) Biomimicry-based strategies for urban heat Island mitigation: a numerical case study under tropical climate
2. Architectureever (2019) Lavasa township | It's Bio-Mimetic history | Biomimicry | India | Architecturever https://architecturever.com/2019/04/08/lavasa-township-and-its-bio-mimetic-history/. Accessed December 8, 2021
3. Arena AP, Funes MN, Henderson GR (2015) Análisis energético de aleros fotovoltaicos instalados en el edificio de la UTN Facultad Regional Mendoza UTN Facultad Regional Mendoza. In: Encuentro de Investigadores y Docentes de Ingeniería. Los Reyunos, San Rafael, Mendoza,Argentina
4. Austin Ortega D, Jiménez U, Mora D, Chen Austin M (2021) Influence of arborization in building energy consumption and thermal comfort: a numerical study in tropical climate. In: 19th LACCEI international multi-conference for engineering, education, and technology: "Prospective and Trends in Technology and Skills for Sustainable Social Development" "Leveraging Emerging Technologies to Construct the Future." https://doi.org/10.18687/LACCEI 2021.1.1.396
5. Badarnah Kadri L (2012) Towards the LIVING envelope: biomimetics for building envelope adaptation. Delft University of Technology. https://doi.org/10.4233/UUID:4128B611-9B48-4C8D-B52F-38A59AD5DE65
6. Barba L, Ruiz C, Rodríguez D, Perén JI (2020) Plan de movilidad urbana sustentable de la Universidad de Panamá (PLAMUP): ETAPA 1. SusBCity 2(1):50–53. https://revistas.up.ac.pa/index.php/SusBCity/article/view/1170
7. Blanco E, Zari MP, Raskin K, Clergeau P (2021) Urban ecosystem-level biomimicry and regenerative design: linking ecosystem functioning and urban built environments. Sustainability 13(1):404. https://doi.org/10.3390/SU13010404
8. Borghesi S, Vercelli A (2003) Sustainable globalisation. Ecol Econ 44(1):77–89. https://doi.org/10.1016/S0921-8009(02)00222-7
9. Boya C, Ardila-Rey J (2020) A method for weather station selection based on wavelet transform coherence for electric load forecasting. IEEE Access XX:1–8. https://doi.org/10.1109/ACCESS.2020.3035022
10. Brown JD (2019) Singapore summit—Biophilic cities. https://www.biophiliccities.org/singapore-summit-reflections
11. Brundtland GH (1987) Report of the world commission on environment and development: our common future ('The Brundtland Report'). https://doi.org/10.9774/gleaf.978-1-907643-44-6_12
12. Castañeda Á, Ocampo G, Sánchez K, Perén J (2018) Movilidad urbana en el Campus Central de la Universidad de Panamá: Caso de la Facultad de Arquitectura y Diseño y la Facultad de Ciencias de la Educación. Revista de Iniciación Científica 4:84–91. https://doi.org/10.33412/REV-RIC.V4.0.1826
13. Chen Austin M, Arnedo L, Student E, Yuil O, Mora D (2021) Energy consumption influenced by occupant behavior: A study in residential buildings in Panama. In: 9 Th LACCEI international multi-conference for engineering, education, and technology: "Prospective and Trends in Technology and Skills for Sustainable Social Development" "Leveraging Emerging Technologies to Construct the Future", no 1. https://doi.org/10.18687/LACCEI2021.1.1.337
14. Chen Austin M, Castillo M, De Mendes Da Silva Á, Mora D (2020) Numerical assessment of bioclimatic architecture strategies for buildings design in tropical climates: a case of study in Panama. In: E3S web of conferences, vol 197, pp 1–10. https://doi.org/10.1051/e3sconf/202019702006
15. Chen Austin M, Chung-Camargo K, Mora D (2021) Review of zero energy building concept-definition and developments in Latin America: a framework definition for application in Panama. Energies 14(18):5647. https://doi.org/10.3390/EN14185647

16. Chen Austin M, Garzola D, Delgado N, Jiménez JU, Mora D (2020) Inspection of biomimicry approaches as an alternative to address climate-related energy building challenges: a framework for application in Panama. Biomimetics 5(3):40. https://doi.org/10.3390/biomimetics5030040
17. Cheshmehzangi A, Dawodu A (2020) Passive cooling energy systems: holistic SWOT analyses for achieving urban sustainability. Int J Sustain Energ. https://doi.org/10.1080/14786451.2020. 1763348
18. Cole RJ (2011) Regenerative design and development: current theory and practice 40(1), 1–6. https://doi.org/10.1080/09613218.2012.617516
19. Dakhia K, Berezowska-Azzag E (2010) Urban institutional and ecological footprint: a new urban metabolism assessment tool for planning sustainable urban ecosystems. Manag Environ Quality: Int J 21(1):78–89. https://doi.org/10.1108/14777831011010874/FULL/PDF
20. David Castro-Gómez C (2012) Mega crecimiento urbano de la ciudad de Panamá y su impacto sobre el hábitat y la vivienda popular. http://biblioteca.clacso.edu.ar/gsdl/collect/clacso/index/ assoc/D5531.dir/gthi2-4.pdf
21. de Costa Trindade Amorim MC, Dubreuil V (2017) Intensity of urban heat islands in tropical and temperate climates. Climate 5(4). https://doi.org/10.3390/CLI5040091
22. De Freitas CM, Schütz GE, De Oliveira SG (2007) Indicadores de sustentabilidade ambiental e de bem-estar em perspectiva ecossistêmica na Região do Médio Paraíba, Rio de Janeiro, Brasil. Cadernos de Saude Publica 23(SUPPL. 4):513–528. https://doi.org/10.1590/S0102-311X2007001600012
23. de Mola UL, Ladd B, Duarte S, Borchard N, La Rosa RA, Zutta B (2017) On the use of hedonic price indices to understand ecosystem service provision from urban green space in five Latin American megacities. Forests 8(12):1–15. https://doi.org/10.3390/f8120478
24. Dicks H, Bertrand-Krajewski J.L, Ménézo C, Rahbé Y (2021) Philosophy of 6 engineering and technology, vol 36, pp 978–981. https://doi.org/10.1007/978-3-030-52313-8_14. hal-03125939
25. DiNardo K (2019) The green revolution spreading across our rooftops—The New York Times. https://www.nytimes.com/2019/10/09/realestate/the-green-roof-revolution.html. Accessed 8 Dec 2021
26. Dodón A, Quintero V, Chen Austin M, Mora D (2021) Bio-inspired electricity storage alternatives to support massive demand-side energy generation: a review of applications at building scale. Biomimetics 6(3):51. https://doi.org/10.3390/BIOMIMETICS6030051
27. Durai Prabhakaran RT, Spear MJ, Curling S, Wootton-Beard P, Jones P, Donnison I, Ormondroyd GA (2019) Plants and architecture: the role of biology and biomimetics in materials development for buildings. Intell Build Int 11(3–4):178–211. https://doi.org/10.1080/17508975. 2019.1669134
28. Eastgate Centre—AskNature (n.d.) https://asknature.org/idea/eastgate-centre/. Accessed 31 Aug 2020
29. Ew ABBR (2019) El futuro de la red eléctrica en la próxima era de movilidad eléctrica, pp 30–37. https://new.abb.com/news/es/detail/49212/el-futuro-de-la-red-electrica-en-la-proxima-era-de-movilidad-electrica
30. Ferwati MS, Alsuwaidi M, Shafaghat A, Keyvanfar A (2019) Employing biomimicry in Urban metamorphosis seeking for sustainability: case studies. Archit City Environ 14(40):133–162. https://doi.org/10.5821/ace.14.40.6460
31. Galan J, Perrotti D (2019) Incorporating metabolic thinking into regional planning: the case of the Sierra Calderona strategic plan. Urban Plan 4(1):152–171. https://doi.org/10.17645/UP. V4I1.1549
32. García-Fuentes M, de Torre C (2017) Towards smarter and more sustainable regenerative cities: the REMOURBAN model. Entrep Sustain Issues 4(3):328–338. https://doi.org/10.9770/JESI. 2017.4.3S(8)
33. Goldstein, B., Birkved, M., Quitzau, M. B., & Hauschild, M. (2013). Quantification of urban metabolism through coupling with the life cycle assessment framework: concept development and case study. Environ Res Lett 8(3):035024. https://doi.org/10.1088/1748-9326/8/3/035024
34. Grupo FARO Universidad Tecnológica Indoamérica GIZ Ecuador (2020) Método para evaluar espacios peatonales urbanos y su aplicación en Ambato, Ecuador. https://www.bivica.org/file/ view/id/5720

35. Guerra M, Pérez A, Arauz S, Arosemena A, Perén J (2019) Caracterización del flujo peatonal en espacios de transición: Caso Estación Vía Argentina y Piex. Revista de Iniciación Científica 5(2):45–51. https://doi.org/10.33412/REV-RIC.V5.2.2503
36. Gunther EA (2021) Vietnam rooftop solar records major boom as more than 9GW installed in 2020—PV Tech. https://www.pv-tech.org/vietnam-rooftop-solar-records-major-boom-as-more-than-9gw-installed-in-2020/. Accessed 8 Dec 2021
37. Hakim, Endangsih T (2020) Evaluation of environmental performance using the Green City index in Depok City, Indonesia. J Phys: Conf Ser 1625(1). https://doi.org/10.1088/1742-6596/1625/1/012001
38. Han Y, Taylor JE, Pisello AL (2015) Toward mitigating urban heat island effects: investigating the thermal-energy impact of bio-inspired retro-reflective building envelopes in dense urban settings. Energy Build 102:380–389. https://doi.org/10.1016/j.enbuild.2015.05.040
39. Hoque S, Iqbal N (2015) Building to net zero in the developing world. Buildings 5(1):56–68. https://doi.org/10.3390/buildings5010056
40. Hosseini SM, Mohammadi M, Rosemann A, Schröder T, Lichtenberg J (2019) A morphological approach for kinetic façade design process to improve visual and thermal comfort: review. Build Environ 153:186–204. https://doi.org/10.1016/J.BUILDENV.2019.02.040
41. Huang L, Wu J, Yan L (2015) Defining and measuring urban sustainability: a review of indicators. Landsc Ecol 30(7):1175–1193. https://doi.org/10.1007/s10980-015-0208-2
42. Hurtado S (2020) Materiales descontaminantes para la purificación del aire en el sector de la construcción. https://repository.upb.edu.co/handle/20.500.11912/5579
43. Iberdrola (n.d.) Green corridors, how to take care of the environment in cities? https://www.iberdrola.com/sustainability/green-corridor. Accessed 8 Dec 2021
44. Kakoulaki G, Kougias I, Taylor N, Dolci F, Moya J, Jäger-Waldau A (2021) Green hydrogen in Europe—A regional assessment: Substituting existing production with electrolysis powered by renewables. Energy Conv Manag 228:113649. https://doi.org/10.1016/J.ENCONMAN.2020.113649
45. Kitchin R, Lauriault TP, McArdle G (2015) Knowing and governing cities through urban indicators, city benchmarking and real-time dashboards. Reg Stud Reg Sci 2(1):6–28. https://doi.org/10.1080/21681376.2014.983149
46. Kwok YT, Lai AKL, Lau KKL, Chan PW, Lavafpour Y, Ho JCK, Ng EYY (2017) Thermal comfort and energy performance of public rental housing under typical and near-extreme weather conditions in Hong Kong. Energy Build 156:390–403. https://doi.org/10.1016/J.ENBUILD.2017.09.067
47. Leaves Remove Pollution—Biological Strategy—AskNature (n.d.). https://asknature.org/strategy/leaves-remove-pollution/. Accessed 8 Dec 2021
48. Lee M, Yoe H (2019) WiBiA: wireless sensor networks based on biomimicry algorithms. Int J Comput Intell Syst 12(2):1212–1220. https://doi.org/10.2991/IJCIS.D.191029.001
49. De León L, Chen Austin M, Carpino C, Mora D (2021) Towards zero energy districts developments base on bioclimatic strategies: a numerical study in a developing country. In: E3S web of conferences, vol 312, p 02017. https://doi.org/10.1051/E3SCONF/202131202017
50. Lin BS, Lin YJ (2010) Cooling effect of shade trees with different characteristics in a subtropical urban park. HortScience 45(1):83–86. https://doi.org/10.21273/hortsci.45.1.83
51. López M (2017) Envolventes arquitectónicas vivas que interactúan con su entorno naturalizando el diseño. https://core.ac.uk/download/pdf/153484217.pdf
52. Maryam Robati SM, Monavari HM (2014) Urban environment quality assessment by using composite index model. Environ Prog Sustain Energy 33(3):676–680. https://doi.org/10.1002/ep
53. Merchant I, Chen Austin M, Mora D (2021) Estimation of the radiative sky cooling potential through meteorological data: a case study in tropical climate. In: E3S web of conferences, vol 312, p 02008. https://doi.org/10.1051/E3SCONF/202131202008
54. MiAmbiente inaugura primer biofiltro en Panamá (n.d.). https://www.laestrella.com.pa/nacional/190517/primer-panama-inaugura-biofiltro-miambiente. Accessed 8 Dec 2021

55. MiBus planning major purchase of electric vehicles Newsroom Panama (n.d.). https://www.newsroompanama.com/environment/mibus-planning-major-purchase-of-electric-vehicles. Accessed 8 Dec 2021

56. Millennium Ecosystem Assessment (2005) Ecosystems and human well-being: synthesis. London

57. Mondal S, Sanyal A, Brahmachari S, Bhattacharjee B, Mujumdar PD, Raviteja J et al (2017) Utilization of constrained urban spaces for distributed energy generation—Development of Solar Paved Pedestrian walkway. Energy Procedia 130(October 2018):114–121. https://doi.org/10.1016/j.egypro.2017.09.406

58. Monitor Deloitte (2019) Ciudades energéticamente sostenibles: la transición energética urbana a 2030. https://www2.deloitte.com/es/es/pages/strategy/articles/ciudades-energeticamente-sostenibles.html

59. Moreno A, Chen Austin M, Mora D (2021) A parametric study of implementing green roofs to improve building energy performance in tropical climate. In: E3S web of conferences, vol 312, p 02004. https://doi.org/10.1051/E3SCONF/202131202004

60. Musango JK, Currie P, Robinson B (2017) Urban metabolism for resource efficient cities: from theory to implementation. www.sustainabilityinstitute.net

61. Pederson-Zari M (2018) Regenerative Urban design and ecosystem biomimicry—Maibritt Pedersen Zari—Google books. https://doi.org/10.4324/9781315114330

62. Petriccione L, Fulchir F, Chinellato F (2021) Applied innovation: technological experiments on biomimetic facade systems and solar panels. Techne-J Technol Archit Environ (2):82–86. https://doi.org/10.13128/techne-10687

63. Putra SA, Trilaksono BR, Riyansyah M, Laila DS (2021) Multiagent architecture for bridge capacity measurement system using wireless sensor network and weight in motion. IEEE Trans Instrum Meas 70. https://doi.org/10.1109/TIM.2020.3031126

64. Quintero A, Zarzavilla M, Tejedor-Flores N, Mora D, Chen Austin M (2021) sustainability assessment of the anthropogenic system in Panama City: application of biomimetic strategies towards regenerative cities. Biomimetics 6(4):64. https://doi.org/10.3390/BIOMIMETICS6040064

65. Quispe VSC (2019) 7 Modelo de optimización combinatoria para bioflujos del transporte. Libros Universidad Nacional Abierta y a Distancia, pp 163–184. https://hemeroteca.unad.edu.co/index.php/book/article/view/4066

66. Ruta Eléctrica entre Panamá y Costa rica sobre rueda – Secretaría Nacional de Energía (n.d.). https://www.energia.gob.pa/ruta-electrica-entre-panama-y-costa-rica-sobre-rueda/. Accessed 8 Dec 2021

67. Sakai S, Nakamura M, Furuya K, Amemura N, Onishi M, Iizawa I et al. (2012) Sierpinski's forest: new technology of cool roof with fractal shapes. Energy Build 55:28–34. https://doi.org/10.1016/J.ENBUILD.2011.11.052

68. Sari DP (2021) A review of how building mitigates the Urban heat Island in Indonesia and tropical cities. Earth 2(3):653–666. https://doi.org/10.3390/earth2030038

69. Schepers P, Helbich M, Hagenzieker M, de Geus B, Dozza M, Agerholm N et al (2021) The development of cycling in European countries since 1990. Euro J Transp Infrastruct Res 21(2):41–70. https://doi.org/10.18757/EJTIR.2021.21.2.5411

70. Shathy ST, Reza MIH (2016) Sustainable cities: a proposed environmental integrity index (EII) for decision making. Front Environ Sci 4(DEC):1–12. https://doi.org/10.3389/fenvs.2016.00082

71. Sheikh WT, Asghar Q (2019) Adaptive biomimetic facades: enhancing energy efficiency of highly glazed buildings. Front Archit Res 8(3):319–331. https://doi.org/10.1016/j.foar.2019.06.001

72. Swanwick C (2009) Society's attitudes to and preferences for land and landscape. Land Use Policy 26(SUPPL. 1):S62–S75. https://doi.org/10.1016/J.LANDUSEPOL.2009.08.025

73. Tecnología mexicana empleada para purificar el aire de las grandes ciudades con microalgas I Mano Mexicana (n.d.). https://manomexicana.com/p/tecnologia-100-mexicana-para-purificar-el-aire-con-microalgas. Accessed 8 Dec 2021

74. Webb M (2021) Biomimetic building facades demonstrate potential to reduce energy consumption for different building typologies in different climate zones. Clean Technol Environ Policy (0123456789). https://doi.org/10.1007/s10098-021-02183-z
75. What Is Biomimicry?—Biomimicry Institute (n.d.). https://biomimicry.org/what-is-biomimicry/. Accessed 21 May 2021
76. Wolch JR, Byrne J, Newell JP (2014) Urban green space, public health, and environmental justice: the challenge of making cities 'just green enough.' Landsc Urban Plan 125:234–244. https://doi.org/10.1016/J.LANDURBPLAN.2014.01.017
77. Yigitcanlar T, Teriman S (2015) Rethinking sustainable urban development: towards an integrated planning and development process. Int J Environ Sci Technol. https://doi.org/10.1007/s13762-013-0491-x
78. Zari MP (2012) Ecosystem services analysis for the design of regenerative built environments. Build Res Inf 40(1):54–64. https://doi.org/10.1080/09613218.2011.628547
79. Zhang X, Skitmore M, De Jong M, Huisingh D, Gray M (2015) Regenerative sustainability for the built environment—From vision to reality: an introductory chapter. J Clean Prod 109:1–10. https://doi.org/10.1016/J.JCLEPRO.2015.10.001
80. Zhang Y (2013) Urban metabolism: a review of research methodologies. Environ Pollut 178:463–473. https://doi.org/10.1016/J.ENVPOL.2013.03.052
81. Zhang Y (2018) Urban metabolism. In: Encyclopedia of ecology, pp 441–451. https://doi.org/10.1016/B978-0-12-409548-9.10756-0

Pho'liage: Towards a Kinetic Biomimetic Thermoregulating Façade

Lise Charpentier, Estelle Cruz, Teodor Nenov, Kévin Guidoux, and Steven Ware

Abstract An adaptive shading device is designed using biomimetics as a tool to optimize thermal comfort and help reduce energy consumption. Inspired by *nyctinastic* movements, ArtBuild's Lab (AB Lab)—a transdisciplinary research laboratory created within ArtBuild's architectural studio—began developing autonomous biomimetic façades in 2015 with the aim of reducing energy consumption in buildings and in particular, mass timber buildings, whose thermal inertia is low. The research project initially mimicked the mechanics and behaviour of stomata cells found abundantly in the plant species, drawing inspiration from the asymmetrical cell wall thickness to activate movement, and then moved on to developing prototypes for solar protection devices whose thermal actuation, shape memory, and geometry combine to enable them to echo the nastic movements described by Darwin. AB Lab's team employs thermobimetals (TBMs)—composite metal alloys that react to temperature variations—to induce nastic movements in their shading devices. By exploiting the differential expansion coefficients of these alloys, the architects were able to shape the solar protection devices to cast measured shadows. Dubbed *Pho'liage*, the

L. Charpentier · K. Guidoux · S. Ware (✉)
ARTBUILD Architect, 58 Rue du Faubourg Poissonnière, 75010 Paris, France
e-mail: swa@artbuild.com

L. Charpentier
e-mail: lic@artbuild.com

K. Guidoux
e-mail: kgu@artbuild.com

E. Cruz
CEEBIOS, European Centre in Biomimetics, Ceebios SCIC, 62 rue du Faubourg Saint-Martin, 60300 Senlis, France
e-mail: estelle.cruz@ceebios.com

MECADEV, UMR CNRS 7179 - National Museum of Natural History of Paris, UMR 7179 / CNRS-MNHN, 1 avenue du Petit Chateau, 91800 Brunoy, France

T. Nenov
BIO VOIE, Rue de la Grande Louvière 15, 7100 La Louvière, Belgium
e-mail: ten@biovoie.eu

© The Author(s), under exclusive license to Springer Nature Singapore Pte Ltd. 2022
F. L. Palombini and S. S. Muthu (eds.), *Bionics and Sustainable Design*, Environmental Footprints and Eco-design of Products and Processes,
https://doi.org/10.1007/978-981-19-1812-4_12

devices react to heat emanating from the sun. When outside temperatures exceed 25 °C, the TBM blades mimic the petals of a plant, opening as "flowers" to form a vast curtain protecting the building from thermal overload. When the temperature drops, the petals deform once again and the flowers close, allowing light to enter the building. Early versions of the *Pho'liage* prototypes revealed several challenges: the temperature-driven deformation of the bimetal, far from being uniform, often took place too abruptly given the dual conflicting expansion forces of the bimetal alloy surfaces. The very nature of the curvature dynamics was repeatedly reviewed. Lifecycle analysis of protective coatings showed the difficulties in sourcing ecological solutions for the alloys' external longevity. Apart from the basic geometry of the flowers, several designs were explored which integrate curve-line folding and adaptable honeycomb support structures, to enhance the efficiency of the open/close shading ratio. Finally, alternatives were suggested that look at reducing the quantity of TBMs, with the alloys acting as actuators whilst other materials such as specific biopolymers provide the shading function.

Keywords Biomimetics · Kinetic adaptive façades · Thermostatic bimetal (TBMs) · Sustainable solutions · Curve-line folding · Envelope · Thermal comfort · Plant movements

1   Introduction

Urban areas are dominated by horizontal dark surfaces, mostly roofs and pavements, absorbing solar radiation and releasing it in the form of longer wavelengths (infrared) as heat also causes the temperature to rise. This is known as the Urban Heat Island (UHI) [1]. Furthermore, climate change causes the temperature to increase and amplifies overheating within urban landscapes [2]. These heat loads have a major impact on the energy performance of buildings, resulting in more cooling energy demands and an increase in total electricity consumption, which, in turn, imply more CO_2 emissions and costs [3]. In the last decades, European cities have found several solutions to this problem such as increasing vegetation on roofs and façades to create evapotranspiration [4], adding more water surfaces for cooling effects [5], using permeable materials as well as coatings with high albedo on exposed urban surfaces [6] to reduce the amount of short-wave heat that gets accumulated. Today, 55% of the world's population lives in urban areas, a proportion that is expected to increase to 68% by 2050 [7]. As we spend most of our time indoors, buildings are becoming real thermal machines with the objective to maintain climatic conditions favourable to human beings. However, most existing buildings are inefficient in their reaction to environmental conditions, displaying poor autoregulation of energy consumption, leading to excessive carbon emissions.

Building envelopes or facades where interior and exterior interact play an essential role in the drive towards zero-carbon or carbon positive buildings as they are the building's primary energy transmission interfaces where interior and exterior interact.

However, most current facades are under or over-mechanized, which poses several problems from high carbon footprints to maintenance difficulties and high costs. The Arab World Institute in Paris illustrates these problems particularly well. Since then, new ideas have emerged such as the "breathing metal" concept developed by architect Doris Kim Sung in 2011 [8], inspired by the autonomous functioning of the pores of our skin to adapt the body's internal tissues to changes in outside temperatures. There is an urgent need to design and optimize facades to develop new dynamic responses and increased autonomy, ideally eschewing regular human intervention and additional energy input from the grid. Over the past decade, more and more façade systems improving user comfort by regulating daylighting have emerged such as static shading elements with variation in shape and size as well as adaptive skins and shading devices for facades. The use of computational design methods and simulations makes it possible to investigate more complex solutions at a building scale by including environmental data within the design process. Despite the improvements, these trends collectively suggest there is an urgent need for creating, designing, and optimizing adaptive shading systems. However, today, kinetic shading systems that can react to the variable nature of weather conditions are more complex and harder to implement due to multiple factors such as trade-offs between energy consumption and environmental performance improvement, research, production, implementation, and costs.

Living systems have the capacity to respond simultaneously to several external environmental factors such as temperature, humidity, or light while optimizing energy consumption. Indeed, over 3.8 billion years of evolution, living systems have been optimized regarding matter, energy, and information flows and sorted out sustainable systems in terms of energy and water management, material production, information processing as well as the organization and collective intelligence efficiency [9]. Biomimetics, the interdisciplinary cooperation of biology and technology or other fields of innovation, aims to solve practical problems through the functional analysis of biological systems and their abstraction into models for the solution [10].

In light of this, the lab team sets out to test a biomimetic approach in response to thermoregulation design challenges, and analyses biomimetics as a tool to develop better building envelopes. In the first section, the following paper will examine how energy requirements are met in the built environment at the building scale. We will specifically look at the role and importance of facades in architecture and explain how autonomous devices and adaptive envelopes can participate in optimizing thermal comfort for the user as well as reduce energy demands. We will describe and show how a biomimetic approach can be used at the architectural scale to respond to these needs. In the second part, we will explain the journey towards developing a biomimetic thermoregulating envelope using a biomimetic approach as a framework. We will describe the different steps behind the design process as well as the challenges and difficulties faced by the team. In the third part, we will comment on the methodology, which helped imagine a new concept and prototype to better address environmental challenges. Lastly, we will reflect on the importance of an in-depth qualitative evaluation of the new design which we refer to today as *Pho'liage*.

2 Setting the Context

2.1 Meeting Energy Requirements

2.1.1 Energy Requirements in Building Regulations

Buildings and their construction are responsible for approximately 40% of EU energy consumption and 36% of greenhouse gas emissions, making them the single largest energy consumer in Europe [11]. In 2012, the building sector in France was responsible for 44% of the country's final energy consumption and emitted 18% of the country's greenhouse gases. The building sector consumes the most energy of all identified sectors, ahead of transport, industry, steel, and agriculture.

In recent years, successive building regulations specifically address thermal issues and have made it possible to significantly reduce energy consumption. In 2004, France created the first climate plan with the aim of stabilizing its greenhouse gas emissions at 1990 levels over the period 2008–2012. In 2009, France created the Grenelle law (law n° 2009–967 of August 3, 2009), which lists commitments in terms of limiting climate change, preservation of biodiversity, maintenance and development of ecosystems and natural environments, reduction of health risks from the environment, and the establishment of an ecological democracy through better public information. The Grenelle law is set to "divide by 4 the greenhouse gas emissions from the 1990 level by 2050". Furthermore, the EU has established a legislative framework that includes the Energy Performance of Buildings Directive 2010/31/EU (EPBD) [12] and the Energy Efficiency Directive 2012/27/EU, to boost the energy performance of buildings. Both directives promote policies that are set to help achieve a highly energy-efficient and decarbonized building stock by 2050. They were amended as part of the Clean Energy for All Europeans package in 2018 and 2019, introducing new elements sending a strong political signal on the EU's commitment to "modernize the building sector in light of technological improvements and to increase building renovations". From 31 December 2020, EU countries must set cost-optimal minimum energy performance requirements for new buildings, for existing buildings undergoing a major renovation, and for the replacement or retrofit of building elements like heating and cooling systems, roofs, and walls. Additionally, all new buildings must be nearly zero-energy buildings (NZEB) and EU countries must draw up lists of national financial measures to improve the energy efficiency of buildings. In addition to these requirements, under the Energy Efficiency Directive (2012/27/EU), national governments are recommended to only purchase buildings that are highly energy efficient. The Commission has established a set of standards and accompanying technical reports to support the EPBD called the energy performance of buildings standards (EPB standards).

2.1.2 Building Facades: Primary Energy Transmission Interfaces

Over the years, building facades have played an active role in the evolution of these increasingly stronger requirements. Indeed, façades are responsible for more than 40% of total energy loss in the winter, as well as overheating in the summer [13], making the employment of air conditioning systems inevitable in the provision of adequate internal comfort for users and building occupants. While the passive housing concept is often achieved by using high-performance insulation materials in façades, the overuse of insulators can lead to an increased load on ventilation and humidity regulation systems, which together amount to additional costs and significant space losses. "Ventilation units consume more than 2% of all electricity in the EU and are amongst the biggest consumers of indoor electricity, after heating, cooling, and lighting" [14]. With an accurate design of façade details and efficient solar shading systems, office buildings in Europe could function efficiently without high energy-consuming cooling systems.

By reducing the use of ventilation systems through alternative solutions, Europeans could save "approximately 1300 PJ in energy use each year by 2025. This is equivalent to the annual gross energy consumption of Austria or Greece" [15]. Following these premises, in the EU, the energy consumption of building construction could potentially be reduced by 10% for oil consumption (approximately 41 m tons) and around 111 m tons for CO_2 production per year [11]. The use of adaptive shading systems with integrated PVs could contribute to reducing the energy demand of buildings and solve the challenging climate change and pollution problems. Space cooling systems (SC) account for an important part of the average European Union household's air conditioning energy consumption (about 5%) and are particularly present in the service sector (approximately 13%) [16]. The European space cooling market is characterized by a huge potential for growth, with the percentage of the surface areas in the service sector being almost ten times higher (30%) than in the residential buildings (4%) [17]. Given the expressed desire by the European markets for increased comfort standards, surface cooling applications are expected to increase, especially within the residential sector [18].

Further possible reasons for the future rise of (SC) applications in Europe are global warming and modern architectural designs with larger glazing areas [19]. This anticipated rise in cooling energy consumption creates a self-nurturing negative feedback loop and amplifies climate change significantly. The (Fig. 1) illustrates the typical constituents of the consumption (for a typical/model office building) divided into heating, cooling, ventilation, lighting, and miscellaneous when using conventional passive façade technologies.

These increasingly stronger requirements call for innovative solutions that can reduce all aspects of energy consumption, including the transformation of facades and the avoidance of high-energy systems such as (SC).

"Envelope" and "skin" are widely used terms in architecture to qualify the roof and the façades. Building envelopes display a wide diversity of architectural expression, building materials, and dimensions spanning the recorded history of architecture. Their strength, aesthetics, and porosity have adapted alongside our cultural evolution

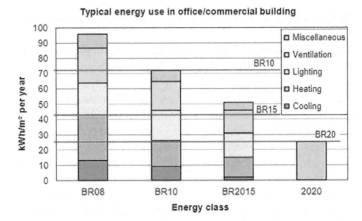

Fig. 1 Changing energy requirements in Denmark where the demands (red lines) are expressed as a limit for total primary energy consumption in kWh/m² per year, with a limit of 25 kWh/m² per year in 2020. The figure also illustrates the typical constituents of the consumption (for a typical/model office building) divided into heating, cooling, ventilation, lighting, and miscellaneous when using conventional passive façade technologies. The 2015 column illustrates how much the energy needs can be reduced by using traditional passive facades in a typical office building, i.e., the requirements can hardly be reached without introducing some form of sustainable energy supply, e.g., PV panels. The last column in the figure shows the great challenge of the 2020 energy requirement, which calls for dynamic façade technologies or other significant measures, e.g., local sustainable energy production. Energy requirements (shown by the horizontal lines) in the Danish building regulations from 2008 till 2020. The first three columns illustrate the typical constituents of the total energy consumption in offices or commercial buildings when using passive façade technologies. The last column shows the 2020 energy demand, which can only be met by dynamic technologies (or by introducing sustainable energy supply

and have responded to environmental constraints [20]. They are considered complex systems since their designers must consider environmental realities, expected levels of human well-being, as well as technical, aesthetic, and financial requirements. Solutions must overcome contradictory requirements such as limiting overheating while providing natural lighting [21].

From a technical standpoint, building envelopes act as barriers that simultaneously control several environmental factors from heat to light to humidity, airborne pollution, acoustics, and mechanical stress [22, 23]. The façade's performance highly influences the building's total energy consumption as it impacts directly the building's exposure to solar gain, a determinant factor on the energy loads required by internal regulation systems—e.g., heating, cooling, and ventilation [24]. Since planning policies frequently encourage urban densification to offset urban sprawl, encouraging renovation rather than the construction of new buildings, façade performance has become a primary lever in the reduction of building energy consumption [21]. Indeed, most of the current stock of European buildings was built between the end of the Second World War and the first oil crisis of 1973 [25]. During this period, building energy optimization was not a major concern since the first building thermal regulations appeared in Europe for new construction in the mid-1970s as a response to the oil crisis [26, 27]. The improvement of the building envelope yielded a significant potential for environmental improvement as highlighted by the increasing number of research programs for the development of efficient retrofitting building systems [28].

2.2 Optimizing Thermal Comfort

2.2.1 The Importance of Adaptive Façade Systems

French thermal regulation RT2012 is one of the measures adopted from the "Grenelle Law" and aims to limit the primary energy consumption of new buildings to a maximum of 50 kWhEP/(m^2) on average. In addition to a maximum consumption requirement, the RT2012 also aims to encourage all building sectors to make technological and industrial changes and to oblige designers to opt for bioclimatic architecture. Indeed, the RT2012 imposes a minimum of 17% or 1/6 of glazed surfaces in buildings [29]. The glazed part of the façade gives by far the largest contribution to the transmission heat losses through the building envelope [r]. However, the glazing also contributes significantly to the heating of the building during the colder seasons when the sun's energy is important through the increased use of solar gain and daylight. To both reduce transmission losses and increase the use of solar energy and daylight façades must adapt to changing external conditions. An ideally designed façade should be interactive and respond intelligently and reliably to the changing outdoor conditions and the occupants' needs. The [30] (2020 Environmental Regulation) that will come into effect in January 2022 for community housing and in 2023 for commercial buildings will among other things, guarantee thermal comfort in the event of strong heatwaves [30].

Loonen defines a Climate Adaptive Building Shell (CABS) as a system that "has the ability to repeatedly and reversibly change its functions, features or behaviour over time in response to changing performance requirements and variable boundary conditions. By doing this, the building shell effectively seeks to improve overall building performance in terms of primary energy consumption while maintaining acceptable thermal and visual comfort. CABS has been classified in to categories based on the bioinspiration approach they deploy (phototropism and heliotropism of plants), the physical aspects of their interface with the environment (blocking, filtering, converting, collecting, or storing energy), the duration of their operating cycles (seconds, minutes, hours, diurnal and seasons), their scale of adaptation to external stimuli (micro-scale and macro-scale), and monitoring systems which control their operation (extrinsic and intrinsic).

2.2.2 Biomimetics as an Opportunity in Architecture

"Biomimetics", "bioinspiration", "biomimicry" are interdisciplinary approaches based on the integration of technology and biology, by transferring living systems' principles into a technological solution [9, 31]. Biomimetics refers to "the interdisciplinary cooperation of biology and technology or other fields of innovation with the goal of solving practical problems through the functional analysis of biological systems, their abstraction into models and the transfer into and application of these models to the solution" [32]. This chapter mostly uses the term "biomimetics" since the *Pho'liage* façade system results from a biomimetic approach rather than bioinspiration or a biomimicry design process.

This approach has inspired innovation in several fields from aeronautics [33, 34], biomedical [35, 36] materials [37, 38] to architecture [39, 40] over the past decades. Indeed, interest in living organisms has been strengthened both by the rise in environmental awareness and advances in technology. This progress allows us to understand life at all levels from molecular to ecosystems, and then to replicate these natural principles and processes with novel technologies [41, 42]. This crosscutting approach is poised to play a major role in solving systemic problems related to energy, health, transports, food security, and creating economic and social value [43, 44].

Since time immemorial, architects have explored the links between Nature and architecture [45, 46]. Architectural trends and movements ranging from Art Nouveau to Ecological, from "biomorphism" to "vernacular" and "bioclimatic", all these compounded sensitivities have driven the promotion of nature-based design. At the crossroads between these different approaches, biomimetics is an opportunity to shift from fossil-fuel-based architecture to regenerative design by translating living systems and ecosystems properties into suitable solutions for the building context [47, 48]. Several examples of building systems or facades illustrate biomimetics in architecture. On the one hand, some designs mimic the efficiency of biological systems in order to optimize building energy consumption [49, 50] or to lighten the building structure [44, 51], etc. On the other hand, some designs mimic ecosystemic functions, which enable buildings to better integrate with ecological cycles, which, in turn, reduces the pressure exerted by urban areas on ecosystems at large [47–49].

Thermoregulation is a contemporary architectural challenge well-explored among biomimetic envelopes. Literature reviews have counted numerous well-documented proofs of the concept of adaptive and non-adaptive biomimetic envelopes [51, 52]. Adaptive building envelopes can adapt their shape due to an intrinsic or extrinsic control. Intrinsic control involves self-adjusting of the façade systems or components while extrinsic control implies façade modification after information retrieving and processing by building sensors [53]. Figure 2 illustrates the diversity of adaptive biomimetic building envelopes such as *HygroSkin*—a climate-responsive pavilion designed by ICD (Fig. 2a), the Thematic Pavilion—an adaptive facade that adapts to the building envelope the bio-inspired *Flectofin* system [54, 55] (Fig. 2b), Bloom—an adaptive material inspired by adaptation mechanisms of living systems [56] (Fig. 2c). Figure 3 presents an overview of non-adaptive biomimetic building envelopes such

Fig. 2 Adaptive biomimetic building envelopes. **a** HygroSkin. © ICD/University of Stuttgart, **b** the thematic pavilion, **c** Bloom. © DO SU Studio Architecture

Fig. 3 Non-adaptive building envelopes. **a** Sierpinski Forest, licence CC-BY-SA Estelle Cruz, **b** the Esplanade Theatre Singapore Art Centre, licence CC-SA, **c** the Nianing Church. © Regis L'Hostis

as the envelope *Sierpinski Forest*—a sun-shading façade component inspired by the fractal geometry of trees [57, 58] (Fig. 3a), the Esplanade Theatre Singapore Art Centre—a shading cladding of a double roof dome inspired by the skin of the durian fruit for thermal regulation [59, 60] (Fig. 3b), and the *Nianing Church*—building inspired by the ventilation system of termites' mounds for passive ventilation and thermal regulation [61, 62] (Fig. 3c).

3   Developing a Biomimetic Envelope

The following section illustrates the development of the biomimetic shading device *Pho'liage* and how an investigation of biological role models can lead to the creation of a functional prototype and the possibility of design variations.

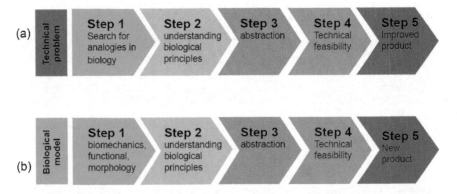

Fig. 4 Biomimetic design process. **a** Technology pull, **b** biology push. Adapted from ISO standard 2015:18,458

3.1 From Biological Movement to Adaptive Shading System

3.1.1 Biomimetics as a Design Framework

Two main approaches exist in biomimetics as defined by the ISO standard 18,458: "technology pull" or "biology push". The ISO has provided the two following definitions: the technology pull process is a "biomimetic development process in which an existing functional technical product is provided with new or improved functions through the transfer and application of biological principles". The biology push process is a "biomimetic development process in which the knowledge gained from basic research in the field of biology is used as the starting point and is applied to the development of new technical products" [9]. The five main steps are presented in Fig. 4 for each approach.

To support biomimetic design processes, more than 43 methods and tools have been developed across all fields [63]. Some of them were specifically proposed to support designers in applying biomimetic in architecture, such as Genius of Biome [64], BioGen [65], ESA—Ecosystem Services Analysis [64].

Pho'liage mainly results from the technology pull approach since the design team first identified the technical challenges—thermal and light regulation through the building facade, and then searched for relevant adaptive strategies within the kingdom of plants. Architect Steven Ware's biology background had a significant impact on the design process.

3.1.2 Initial Concepts

The team explored initial concepts involving direct use of thermostatic bimetal bands with incisions cut parametrically into the material. This approach was chosen as appropriately frugal given the cost of TBM materials and the difficulty of tooling the

Fig. 5 Conceptual rendering of the façade of the University Hospital CHU Nantes, France featuring bands of thermo-bimetal into which incisions 'free' the memory shape to provide shading. © ArtBuild

bands. It also effectively employed the factory curved shape memory, as the bands of TMB were stretched flat from their initial factory roll configuration, and the incision would "free" the bands according to the shape of each cut, allowing them to take up their initial memory shape. As the curved shapes heat up in the sun, they go back to the "lay flat" position and thus close the gaps made by the incisions, protecting the façade from solar gain (Fig. 5).

Thermostatic bimetals (TBMs) are alloys that expand with rising temperature and contract with falling temperature. In ancient civilizations, wheel rims were heated before fitting and shrunk on cooling, thus holding the wheel firmly together. Compound materials made from layers of different materials with different properties that, once combined, produce a material with new properties, are also known from early times. However, engineers began to exploit the possibilities offered by compound materials incorporating layers with different thermal expansion rates in the eighteenth century. Today, TBM continues to provide a source of inspiration for inventors seeking novel ideas and applications. Manufacturers have developed and improved manufacturing processes to produce TBM at an industrial scale from hot rolling cladding methods to cold cladding processes. Besides, research work continuously is made to explore TBM properties. ArtBuild's team has explored and carried out several experiments with this specific material since 2015. Figures 6a–d successively explain how TBM operates.

Thermostatic bimetal bending is directly proportional to the difference in the coefficient of expansion and the temperature change of the component strips, and inversely proportional to the thickness of the combined strips. The amount of bending is also affected by the ratio of the moduli of elasticity of the two strips and by their thickness ratio [64].

Practical tests with both thermo bi-plastics and thermo bi-metals led the team to focus on triangular shapes, which best-expressed curving responses to heat changes as well as rigidity, anticipating the needs for the elements of the facade to be positioned on the exterior of the building envelope to be effective, all the while subjected to wind loads and pollution, one side of the triangle assumed to provide a stable fixing to the support structure.

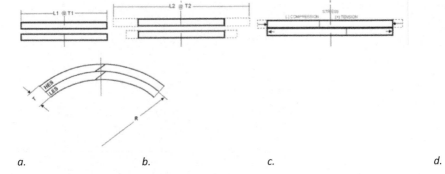

a.   b.   c.   d.

Fig. 6 **a** Two metal strips of identical length (at a given temperature) having high and low coefficients of thermal expansion. **b** When the temperature rises, their relative lengths will change. **c** When the strips are bonded together, and the temperature is raised, the high expansion strip will be under compression and the low expansion strip will be under tension. **d** These forces produce a moment that causes the element to bend in a uniform arc

The form-finding process followed several steps. Based on the studies of existing solutions together with the available literature [64, 65], the team conducted a detailed feasibility study using the manufacturer's data for the use of TBM material. A reference for thermostatic bimetal produced by Engineered Material Solutions® [64] was selected for its temperature activation range and relative mechanical resilience vs action span. A theoretical model was then created using the finite element method (FEM). In combination with wind-load calculations and the applied thermostatic theory of Antoine Yvon Villarceau [61], the first 3D parametric models were built on Rhinoceros® and Grasshopper®. Those 3D models helped to visualize and calculate the open/closed optimal ratio induced by a temperature change, before engaging in the physical testing as the next step of the R&D process (Fig. 7).

The team then tested a prototype of assembled individual triangular "petals", which made more efficient use of the TBM material. The resulting "flowers" or petal

Fig. 7 First 3D parametric models were built on Rhinoceros® and Grasshopper®. © ArtBuild

clusters are connected by stainless steel cables to be mounted, once again, on the exterior of the façade. This configuration led to more specific studies on the detailing of attachment components given the need to both hold the TBM petals in place, all the while allowing for their movement when required.

3.2 Developing the First Prototype for Application

3.2.1 Studying Plant Movements

Charles Darwin's publication *The Power of Movement in Plants* in 1880 saw an extension of his work on the theory of natural selection to focus on the way plants react to external stimuli, introducing the reader to the behaviour of certain species of the flowering plant genus Medicago: one during the day ("leaves during the day"), the other at night ("leaves asleep at night"). The movements he evokes are known collectively as nyctinasty and have often been a source of inspiration for scientists, photographers, artists, and architects. The movement that Darwin evokes concerns a non-linear response known as nyctinasty. The term nyctinastic (from Greek *nux, nukt-* "night" + *nastos* "pressed") is used to describe the closure of leaves during the night, nastic movements occur in leaves, flowers, cotyledons, or branches in response to the arrival of dusk and dawn. These movements are described as *nastic*, or "direction-independent, reversible orientation changes in response to direction-independent stimuli".

Plants' surfaces, like other biological surfaces, feature multifunctional interfaces with their environment [62]. For more than 400 million years, land plants have evolved a wide diversity of biological surface structures capable of regulating environmental factors—e.g. heat, light, water. The properties of these functional surfaces, for example, the absorption of water or light reflection can be translated into useful models for the development of technical solutions. Figure 8 shows the multifunctional properties of the surfaces of plant leaves and illustrates the highly functional plant surfaces that provide water, light, thermal, mechanical, and air regulation.

Flower petal movements in the living world are mostly induced by a difference in growth rate on each side of the petal. In species that open in the morning primarily as a response to temperature, the inner surface of the petal rapidly expands when the

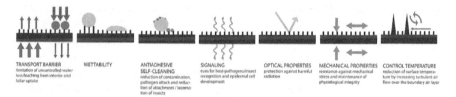

TRANSPORT BARRIER — limitation of uncontrolled water loss/leaching from interior and foliar uptake

WETTABILITY

ANTIADHESIVE SELF-CLEANING — reduction of contamination, pathogen attack and reduction of attachment / locomotion of insects

SIGNALING — cues for host-pathogens/insect recognition and epidermal cell development

OPTICAL PROPIERTIES — protection against harmful radiation

MECHANICAL PROPIERTIES — resistance against mechanical stress and maintenance of physiological integrity

CONTROL TEMPERATURE — reduction of surface temperature by increasing turbulent air flow over the boundary air layer

Fig. 8 Koch, K., Bhushan, B., and Barthlott, W., Multifunctional surface structures of plants: an inspiration for biomimetics. Progress in Materials Science, 2009, 54, pp. 137–178

Table 1 Number of optimized design versions studied with technical characteristics. © ArtBuild

Number of TBM petal-like structures	Perforated	Imperforated	Materials	Maximum filed level (dB)
2	V1	V1	TBM	73.3
3 (trilobal)	**V2**	**V2**	**TBM**	**66.33**
4	V3	V3	TBM	74.66
8	n/a	V4	TBM	70.33
8 + 4	n/a	V5	TBM/PLA	74

temperature rises due to the release of phytohormone/metabolites, notably absent on the outer surface. Cooling results in faster growth on the outer surface. The study of the nastic movement phenomena by the research team led to the development of several bio-inspired TBM-made prototype propositions (Table 1). The change in the TBM curving shape is memory-shaped, which makes possible the long-term use of our optimized design without any maintenance.

3.2.2 Optimizing Shape and Form

Based on a combination of different shapes and numbers for petals, the team tested various configurations [66] considering the physical properties of the material, which were not available through the TBM catalogue literature. This modelling was especially aimed at minimizing the loss and waste of material during the punch press industrial process, as well as locating the critical actuating surfaces in areas when sunlight masking would be enhanced or avoided.

The following key elements came to the fore in the design of subsequent prototypes:

- Optimizing the functional area of the TBM actuator and reducing the masking effect.
- Optimizing the open/fold ratio.
- Reducing the sound made by sudden distortion (Field Level in dB) of the TBM elements when heated and cooled.

As a result of this study, the perforated design of the V2 thermostatic bimetal element was chosen as the best candidate for further research given the specific application of the TBM component as the actuator of the shading device (see Figs. 8 and 9). Lab test results supported the trilobal petal-like TBM actuator design that reduced material waste, increased the open/close ratio efficiency, reduced sound emissions during the activation process, and showed slower 'closing time' during the cooling process leading to increased solar shading performance (Fig. 10).

Once the ground components were defined in terms of material specification and general shape, it was possible to create a larger mock-up model (see Fig. 11).

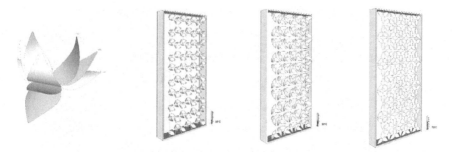

Fig. 9 V2 imperforated prototype. Change in shape within temperature variation. © ArtBuild

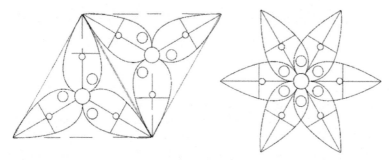

Fig. 10 Perforated prototype (Left: Photo of the folded position, Right: CAD of the open position). © ArtBuild

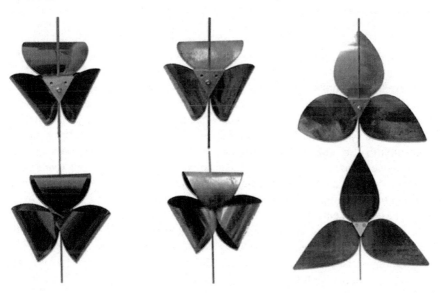

Fig. 11 Section of a mock-up model after outdoor anti-corrosion tests of 36 months (Left: painted TBM petals, Right: unprotected TBM petals). © ArtBuild

a.   b.

Fig. 12 Outdoor test aperture. Comparison of the movement of painted and unpainted TBM elements during the summer after being exposed for 10 s (**a**) and for 1 min on the left (**b**). © ArtBuild

The research team then focused on corrosion protection to the TBMs, all the while seeking to optimize the reactivity of the petal curving dynamics. A combination of decorative, Cr-8VI-free coatings was chosen, the chemical properties acting as an anti-corrosion protection of the TBMs while the dark colour and matt finish enable the petals to efficiently absorb the sun's heat leading to faster actuation responses. The flexibility of the TBMs related to the overall memory-shaped performance was not compromised with the application of the coatings (Fig. 12).

Having produced several prototypes at various stages of the R&D, the team tested different assembly configurations of the TBM actuator and PV cell subcomponents to explore a cost-effective solution with the highest possible efficiency and reactivity. The team continues to work on this concept through theoretical models and lab tests. The renewable electricity production combined with the dynamic memory-shaped movement of the TBM actuator distinguishes this patent-pending technology from other existing solutions (see Figs. 13 and 14). As mentioned previously, the solar shading device works without motorization or maintenance needs in the timeframe of the warranty period. This unique passive kinetic device is activated and controlled fully by solar radiation. The PV cells induce important heat loss during energy production cycles, mostly because of the Joule heating effect as process output. The positive interaction of both photo-sensitive (PV) and thermo-reactive (TBM) materials' heat exchange is beneficial to the overall energy efficiency of the device resulting from the improvement of the actuator reactivity.

The PV components are connected to a resistance element using conducting cables and could be expected to improve the overall performance of the system through the increased speed of the opening and closure of the flowers. The use of PVs questions the autonomous nature and the passivity of the system, which switches from being a low-tech system to a high-tech one. Consequently, the team decided not to continue research on the integration of PV components.

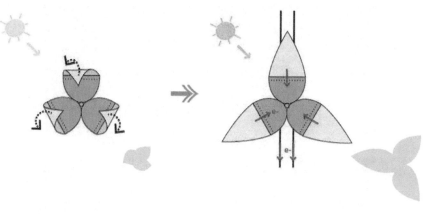

Fig. 13 TBM actuator (in blue) and PV cell (in white); the two supporting/conductive cables are illustrated in black. © ArtBuild

Fig. 14 The PV components (in grey) are physically related to resistance (green colour) through conductors (blue/red colour) trapped in between the LES and HES TBM layers. © ArtBuild

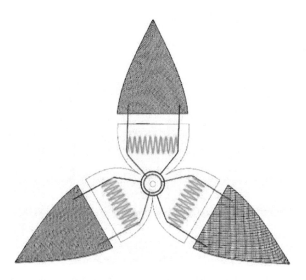

3.2.3   Implementation and Limits of the Prototype

ArtBuild was selected to be the lead architects of the new headquarters of the CIRC—IARC, the International Agency for Research on Cancer in Lyon, France, which is planned to be completed by the summer of 2022. The bio-inspired shading system *Pho'liage* will be installed on the façade to the circular courtyard, forming a vast curtain between the two urban windows and covering 270 degrees of the building's round patio (Fig. 15). The shading device, composed of 1523 flowers and 82 cables, will extend from the second floor to the fifth floor (Fig. 16). The TBM flower shading elements are designed to be fastened to a support structure of tensioned metallic cables using custom-made fixings. The cables matrix creates a vast mesh with one

Fig. 15 CIRC—IARC, International Agency for Research on Cancer in Lyon, view of the courtyard. © ArtBuild

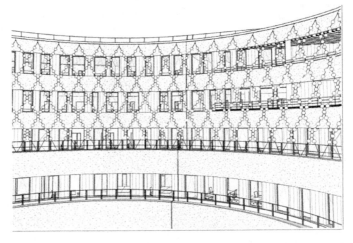

Fig. 16 Layout of the TBM flowers on the cables. Facade of the CIRC in the courtyard. © ArtBuild

flower per crossover and three flowers in between (Fig. 16). Each cable features a turnbuckle mechanism and can be removed entirely should the shading system need maintenance or replacement.

The implementation of *Pho'liage* on the CIRC—IARC has helped the team better understand the challenges behind designing building envelopes with complex geometries and the issues linked to translating the prototype into real-life architectural projects (Fig. 17).

In September 2020, the lab team carried out an analysis and critique of the *Pho'liage* project, taking into account the difficulties of the implementation of the

Fig. 17 Pho'liage fixing elements on the roof of the CIRC. © ArtBuild

CIRC. Indeed, the team reflected on and questioned the entire design process from reconsidering the problem statement to looking at the implementation of the shading system in a real project.

Several issues were raised concerning the cost, environmental impact, and efficiency of the system. Measuring the overall performance of the device includes measuring both ease of use and operation to understand how it responds to the initial problem statement: improving the user's comfort while reducing energy demand.

- The use of TBM poses economic concerns as TBM is expensive, on the CIRC, the cost of one flower is 55 euros.
- Sourcing of the TBM has proved to be difficult as it can only be supplied by a small number of companies manufacturing special alloys. The companies and their teams of specialized engineers have been very difficult to contact.
- TBM is an alloy thus it cannot be recycled which raises environmental concerns.
- The curled (closed position)/uncurled (open position) ratio isn't efficient enough as the surface deployed when the flower is open is small, thus creating a small amount of shade.
- Although some work has been made to improve acoustic disturbance, the system still emits a sudden metallic snapping sound when coming back to its initial curled position.
- The system cannot be turned on or off by the user. According to LEAMAN et BORDASS's study [66] there is increasing evidence that occupants prefer to have some level of personal control of their local indoor environment [67].
- The cable mesh structure supporting the flower elements isn't ideal as the orientation of the flowers cannot be controlled and doesn't suit free form or irregularly shaped buildings.

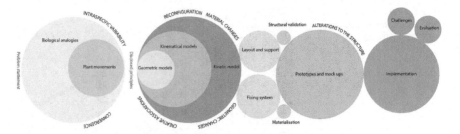

Fig. 18 Different stages for designing an improved version of Pho'liage. Diagram adapted from Shleicher, S. Bio-inspired Compliant Mechanisms for Architectural Design (2016), p. 68

4   Addressing Environmental Challenges

This section shows the process behind the development of a second improved version of the shading device *Pho'liage* aiming to both address an environmental challenge and to improve its operation. In this context, the team focused on the fact that thermostatic bimetal isn't environmentally efficient as it is an alloy, thus difficult to recycle and costly in this specific field of application.

4.1   *Optimizing* Pho'liage

4.1.1   Research Method

Concerning the method, the team of architects and designers identified six distinct stages behind the development of *Pho'liage:* research, problem formulation, design process, prototyping, implementation, and evaluation. The below table describes the steps for each consecutive phase, and the diagram in Fig. 18 details steps 2 and 3.

4.1.2   Exploring Biological Role Models for Deployment Mechanisms

Improving the open/closed ratio and reducing the amount of TBM were identified as the two main technical challenges for designing a new improved version of the *Pho'liage* flower. The team investigated biological role models in the scientific literature in order to understand how plant species take advantage of mechanical, compositional, and structural gradients to perform mobility with minimal energy use. We explored a specific category of reversible plant movement linked to deployment or triggering movement, which are known as *elastic instabilities and snapping motions* (see Fig. 19). The figure below shows the different categories within nastic movements. After in-depth research in the available scientific literature, two

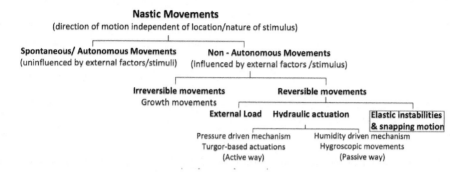

Fig. 19 Classification of nastic plant movements from Saad Elghazi Y., Hamza N., Dade-Robertson M. Responsive Plant-inspired skins: a review. (Studied plant movement in red)

biological role models—the Venus flytrap (Dionaea muscipula) and the Aldrovanda vesiculosa—were isolated and analysed (see Table 2).

According to Schleicher, plants' flexibility represents a compliant mechanism that reduces the number of mechanical parts by an integrative design (Schleicher 2016). Taking this into consideration, the team made several paper mock-ups to analyze curling and folding mechanisms in order to test various deployment movements mimicking the elastic snapping motions found in plants. AB Lab's team explored the concept of curved line folding in the scientific literature and studied the fact that in pliable systems the deformation of individual parts is constrained by their neighbouring elements. Curve-line folding is a type of folding technique defined as the act of folding a flat sheet of material along a curved crease pattern in order to create a 3D shape [65]. It uses the combination of folding (plastic deformation) and bending (elastic deformation). In other words, when one surface area is bent, the forces and movements are transmitted through the curved creases to the adjacent surface areas, which results in a folding motion. Vergauwen researched this approach

Table 2 Description of method behind redesign of *Pho'liage*. © ArtBuild

Phase	Description
I. Research	Documentation and understanding of the context, formulation of a problem
II. Design process	Using the biomimicry methodology: abstracting a technical problem, transposing to biology, identifying potential biological models, selecting biological models of interest, abstracting biological strategies, transposing to technology
III. Concept development and prototyping	Refining shape, form, material, and finishes, fabrication of the prototype
IV. Implementation	Realization of the project on a building, the challenges, and difficulties
Evaluation and impact	Assessment of the system through multiple criteria

Fig. 20 Curve line folding principles © ArtBuild

and discussed its application in the context of kinetic shading systems in architectural engineering. The researcher explains that the elastic deformations that occur when a flat sheet is forced into a curved shape can produce an interesting transformation process that could be used for the development of a new type of deployable structure.

The practical mock-ups and the exploration of the curve-line fold concept have shown that the employment of elasticity within a structure facilitates not only the generation of complex geometries but also enables the creation of elastic kinetic structures. Indeed, compliant mechanisms combine strength with elasticity and gain some of their mobility from flexible members' deflection rather than only from movable joints (Fig. 20).

4.1.3   Model Making and Testing

The above research enabled the team to start designing a deployable surface. To create an autonomous shading element, the team focused on finding a trigger to induce movement in order to open and close the flower. Previous investigations using TBM in the initial concept proved to the team that the material could be used as an actuator force. AB Lab subsequently attempted to create a new hybrid shading element consisting of two distinct parts: a flat deployable surface and an actuator to create a folding motion. The initial mock-ups involved paper surfaces with curved creases and a TBM actuator causing the flower to deform and bend as it furled and unfurled.

The architect/designer team then conducted a series of tests to find the best shape for the deployable surface (Fig. 21). A triangular shape proved to be optimal as the actuator area/polymer surface area ratio is highly efficient, with a small actuator's surface and a large deployable surface. Furthermore, it enabled a convenient tiling pattern when juxtaposed. Once the team agreed on the shape, multiple curve-line paper mock-ups were made to test different curvatures for each crease with various radii (Table 3). Manual tests were undertaken to determine which design required the least pressure and force for deformation and bending to occur. A series of tests with weights were conducted (Fig. 22). The opening and closure of each paper/TBM mock-up were tested by applying temperatures of up to 65 °C directly to the TBM actuator with a heat gun (see Fig. 23). The actuator was prestressed into a curved shape and then mechanically fastened to the paper surface. Following the heating

Fig. 21 Exploration of shape and curve line folding principles. © ArtBuild

phase, the TBM component deforms and lays flat, resulting in the flower's opening (Figs. 23 and 24).

4.1.4 Layout and Support Structure

The team then worked on a support structure to position each individual flower element. A honeycomb matrix made from assembled folded aluminium strips was retained, supporting the flowers and enabling the precise orientation of each flower with respect to wind loads and solar movements. Each flower is screwed onto an angle of the hexagonal shape, presenting its TBM actuator to the sun (Fig. 25). This support structure presents several advantages compared to the initial cable system, including greater resistance to environmental conditions and the ability to cover irregularly shaped facades (Fig. 26).

Table 3 Describes the two selected biological role models. © ArtBuild

Name of selected biological models	Description of biological model	Image of the studied organism
Venus flytrap (Dionaea muscipula)	The leaves of the Venus flytrap snap shut and trap prey within milliseconds by turning physical signals into electrical signals. These carnivorous plants rely on nutrients from small prey animals when growing in nutrient-poor soil. When an unsuspecting prey brushes up against two touch-sensitive hairs on the inside of the trap-shaped leaves, the trap snaps shut, ensnaring the prey for later digestion. The touch-sensitive hairs, known as trigger hairs, signal trap closure using sodium-activated action potentials (APs)	*Trap of Dionaea muscipula. Image from Wikipedia Commons*
Aldrovanda vesiculosa	The trap consists of two lobes connected with hinge zones to a midrib. While the lobes and the central area are of higher stiffness, the hinge zone is more flexible and describes a curved fold line (Fig. 3). After being triggered by prey, a change of turgor pressure leads to a small change in the bending curvature of the midrib, which is amplified by a curved line folding mechanism and leads to a complete closure of the trap (Poppinga et al. 2016)	*Aldrovanda vesiculosa.* Stem with leaves with orbicular lamina (trap). Image from Wikipedia commons

Fig. 22 Testing various curvatures with weights. © ArtBuild

Fig. 23 Test of flower deformation with a heating source (TBM and paper). TBM surface temperature: **a** 20 °C, **b** 30 °C, **c** 45 °C, **d** 65 °C. © ArtBuild

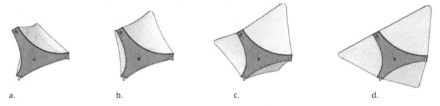

Fig. 24 Drawing of flower aperture depending on heating. TBM actuator (dark grey) deployable surface (light grey). TBM surface temperature: **a** 20 °C, **b** 30 °C, **c** 45 °C, **d** 65 °C. © ArtBuild

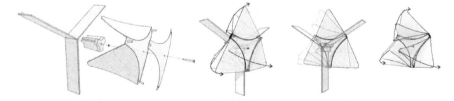

Fig. 25 **a** Exploded view. **b** Mechanism and operation. © ArtBuild

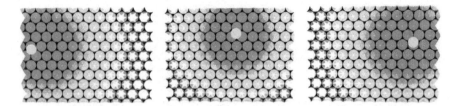

Fig. 26 Opening and closure of the flower element according to sun position. Red represents surface temperature above 65 °C, orange and yellow temperatures between 30 and 40 °C and blue below 21 °C. © ArtBuild

4.1.5   Material Research

According to Shleicher, materials for bio-inspired compliant mechanisms need to be chosen to maximize flexibility rather than stiffness. Stiffness and strength are not the same, and it is possible to make a structure both flexible and strong. A structure is described as demonstrating low stiffness when a small load can cause a large deflection. Yet this reveals nothing about its strength. "Strength" is a material property that specifies the stress that a structure can withstand before failure. The deflection of a material under a given load, which appears before a structure fails, is dependent on its stiffness and rigidity. A structure's flexural rigidity, however, results from its material properties and its geometry. We can say that there are two means to effectively maximize the flexibility of a structure: either by changing its material settings or by modifying its shape.

Material choice is of primary importance. It can be said that materials with a higher strength-to-modulus ratio can undergo much larger deflections and achieve smaller bending radii before undergoing plastic deformation and failure than those with a lower ratio. High values can be found among plastics and polymers. This material group is particularly well suited for compliant mechanisms and elastic kinetic structures.

Translating the paper mock-ups to plastic prototypes has proven to be difficult as polymers react differently to paper. The properties influencing material choice include:

- Low environmental impact.
- Elastic: flexible and bendable.
- Durable: resistant to outdoor environment and weather (heat, light, wind, water).
- Secure: comply with French fire regulations.

The team ran a second series of tests using TPE (thermoplastic elastomers) to find the right material with the right properties (Table 4). The different models were 3D modelled, and the team noted that the models were too soft and flexible, especially around the crease. Rhinoceros software was used to create models with greater material thickness around the crease and thinner zones around the edges of the piece, as in Fig. 27.

Table 4 Different curvatures of the folding crease (*r: Radius). Circles for tests 1, 2, 3 have the same centre. © ArtBuild

Test 1 r = 68 mm	Test 2 r = 66 mm	Test 3 r = 64 mm
Test A r = 46 mm	Test B r = 51 mm	Test C r = 68 mm

AB Lab collaborated with SIPLAST, a plastics processing company based near Saint Etienne in France, which develops biopolymers. The aim of the collaboration was to create a suitable biopolymer with the right properties for the deployable surface of the flower. AB Lab specifically requested a biopolymer material, thus produced from renewable biomass sources, such as vegetable fats and oils, corn starch, straw, woodchips, sawdust, recycled food waste. SIPLAST tested multiple molecular compositions for different materials in order to find the right balance for the required biopolymer. Table 5 shows the different test results obtained according to temperature and material combination. SIPLAST's engineers designed a special custom-made mould in order to create a triangular pliable surface with creases out of biopolymer. After various tests, they decided it was best to remove some materials in

Table 5 Thermoplastic elastomers properties

Material name	Description	Properties				Environmental impact
TPE	Highly versatile thermoplastic elastomers (TPEs) suitable for most environments and processes	**Flexibility**	**Resistance**	**Aesthetics**	**Security**	Can be supplied as biosourced or recycled
		Highly chemically compatible with other polymers	Excellent UV resistance	Surface can be mat or glossy	Fire resistant	
		Elastic memory	Mechanical properties			
		From soft to semi-rigid (from 40 shore A to 80 shore A)				

Fig. 27 3D models using Rhinoceros and 3D prototyping with varying thicknesses. © ArtBuild

the middle in order to reduce TBM's effort (Figs. 28 and 29) and created the appropriate tooling for injecting polymer. The team compared the paper and biopolymer prototypes, noting that the operation of the flower with cut-outs in the centre was close to that of the paper mock-up. Some trials were made to connect the TBM actuator to the biopolymer surface, according to different points of contact (Fig. 30) (Table 6) (Fig. 31).

4.2 Opportunities and Prospective Outlook

As outlined by this case of study and several comparative analyses [cruz] [kuru] [al-oba], the building energy-saving potential of biomimetic building envelopes is promising.

Crossover initiatives linking research bodies with public procurement bodies could lead to new developments in biomimetic design. Public building projects notoriously suffer from limited time frames, and design application progress requiring

Fig. 28 The final shape, ready to be produced. © ArtBuild

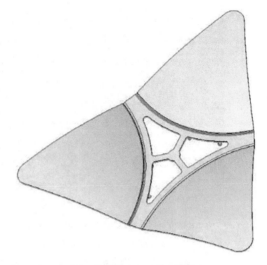

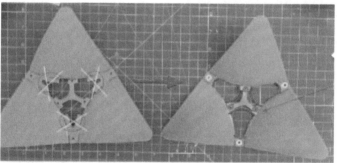

Fig. 29 The final shape ready to be produced. © ArtBuild

Table 6 Operation of the flower within different material compositions © ArtBuild

		Material	Temperature stability (°C)	Hardness/flexibility	Observation
1	A	40% SEBS and 60% PP SEBS charged 30% short wood fibre + red dye	135	Good flexibility A+	Operates well with changes on the hinge
2	B	50% SEBS and 50% PP SEBS charged 30% short wood fibre + green dye	135	Good flexibility A++	Operates well with changes on the hinge
3	C	70% SEBS and 30% PP SEBS charged 50% long wood fibre	135	Very good flexibility	Operates well with changes on the hinge but fragile hinge when fibre

a.   b.

Fig. 30 Aperture tests by SIPLAST. **a** Open flower, **b** folded flower. © ArtBuild

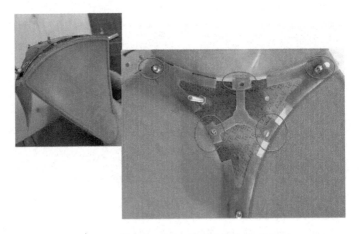

Fig. 31 Points of contact between biopolymer and TBM. © ArtBuild

knowledge generation, the creation of structuring tools, and biological data mining, may be accelerated through biomimetics.

Selecting and abstracting the appropriate biological model for a biomimetic solution is intricate. To help focus a research project, even trained biomimetic practitioners need a preselection of biological scales or organism groupings. This approach has been shown to stimulate co-discoveries, a by-product beneficial for technological breakthroughs and contribution in biological data. It would seem productive to apply this multidisciplinary work specifically with a focus on several taxonomic groups at a time and to assess the effects of hybridization of biological strategies on the design of a biomimetic envelope element with multi-regulation targets and specifications.

As seen in this study, the methodologies and tools used in the bioinspiration design process are diverse, and yet the number of projects in the literature reaching a TRL of

6 is low. Despite a high potential for product development, the implementation of Bio-BS elements in practice is challenging. During abstraction and technical feasibility steps, designers must consider both market specifications, and retrieve feedback and experience from the users afterward, allowing scalable and repeatable models, and avoiding successful but unique-application biomimetic designs.

In addition, we suggest that addressing multi-regulation requires mechanisms in the early stage of the design process, assisted with data exploration and structuring tools. Further research from the authors is ongoing and focuses on the development of tools to access biological data during the design process, helping combine different biological strategies.

ArtBuild is currently reflecting on upcoming potential research topics. AB Lab's overall research on biomimetic facades gives an overview of how an architectural studio can link the different prisms of architectural design together with basic biological research into one coherent narrative. The final rendered images (Fig. 32) showing the biomimetic facade on a utopic project glass tower are part of a larger video and hopefully a source of inspiration, which might spark new ideas and encourage the next generation of researchers to contribute in their own way. Prototypes of *Pho'liage* will be shown in various exhibitions in cities across Europe in 2022.

Fig. 32 Prototype of *Pho'liage* showing different polymers. © ArtBuild

Fig. 33 Render extract from video representing Pho'liage on a circular building facade view from the interior. **a** Open position where the flowers are deployed, **b** closed position. © ArtBuild

The authors hope that this paper provides a comprehensive summary and helpful considerations resulting from this complex, multi-variable challenge, as well as a stimulus for further research on this topic. The chapter also outlines the challenges, issues, and difficulties linked to the implementation of the project and the practical use of the biomimetic approach. The reflections on the methodology show an attempt to design a suitable biomimetic working process that acts as a framework for AB Lab and which could be reused to frame future projects (Fig. 33).

It is interesting to note that engineers still tend to apply a traditional way of thinking to their design of kinetic structures, which prioritizes rigid-body mechanics. Bioinspired mechanisms and clever use of elastic material behaviour and structural instabilities make compliant mechanisms a highly suitable solution for autonomous shading systems but there is still a lot to be explored. Designers could investigate, for example, how climatic requirements may best transform the shapes and tilling symmetries of the shading systems. Material scientists and engineers could devote themselves to increasing the durability of elastic materials in the most effective way.

References

1. Mitigation of the heat island effect in urban New Jersey (2015) Taylor & Francis. https://www.tandfonline.com/doi/abs/https://doi.org/10.1016/j.hazards.2004.12.002

2. Chapman S, Watson JEM, Salazar A, Thatcher M, McAlpine CA (2017) The impact of urbanization and climate change on urban temperatures : a systematic review. Landscape Ecol 32(10):1921–1935. https://doi.org/10.1007/s10980-017-0561-4
3. Ortiz L, González JE, Lin W (2018) Climate change impacts on peak building cooling energy demand in a coastal megacity. Environ Res Lett 13(9):094008. https://doi.org/10.1088/1748-9326/aad8d0
4. Gago EJ, Roldan J, Pacheco-Torres R, Ordóñez J (2013) The city and urban heat islands: a review of strategies to mitigate adverse effects. Renew Sustain Energy Rev 25:749–758
5. Gunawardena KR, Wells MJ, Kershaw T (2017) Utilising green and bluespace to mitigate urban heat island intensity. Sci Total Environ 584–585(April):1040–1055 (Elsevier B.V)
6. Pacheco-Torgal F, Labrincha J, Cabeza L, & Granqvist CG (2015) Eco-efficient materials for mitigating building cooling needs: design, properties and applications. Elsevier Ltd., pp 1–9
7. 68% of the world population projected to live in urban areas by 2050, says UN I UN DESA I United Nations Department of Economic and Social Affairs (2018). United Nations Department of Economic and Social Affairs. https://www.un.org/development/desa/en/news/population/2018-revision-of-world-urbanization-prospects.html
8. Bloom (n.d.) DOSU studio. https://www.dosu-arch.com/bloom. Accessed 7 Dec 2021
9. ISO 18458:2015—Biomimetics—Terminology, concepts and methodology
10. Chayaamor-Heil N, Guéna F, Hannachi-Belkadi N (2018) Biomimétisme en architecture. État, méthodes et outils. Cahiers de La Recherche Architecturale, Urbaine et Paysagère, 1. https://doi.org/10.4000/craup.309
11. European solar-shading organization (2006) Energy saving and CO_2 reduction potential from solar-shading systems and shutters in the EU-25. http://www.buildup.eu/sites/default/files/ESC ORP-EU25_p2528.pdf
12. Energy performance of buildings directive—European Commission (2021) Energy—European Commission. https://ec.europa.eu/energy/topics/energy-efficiency/energy-efficient-buildings/energy-performance-buildings-directive_fr
13. De Groote M, Volt J, Bean F (2017) Is Europe ready for the smart buildings revolution?—Mapping smart-readiness and innovative case studies. Buildings Performance Institute Europe (BPIE). http://bpie.eu/wp-content/uploads/2017/02/STATUS-REPORT-Is-Europe-ready_FINAL_LR.pdf
14. European Commission, Ventilation units: energy savings. https://ec.europa.eu/info/energy-climate-change-environment/standards-tools-and-labels/products-labelling-rules-and-requirements/energy-label-and-ecodesign/energy-efficient-products/ventilation-units_en
15. EUROSTAT (2018) Electricity production, consumption and market overview: net electricity generation, EU-28, 1990–2016. https://ec.europa.eu/eurostat/statistics-explained/index.php/Electricity_production,_consumption_and_market_overview
16. Pezzutto S, Fazeli R, De Felice M, Sparber W (2016) Future development of the air-conditioning market in Europe: an outlook until 2020. In: WIREs (Wiley interdisciplinary reviews): energy and environment, vol 5, no 6, pp 649–669
17. Shanks K, Nezamifar E, Impacts of climate change on building cooling demands in the UAE. In: Proceedings of the SB13 Dubai: advancing the Green Agenda technology—Practices and policies, Dubai, UAE. Accessed 8–10 Dec 2013
18. Sandak A, Sandak J, Brzezicki M, Kutnar A (2019) State of the art in building façades. Environ Footprints Eco-Des Prod Process 1–26. https://doi.org/10.1007/978-981-13-3747-5_1
19. Cruz E (2017) Biomimetic solutions to design multi-functional envelopes CEEBIOS, European centre of excellence in biomimicry of the city of Senlis
20. MH-OR, Technology roadmap: energy efficient envelopes. http://www.iea.org/publicationsand undefined2012
21. Badarnah L (2012) Towards the living envelopes: biomimetics for building envelopes
22. COST Action TU1403—Adaptive facades network—Webpage of COST action TU1403. http://tu1403.eu/. Accessed 12 Mar 2020
23. AP-E. and Buildings and undefined 2016, Forty years of regulations on the thermal performance of the building envelope in Europe: achievements, perspectives and challenges. Elsevier.

Accessed 30 Nov 30 2021. https://www.sciencedirect.com/science/article/pii/S03787788163 05400

24. Nemry F, Uihlein A, Colodel CM., Wetzel C, Braune A, Wittstock B, & Frech Y (2010) Options to reduce the environmental impacts of residential buildings in the European Union—Potential and costs. Elsevier. https://www.sciencedirect.com/science/article/pii/S03787788100 00162. Accessed 23 Nov 2020

25. Vincent JFV, Bogatyreva OA, Bogatyrev NR, Bowyer A, Pahl AK (2006) Biomimetics: its practice and theory. J R Soc Interface 3(9):471–482. https://doi.org/10.1098/rsif.2006.0127

26. Objectifs et intérêts de la RT (2012). Bureau d'études thermiques RT2012 en ligne, 2018. https://www.e-rt2012.fr/explications/generalites/objectifs-interets-rt-2012/. Accessed 7 Dec 2021

27. RE 2020 (2021) IFPEB. https://www.ifpeb.fr/neutralite-carbone/re-2020/

28. ISO/TC 266 (2015) Biomimetics, ISO 18458:2015, vol 2015

29. Ward TA, Rezadad M, Fearday CJ, Viyapuri R (2015) A review of biomimetic air vehicle research: 1984–2014. Int J Micro Air Veh 7(3):375–394. https://doi.org/10.1260/1756-8293.7.3.375

30. Siddiqui NA, Asrar W, Sulaeman E (2017) Literature review: biomimetic and conventional aircraft wing tips. Researchgate.net 4(2). https://doi.org/10.15394/ijaa.2017.1172

31. Sheikhpour M, Barani L, AK-J. of C. Release, and undefined 2017, Biomimetics in drug delivery systems: a critical review. Elsevier. https://www.sciencedirect.com/science/article/pii/S0168365917301372. Accessed 26 Nov 2021

32. Barthelat F (2015) Full critical review architectured materials in engineering and biology: fabrication, structure, mechanics and performance. Taylor Fr 60(8):413–430. https://doi.org/10.1179/1743280415Y.0000000008

33. von Gleich A (2010) Potentials and trends in biomimetics. Springer

34. Pawlyn M (2011) Biomimicry in architecture

35. Knippers J, Speck T, Nickel KG (2016) Biomimetic research: a dialogue between the disciplines, pp 1–5

36. Sharma S, Sarkar P (2019) Biomimicry: exploring research, challenges, gaps, and tools. Smart Innov Syst Technol 134:87–97. https://doi.org/10.1007/978-981-13-5974-3_8

37. Lepora NF, Verschure P, Prescott TJ (2013) The state of the art in biomimetics. Bioinspir Biomim 8(1):013001. https://doi.org/10.1088/1748-3182/8/1/013001

38. Snell-Rood E (2016) Interdisciplinarity: bring biologists into biomimetics. Nature 529(7586):277–278. https://doi.org/10.1038/529277a

39. Biomimetic technology market report, industry analysis, and forecast. https://bisresearch.com/industry-report/biomimetic-technology-market.html. Accessed 29 Jul 2020

40. Gruber P (2011) Biomimetics in architecture [Architekturbionik]

41. Zari MP, Biomimetic design for climate change adaptation and mitigation. https://doi.org/10.3763/asre.2008.0065

42. Gruber P, Jeronimidis G (2012) Has biomimetics arrived in architecture? Bioinspir Biomim 7(1):1–3. https://doi.org/10.1088/1748-3182/7/1/010201

43. McCafferty DJ, Pandraud G, Gilles J, Fabra-Puchol M, Henry PY (2018) Animal thermoregulation: a review of insulation, physiology and behaviour relevant to temperature control in buildings. Bioinspiration Biomimetics 13(1). https://doi.org/10.1088/1748-3190/aa9a12

44. Badarnah L, Lidia (2017) Form follows environment: biomimetic approaches to building envelope design for environmental adaptation. Buildings 7(2):40. https://doi.org/10.3390/buildings7020040

45. McCafferty DJ, Pandraud G, Gilles J, Fabra-Puchol M, Henry P-Y (2017) Animal thermoregulation: a review of insulation, physiology and behaviour relevant to temperature control in buildings. Bioinspir Biomim 13(1):011001. https://doi.org/10.1088/1748-3190/aa9a12

46. Blanco E, Zari MP, Raskin K, Clergeau P (2021) Urban ecosystem-level biomimicry and regenerative design: linking ecosystem functioning and urban built environments. Sustainability 13(1):404. https://doi.org/10.3390/su13010404

47. Zari MP (2015) Ecosystem services analysis: mimicking ecosystem services for regenerative urban design. Int J Sustain Built Environ 4(1):145–157. https://doi.org/10.1016/j.ijsbe.2015.02.004

48. Kuru A, Oldfield P, Bonser S, Fiorito F (2019) Biomimetic adaptive building skins: energy and environmental regulation in buildings. Energy Build 205:109544. https://doi.org/10.1016/j.enbuild.2019.109544
49. Loonen R et al (2015) Design for façade adaptability: towards a unified and systematic characterization. In: 10th conference advanced building skins
50. One Ocean, Thematic Pavilion EXPO 2012/soma I ArchDaily. https://www.archdaily.com/236979/one-ocean-thematic-pavilion-expo-2012-soma/. Accessed 14 Mar 2020
51. Singapore Arts Center—Esplanade theatres on the bay—E-architect. https://www.e-architect.co.uk/singapore/singapore-arts-center. Accessed 14 Mar 2020
52. CHURCH NIANING I IN SITU architecture—Rethinking the future awards. https://awards.rethinkingthefuture.com/gada-winners-2019/church-nianing-in-situ-architecture/. Accessed 14 Mar 2020
53. IN SITU architecture: project. http://www.insitu-architecture.net/en/projets/12404-church.html. Accessed 14 Mar 2020
54. Fortmeyer RM, Linn CD, Kinetic architecture: designs for active envelopes
55. Doris Kim Sung: metal that breathes I TED talk. https://www.ted.com/talks/doris_kim_sung_metal_that_breathes. Accessed 14 Mar 2020
56. Brochures (2021) Auerhammer. https://www.auerhammer.com/en/downloads/brochures.html
57. Pelicaen E (2018) Development of an adaptive shading device using a biomimetic approach: redesign of the South Façade of the Arab World Institute, Master thesis, Brussels Faculty of Engineering, Vrije Universiteit Brussel (VUB). https://www.scriptieprijs.be/scriptie/2018/development-adaptive-shading-device-using-biomimetic-approach
58. Modin H (2014) Adaptive building envelopes, Master thesis, Department of Civil and Environmental Engineering, Chalmers University of Technology, Göteborg (Sweden). http://publications.lib.chalmers.se/records/fulltext/214574/214574.pdf
59. Engineered Material Solutions (Wickeder Group), Thermostatic Bimetal Designer's Guide. https://www.emsclad.com/fileadmin/Data/Divisions/EMS/Header/Bimetal_Desingers_Guide.pdf
60. Villarceau AY (1862) Recherches sur le mouvement et la compensation des chronomètres, 1st edn, Mallet-Bachelier, Paris
61. Koch K, Bhushan B, Barthlott W (2009) Multifunctional surface structures of plants: an inspiration for biomimetics. Prog Mater Sci 54(2):137–178. https://doi.org/10.1016/j.pmatsci.2008.07.003
62. Nicol JF, Humphreys MA (2002) Adaptive thermal comfort and sustainable thermal standards for buildings. Energy Build 34(6):563–572
63. Speck T, Knippers J, Speck O (2015) Self-X materials and structures in nature and technology: bio-inspiration as a driving force for technical innovation. Archit Des 85(5):34–39. https://doi.org/10.1002/ad.1951
64. Vergauwen A, Laet LD, Temmerman ND (2017) Computational modelling methods for pliable structures based on curved-line folding. Comput Aided Des 83:51–63. https://doi.org/10.1016/j.cad.2016.10.002
65. Nenov T (2018) Development of adaptive shading device from thermostatic bimetal components using a biomimetic approach. In: Biovoie working papers
66. In buildings with central control, occupants have to adapt to a particular temperature and may experience this as "uncomfortable". In their opinion, when occupants have control over temperature changes, they find the environment more comfortable. This is confirmed by (FRONTCZAK and WARGOCKI, 2011) who show, through surveys, that giving occupants the possibility to control the indoor environment improves thermal and visual comfort as well as overall building satisfaction
67. Werner S (2016) European space cooling demands. Energy 110(C):148–156
68. Economidou M, Todeschi V, Bertoldi P, D'Agostino D, Zangheri P, Castellazzi L (2020) Review of 50 years of EU energy efficiency policies for buildings. Energy Build 225:110322. https://doi.org/10.1016/J.ENBUILD.2020.110322

Printed in Great Britain
by Amazon